Sterile Processing of Pharmaceutical Products

Sterile Processing of Pharmaceutical Products

Engineering Practice, Validation, and Compliance in Regulated Environments

Sam A. Hout
Mission Viejo
CA, USA

WILEY

This edition first published 2022

Registered Office
John Wiley & Sons, Inc., 111 River Street, Hoboken, NJ 07030, USA

Editorial Office
111 River Street, Hoboken, NJ 07030, USA

For details of our global editorial offices, customer services, and more information about Wiley products visit us at www.wiley.com.

Wiley also publishes its books in a variety of electronic formats and by print-on-demand. Some content that appears in standard print versions of this book may not be available in other formats.

Library of Congress Cataloging-in-Publication Data applied for:
ISBN: 9781119802327

Cover design by Wiley
Cover image: © AlexRaths/Getty Images

FSC
www.fsc.org
MIX
Paper from responsible sources
FSC® C013604

Contents

Preface

There are two ways to manufacture sterile drug products: aseptic manufacturing and sterile fill-finish. First is a process in which the drug product, container, and closure are first subjected to sterilization methods separately, as appropriate, and then brought together (aseptic manufacturing). Second is fill and seal drug product in container and terminally sterilize.

Many complex drug products are not amenable to terminal sterilization, leading to increased demand for aseptic manufacturing and sterile fill-finish capabilities. Aseptic processing is uniquely challenging because it requires careful planning, thoroughly trained personnel, and specialized facilities/equipment to properly execute. Cleanroom facilities and aseptic processes should be designed to minimize contamination risk from materials, equipment, and personnel. Sterile lyophilization requires investment into specialized equipment, facilities, and knowledge.

Drug products that are delivered via the parenteral, ophthalmic, inhaled, or otic route present an increased risk of infection or harm because they bypass many of the body's natural defenses. To ensure patient safety, the FDA requires that drug products delivered via these routes be supplied as sterile products. This designation includes many complex drug products, including ophthalmic suspensions, sterile injectables, lyophilized powders for injection, and aqueous-based aerosols for inhalation.

"Sterile products" refer to products that are going to be administered using an enteral route of administration. The "products" are going to be infused directly into the bloodstream or body tissue, and it is extremely important they must be "sterile." Consider that when a patient takes a tablet orally, their digestive system (acting as part of the immune system) could identify and kill any bacteria that may be present. With a parenteral route of administration, if any contaminants are present, they go right into the bloodstream, bypassing the digestive system.

Regulatory guidance is intended to help manufacturers meet the requirements in the FDA current good manufacturing practice (cGMP) regulations (21 CFR parts 210 and 211) when manufacturing sterile drug and biological products using aseptic processing.

Sterile injectables manufacturing involve the production of vaccines. A vaccine must be rigorously tested before the public gets it. We often think of vaccines as treatments for illness, rather they are preventative measures for specific targeted viruses such as the flu vaccines. People who are well are given vaccines to keep them from getting sick. Therapeutic drugs are developed to cure diseases that already invaded the human body. They are tested and qualified for their efficacy in curing-specific disease indication and proven safety through extensive clinical trials that define their side effects.

Researchers will study the virus and attempt to determine which type of vaccine may work best. There are several kinds of vaccines. Some have a weakened live virus, which triggers a protective immune response but does not cause illness. Some vaccines contain inactive virus that creates an immune response in the body. Other types of vaccines utilize genetically engineered RNA or DNA, which triggers a protein that can prevent the virus from binding to our cells causing illness. The selection of which vaccine route to follow is dependent on experimental results that identify the best possible outcome. Following lab testing to ensure that a vaccine is working on the molecular cellular level, vaccines will undergo animal testing to predict safety for humans. This is followed by clinical trials targeting efficacy and safety for human use.

Phase 1 clinical trials involve a small number of healthy volunteers to test the vaccine for serious side effects. Phase 2 involves smaller studies evaluating efficacy. At this stage dosage levels of the vaccine are evaluated for scheduling of dosages for multiple applications. Safety is constantly monitored by observing clinical trial subjects for any side effects and whether the immune response or antibody accumulation is effective to warrant moving forward to additional clinical studies. In Phase 3, larger field studies are conducted. This involves testing a population, vaccinating some while having a control group, and monitor the effect over time to observe specific effects or difficulties. Common denominators among subjects are carefully evaluated for short-term side effects and dosages levels impact on side effects that are statistically evaluated and documented. If Phase 3 shows that the vaccine is safe and effective, it is approved and registered for commercial use. Following a working vaccine approval by the FDA, there is considerable time to scale up mass-production and distribution in commercial operations.

The main goal of vaccination is to inoculate huge numbers of people to develop immunity in the local community that would protect against larger outbreaks. Testing and monitoring for longer term safety concerns continue even after the vaccine is generally available. Common side effects of vaccines include redness, swelling, and associated pain at the location of injection, low-temperature fever;

severe side effects like allergic reactions are relatively uncommon. Vaccines are developed for their protective benefits, which outweigh the risks.

Vaccine newer biotechnology methods are based on cell culture that could allow for more rapid development. As recently experienced, a pandemic circling the globe (COVID-19), there would be an international effort to find a viable vaccine. Collaboration between the private sector and public research labs would accelerate vaccine development.

Recent approaches to expedite vaccine development utilize unconventional study design such as for the coronavirus vaccine. Published reports on infectious diseases show that instead of traditional Phase 3 trials, volunteers at low risk of developing a severe form of COVID-19 (Younger healthy people without chronic conditions) are pursuing a human challenge study. These subjects could be exposed to the coronavirus, are monitored closely, and given care to prevent severe reactions. This type of study would involve fewer participants but could be done in less time. There are ethics concerns with this type approach as typically clinicians' ethics rules generally forbid deliberately infecting human beings with a virus that might cause serious disease.

Besides vaccine trials that target preventing the disease, testing potentials of therapeutic treatments aim to make sick people well again continue to focus on drug development. For example, the recent development of a sterile injectable antiviral drug Remdesivir that is intended for treating severe cases of coronavirus infection in a hospital setting. There are also drugs that could potentially address the immune system's response to a virus. Patients get sick as the immune system reacts against the virus, whereby they develop excessive inflammation. Therefore, anti-inflammatory drugs that might mitigate the immune system's effects might be considered for clinical trials.

All ingredients of vaccines play necessary roles either in making the vaccine, triggering the body to develop immunity, or in ensuring that the final product is safe and effective. These include:

- Adjuvants that help boost the body's response to vaccine.
- Stabilizers that help keep vaccine effective after manufactured
- Formaldehyde that is used prevent contamination by bacteria during the vaccine manufacturing process.
- Thimerosal that is also used during the manufacturing process but is no longer an ingredient in any vaccine except multi-dose vials of the flu vaccine. Single-dose vials of the flu vaccine are available as an alternative.

The diseases vaccines prevent can be dangerous, or even deadly. Statistically, the chances of a child getting diseases such as measles, pertussis, or another vaccine-preventable disease might be low, and the child might never need the protection vaccines offer. However, one chooses to vaccinate to ensure that the protection vaccines provide is in place if needed. Immunity is the body's way of preventing disease. At birth, the immune system is not fully developed, which can put a child at greater risk for infections. Vaccines reduce risk of infection by working with body's natural defenses to help safely develop immunity to disease.

Vaccines use very small amounts of antigens (toxins) that help the immune system recognize and learn to fight serious diseases. Antigens are parts of germs that cause the body activate immune system. Getting a disease or getting a vaccine can both give future protection from that disease. With the disease humans get sick, but with the vaccine they get that protection.

Mission Viejo, CA, USA
December 2021

Sam A. Hout

Acknowledgments

This book is written with the support of my family, my patient, calm, and loving wife Mona, my role-model daughter Samantha, and my inspirational son Owen. This is also extended to my brothers. A special dedication is to the loving memory of my older brother, Dr. Salim Hout, MD who influenced my upbringing and thought process by building unwavering support, self-confidence, and unconditional love, in addition, the transformational memory of my parents who supported our lives with everlasting compassion and support. I am very grateful to all my colleagues in the United Kingdom, the United States, and California who supported my growth and effort over the last three decades. My friend and colleague Shawn Mossanen, software architect played a major role in helping my understanding of computerized systems logistics, especially ERP chain supply systems. Special thanks go to Steven Short, who contributed to my explanations of specific aspects of sterile processes through our many communications over three years. The lifetime support and guidance of my brotherly friend Mazen Tawil is noted for his loyalty, thoughtfulness, and camaraderie.

Acknowledgments

[illegible]

1

Introduction

This book is engineering driven to highlight basis of design, design criteria, and assumptions to coach and guide professionals involved in learning more about sterile manufacturing engineering and in pharmaceutical and biopharmaceutical manufacturing facilities, including operational requirements. The intent of this book is to present a simplified treatise with some very specific description of aseptic processing techniques and sterilization methods.

Sterile manufacturing engineering and process validation guides (US FDA, Guidance for Industry – Process Validation) are intended to manage risk, introduce Quality by Design (QbD), and ensure continued process reliability, consistency, and repeatability.

Sterile manufacturing guidelines are based on specific elements in the construction and installation of classified clean rooms, critical utility systems such as water for injection (WFI), solution transfer systems, powder transfer systems, and monitoring and controls, including automation of processes. Specialized technologies [1] in design of fill/finish lines of vials, cartridges, bottles for ophthalmic or otic purposes, prefilled syringes (PFSs), or ampoules, which might be filled and sealed aseptically or IV bags that are mostly terminally sterilized. In addition, considerations for personnel and material flow to prevent cross contamination are usually very well defined to separate in and out movements of sterile suites and cascading clean rooms through separate paths, including the use of path-through transfer systems for material flow.

Overall, paying attention to the design, construction, and installation should focus on process flow and employ standardization of equipment and methodologies as much as possible. Building sustainable premises, equipment qualifications, cleaning validation, and sanitization requirements should be all in consideration as part of the design of pharma/biopharma processing. Similarly, process validation and periodic revalidation of manufacturing systems are all important stages and steps that need to be on well-defined schedules to ensure reliability of drug production [2].

Sterile Processing of Pharmaceutical Products: Engineering Practice, Validation, and Compliance in Regulated Environments, First Edition. Sam A. Hout.

The main emphasis of clean room classifications is to quantify particle limits when rooms are in operation (dynamic conditions) for nonviable particles. In addition, microbiological limits are emphasized when classified rooms are in operation for viable particles. Media fills are conducted to certify rooms for aseptic conditions of drug products.

EN ISO 14644 Methodology

Class ISO 146144-1 (Federal Standard 209E)	Average airflow velocity, m/s (ft/min)	Air changes per hour	Ceiling coverage (%)
ISO 8 (Class 100,000)	0.005–0.041 (1–8)	5–48	5–15
ISO 7 (Class 10,000)	0.051–0.076 (10–15)	60–90	15–20
ISO 6 (Class 1,000)	0.127–0.203 (25–40)	150–240	25–40
ISO 5 (Class 100)	0.203–0.406 (40–80)	240–480	35–70
ISO 4 (Class 10)	0.254–0.457 (50–90)	300–540	50–90
ISO 3 (Class 1)	0.305–0.457 (60–90)	360–540	60–100
ISO 1–2	0.305–0.508 (60–100)	360–600	80–100

Microbiological Limits

	Recommended limits for microbial contamination			
Grade	Air sample, cfu/m^3	Settle plate, Ø 90 mm, cfu/4 h	Contact plates, Ø 55 mm, cfu/plate	Glove print, 5 fingers, cfu/glove
A	<1	<1	<1	<1
B	10	5	5	5
C	100	50	25	—
D	200	100	50	—

Monitoring of classified areas is continuous whether the room is in operation or not. Particles are measured through a continuous particle monitoring computerized system under measured conditions of temperature, pressure, and relative humidity. In addition, Environmental Monitoring Systems (EMS) with automated features are fitted with alarms to control alert and action levels in classified rooms limit of operation and regulatory requirements. Microbial counts are sampled and read using settling plates both at the aseptic core and equipment surroundings inside the classified clean room. Locations of these plates and frequency of sampling are based on formal risk analysis, sampling requirements, and validation protocols.

Various technologies are employed in sterile manufacturing such as Isolator technologies, Blow/Fill/Seal, aseptic customized tubing set as needed for product specifications, and terminally sterilized drug products, including sterile injectables (parenteral medicines). Protective personnel garments are specified for each clean room classification to cover hygiene, method of gowning, and required training and certification. General aspects of wall materials of construction, sinks, drains, changing rooms, airlocks (material air lock [MAL]) (anterooms), and air supply and return circulation are major considerations in the design criteria and specifications used in clean room classifications. The use of sanitizers, disinfectants, detergents, and fumigation is closely controlled and specified for room and equipment cleaning.

Moist heat (saturated steam; not superheated steam), radiation, or ethylene oxide gasification might be employed in sterilization of components, active ingredients, excipients, or drug products. Biological and chemical indicators are utilized to confirm sterility during routine production. In aseptic filling, liquid materials are filtered through a 0.22-μm sterile filter. Filter integrity is tested before and after use. Drug aseptic conditions require all filling operations to occur under class A (ISO 5 zone) conditions in the clean aseptic core; similarly, all capping will be under class A conditions. Aseptically filled vials, cartridges or PFSs must pass 100% integrity testing.

To verify sterile products, sterility assurance level of 10^{-6} is achieved by overkill methods and sterility tests must confirm process capability. Pharmaceutical Inspection Cooperation Scheme (PIC/S) conventions highlighted details for validation of aseptic processes, and recommendations for sterility testing. In addition, it covered isolators for aseptic and sterility testing. Other guidelines governing these areas are EN ISO 14644-1, -2, International Society of Pharmaceutical Engineers (ISPE) guidelines, and US FDA aseptic processing guide.

2

Sterilization

Drug product and materials that are used in medical and surgical applications, which are used in health-care facilities also, are made of materials that are heat stable and therefore undergo heat, primarily steam [3], terminal sterilization. An increase in medical devices and components made of materials (e.g. plastics) that require low-temperature sterilization necessitated the use of other means of sterilization such as radiation [4] and ethylene oxide (ETO) gas [5], which has been used for heat- and moisture-sensitive medical devices. Recently, low-temperature sterilization systems (e.g. hydrogen peroxide gas plasma, peracetic acid immersion, ozone) have been developed and are being used to sterilize medical devices. This section reviews sterilization technologies for drug products and components used in healthcare.

Sterilization destroys all microorganisms on the surface of an article or in a drug compounded fluid to prevent disease transmission associated with the use of that item [6]. Inadequately sterilized critical components represent a high risk of transmitting pathogens. The sterility assurance level (SAL) of the product is defined as a single-viable microorganism occurring on a product after sterilization. An acceptable SAL is the probability of a spore surviving where one in one million would be 10^{-6}. SAL is an estimate of lethality of the entire sterilization process [7].

Medical products that have contact with sterile tissues or fluids are considered critical. These items should be sterile when used because any microbial contamination could result in disease transmission. Besides sterile injectable drugs, other items include surgical instruments, biopsy forceps, and implanted medical devices. If these items are heat resistant, the recommended sterilization process is steam sterilization, because it has the largest margin of safety due to its reliability, consistency, and lethality.

Sterile Processing of Pharmaceutical Products: Engineering Practice, Validation, and Compliance in Regulated Environments, First Edition. Sam A. Hout.

Steam Sterilization

Moist heat in the form of saturated steam under pressure is the most reliable. Steam sterilization is nontoxic and sporicidal. Steam sterilization is accomplished in an autoclave, whereby products and components are exposed to direct steam contact at the required temperature and pressure for the specified time. Steam-sterilizing temperatures are between 121.1 °C (250 °F) and 132 °C (270 °F). These temperatures must be maintained for a minimal time to kill microorganisms based on a validated sterilization factor, F_0, and kill rate based on *D*-values for specific products and components needed to be sterilized. *D*-values (time to reduce the surviving population by 90% or 1 log10) allow a direct comparison of the heat resistance of microorganisms. *D*-values can be determined at various temperatures and are used to designate the exposure temperature. At constant temperatures, sterilization times vary depending on the type of item (e.g. metal versus rubber, plastic, items with lumens), porous, or whether the item is wrapped or unwrapped, and the sterilizer type.

Steam cycle is monitored by mechanical, chemical, and biological monitors and indicators. Steam sterilizers usually are monitored using a printout (or graphically) by measuring temperature, the time at temperature, and pressure. Typically, chemical indicators are affixed to the outside and incorporated into the pack to monitor the temperature or time and temperature. The effectiveness of steam sterilization is monitored with a biological indicator containing spores of *Bacillus stearothermophilus*. Positive spore test results are a relatively rare event, which can be attributed to operator error, inadequate steam delivery, or equipment malfunction.

Moist heat destroys microorganisms by the irreversible coagulation and denaturation of enzymes and structural proteins. In support of this fact, the presence of moisture significantly affects the coagulation temperature of proteins and the temperature at which microorganisms are destroyed.

Flash Sterilization

Flash sterilization is a modification of conventional steam sterilization (either gravity, pre-vacuum, or steam-flush pressure-pulse) in which the flashed item is placed in a compartment, or rigid container to allow for rapid penetration of steam.

Flash sterilization is considered acceptable for processing cleaned patient-care items that cannot be packaged, sterilized, and stored before use. It also is used when there is insufficient time to sterilize an item by the preferred package method. Flash sterilization is not recommended for implantable devices; however, flash sterilization may be unavoidable for some devices (e.g. orthopedic screw, plates).

Low-Temperature Sterilization Technologies

ETO has been widely used as a low-temperature sterilant. It has been the most commonly used process for sterilizing temperature- and moisture-sensitive components and medical devices. Due to environmental reasons and exposure limitations, alternative technologies to chlorofluorocarbon that are currently available include 100% ETO; ETO with a different stabilizing gas, such as carbon dioxide or hydrochlorofluorocarbons (HCFC), immersion in peracetic acid, hydrogen peroxide gas plasma, and ozone. In addition, vaporized hydrogen peroxide (VHP), vapor phase peracetic acid, gaseous chlorine dioxide, ionizing radiation, or pulsed light, or combined mixtures of peracetic acid and hydrogen peroxide are being utilized. The challenges concerning these technologies involve the reliability of these systems to fully kill pathogens. This occurs because microorganisms must be in direct contact with the sterilant for inactivation to occur. Several peer-reviewed scientific publications have data demonstrating concerns about the efficacy of several of the low-temperature sterilization processes (i.e. gas plasma, VHP, ETO, peracetic acid).

Ethylene Oxide Gas Sterilization

ETO is a colorless flammable and explosive gas. There are four essential parameters (operational ranges): gas concentration (450–1200 mg/l), temperature (37–63 °C), relative humidity (40–80%) (water molecules carry ETO to reactive sites), and exposure time (one to six hours) that influence the effectiveness of ETO sterilization.

The use of ETO evolved when few alternatives existed for sterilizing heat- and moisture-sensitive products, components, and medical devices. The main disadvantages associated with ETO are lengthy cycle time, cost, and its potential hazards to staff; the main advantage is that it can sterilize heat- or moisture-sensitive medical equipment without deleterious effects on heat-sensitive materials used in medical devices. Acute exposure to ETO may result in irritation. Occupational exposure in health-care facilities has been linked to hematologic changes and an increased risk into human tissue.

ETO is absorbed by many materials. For this reason, following sterilization the item must undergo aeration to remove residual ETO. Guidelines have been promulgated regarding allowable ETO limits for devices, (or active pharmaceutical ingredient [API]) that depend on how the device is used, how often, and how long in order to pose a minimal risk to patients in normal product use. ETO toxicity has been established in a variety of animals. Exposure to ETO can cause eye pain, sore throat, difficulty breathing, and blurred vision. Exposure can also cause dizziness, nausea, headache, convulsions, blisters, and vomiting and coughing. Occupational Safety Health Administration (OSHA) has established a PEL of 1-ppm airborne ETO in the workplace, expressed as a TWA for an 8-hour work shift in a 40-hour work week. The "action level" for ETO is 0.5 ppm, expressed as an 8-hour TWA, and the short-term excursion limit is 5 ppm, expressed as a 15-minute TWA.

The microbicidal activity of ETO is considered to be the result of alkylation of protein, DNA, and RNA. The excellent microbicidal activity of ETO has been demonstrated. ETO inactivates all microorganisms although bacterial spores (especially *Bacillus atrophaeus*) are more resistant than other microorganisms. For this reason, *B. atrophaeus* is the recommended biological indicator. Like all sterilization processes, the effectiveness of ETO sterilization can be altered by lumen length, lumen diameter, inorganic salts, and organic materials.

Hydrogen Peroxide Gas Plasma

Gas plasmas are generated in an enclosed chamber under deep vacuum using radio frequency or microwave energy to excite the gas molecules and produce charged particles, many of which are in the form of free radicals that are capable of interacting with essential cell components (e.g. enzymes, nucleic acids) and thereby disrupt the metabolism of microorganisms. The type of gas used and vacuum are two important variables that can determine the effectiveness of this process. Sterilization chamber is evacuated, and hydrogen peroxide solution is injected vaporized. Hydrogen peroxide vapor diffuses through the chamber (50 minutes), exposes all surfaces of the load to the sterilant, and initiates the inactivation of microorganisms. An electrical field created by a radio frequency is applied to the chamber to create a gas plasma. Microbicidal-free radicals are generated in the plasma. The excess gas is removed and in the final vent stage of the process, the sterilization chamber is returned to atmospheric pressure by introduction of high-efficiency filtered air. The by-products of the cycle (e.g. water vapor, oxygen) are nontoxic and eliminate the need for aeration. The process operates in the range of 37–44 °C and has a cycle time of 75 minutes. Sterilization efficacy is improved by using two cycles with a hydrogen peroxide diffusion stage and a plasma stage per sterilization cycle. Total processing time is reduced from 73 to 52 minutes. The biological indicator used with this system is *B. atrophaeus* spores.

Disinfection and Surface Sterilization Using Peracetic Acid

Peracetic acid is a highly biocidal oxidizer that maintains its efficacy in the presence of organic soil. Peracetic acid removes surface contaminants (primarily protein). Mechanism of action of peracetic acid is its ability to denature proteins, disrupts cell wall permeability, and oxidizes bonds in proteins, enzymes, and other metabolites. Peracetic acid will inactivate Gram-positive and Gram-negative bacteria, fungi, and yeasts in <5 minutes at <100 ppm.

Ionizing Radiation

Sterilization by ionizing radiation, primarily by cobalt 60 gamma rays or electron accelerators, is a low-temperature sterilization method that has been used for a number of medical products (e.g. tissue for transplantation, pharmaceuticals, and medical devices). There are no Food and Drug Administration (FDA)-cleared ionizing radiation sterilization processes for use in health-care facilities. Because of high sterilization costs, this method is an unfavorable alternative to ETO and plasma sterilization in health-care facilities but is suitable for large-scale sterilization of components and API.

Dry-Heat Sterilizers

This method should be used only for materials that might be damaged by moist heat or that are impenetrable to moist heat (e.g. powders, petroleum products, sharp instruments). The advantages for dry heat include the following: it is nontoxic and does not harm the environment; a dry heat cabinet is easy to install and has relatively low operating costs; it penetrates materials; and it is noncorrosive for metal and sharp instruments. The disadvantages for dry heat are the slow rate of heat penetration and microbial killing makes this a time-consuming method. In addition, the high temperatures are not suitable for most materials. The most common time–temperature relationships for sterilization with hot air sterilizers are 170 °C (340 °F) for 60 minutes, 160 °C (320 °F) for 120 minutes, and 150 °C (300 °F) for 150 minutes. *B. atrophaeus* spores' biological indicators should be used to monitor the sterilization process for dry heat because they are more resistant to dry heat than are *Geobacillus stearothermophilus* spores, which are more suitable for moist heat (steam).

There are two types of dry-heat sterilizers: the static-air type and the forced-air type. The static-air type is referred to as the oven-type sterilizer as heating coils in the bottom of the unit cause the hot air to rise inside the chamber via gravity convection. This type of dry-heat sterilizer is slower in heating, requires longer time to reach sterilizing temperature, and is less uniform in temperature control throughout the chamber than is the forced-air type. The forced-air or mechanical convection sterilizer is equipped with a motor-driven blower that circulates heated air throughout the chamber at a high velocity, permitting a more rapid heat transfer.

Filtration

Although filtration is not a lethality-based process, it is used to remove bacteria from thermolabile pharmaceutical fluids that cannot be purified by any other means. In order to remove bacteria, the membrane pore size (0.22 μm) must be smaller than the bacteria and uniform throughout transference of the sterile filtrate into the final container under aseptic conditions.

Microwave

Microwaves are used in medicine for disinfection of soft contact lenses, dental instruments, dentures, milk, and urinary catheters for intermittent self-catheterization. However, microwaves must only be used with products that are compatible (do not melt). Microwaves are radio-frequency waves, which are usually used at a frequency of 2450 MHz. Microwaves are effective microbicide. Microwave oven (2.45 GHz) inactivates bacterial cultures, mycobacteria, viruses, and *G. stearothermophilus* spores within 60 seconds to 5 minutes depending on the challenge organism. Microwaves used for sterilization of medical devices have not been FDA cleared.

Vaporized Hydrogen Peroxide (VHP®)

Hydrogen peroxide solutions have been used as chemical sterilant. VHP is used for the sterilization of medical equipment, isolator in use to produce pharmaceutical products in aseptic filling. VHP is delivered to the reaction site using deep vacuum to pull liquid hydrogen peroxide (30–35% concentration) from a cartridge through a heated vaporizer and then, following vaporization, into the sterilization chamber. VHP offers several appealing features that include rapid cycle time (e.g. 30–45 minutes); low temperature; environmentally safe by-products (H_2O, oxygen), good material compatibility; and ease of operation, installation, and monitoring. VHP has limitations, including cellulose cannot be processed; nylon becomes brittle; and VHP penetration capabilities are less than those of ETO.

Ozone

Ozone has been used for years as a drinking water disinfectant. Ozone consists of O_2 with a loosely bonded third nascent oxygen atom that is readily available to attach to, and oxidize, other molecules. This additional oxygen atom makes ozone a powerful oxidant that destroys microorganisms but is highly unstable (half-life of 22 minutes at room temperature). A sterilization process, which uses ozone as the sterilant, was cleared by FDA. The sterilizer creates its own O_3 sterilant internally from United States Pharmacopeia (USP) grade oxygen and purified water. The sterilant is converted back to oxygen and water vapor at the end of the cycle by passing through a catalyst before being exhausted into the room. The duration of the sterilization cycle is about four hours, and it occurs at 30–35 °C. Microbial efficacy has been demonstrated by achieving a SAL of 10^{-6} with a variety of microorganisms, including a most resistant microorganism, *G. stearothermophilus*.

Formaldehyde Steam

Low-temperature steam with formaldehyde is used as a low-temperature sterilization method. The process involves the use of formalin, which is vaporized into a formaldehyde gas that is admitted into the sterilization chamber. A formaldehyde concentration of 8–16 mg/l is generated at an operating temperature of 70–75 °C. The sterilization cycle consists of a series of stages that include an initial vacuum to remove air from the chamber and load, followed by steam admission to the chamber with the vacuum pump running to purge the chamber of air and to heat the load, followed by a series of pulses of formaldehyde gas, followed by steam. Studies indicate that formaldehyde is a mutagen and a potential human carcinogen, and OSHA regulates formaldehyde. The permissible exposure limit for formaldehyde in work areas is 0.75 ppm measured as an 8-hour TWA.

Gaseous Chlorine Dioxide

Chlorine dioxide is not mutagenic or carcinogenic in humans. As the chlorine dioxide concentration increases, the time required to achieve sterilization becomes progressively shorter. For example, only 30 minutes were required at 40 mg/l to sterilize *B. atrophaeus* spores at 32 °C. No gaseous chlorine dioxide system is FDA cleared.

Vaporized Peracetic Acid

The sporicidal activity of peracetic acid vapor at 20–80% relative humidity and 25 °C was determined on *B. atrophaeus* spores on paper and glass surfaces. Appreciable activity occurred within 10 minutes of exposure to 1 mg of peracetic acid per liter at 40% or higher relative humidity. No vaporized peracetic acid system is FDA cleared.

Infrared Radiation

An infrared radiation is capable of destroying bacterial spores. Some of the possible advantages of infrared technology include short cycle time, low energy consumption, no cycle residuals, and no toxicologic or environmental effects. This may provide an alternative technology for sterilization of selected heat-resistant instruments. FDA has not cleared IR systems for this purpose.

Sterilization Cycle Verification

A sterilization process should be verified before it is put into use. All steam, ETO, and other low-temperature sterilizers are tested with biological and chemical indicators upon installation, when the sterilizer is relocated, redesigned, after major repair and after a sterilization failure has occurred to ensure they are functioning prior to placing them into routine use. Three consecutive empty steam cycles are run with a biological and chemical indicator in an appropriate test package. Each type of steam cycle used for sterilization (e.g. vacuum-assisted, gravity) is tested separately. In a pre-vacuum steam sterilizer, three consecutive empty cycles are also run with a Bowie–Dick test. The sterilizer is not put back into use until all biological indicators are negative and chemical indicators show a correct end-point response.

Biological and chemical indicator testing is also done for ongoing quality assurance testing of representative samples of actual products being sterilized and product testing when major changes are made in packaging, wraps, or load configuration. Biological and chemical indicators are placed in products, which are processed in a full load. When three consecutive cycles show negative biological indicators and chemical indicators with a correct end point response, you can put the change made into routine use. Items processed during the three evaluation cycles should be quarantined until the test results are negative.

Monitoring

The sterilization procedure should be monitored routinely by using a combination of mechanical, chemical, and biological indicators to evaluate the sterilizing conditions and indirectly the microbiologic status of the processed products or components [8]. The mechanical monitors for steam sterilization include the daily assessment of cycle time and temperature by examining the temperature record chart (or computer printout) and an assessment of pressure via the pressure gauge, including DART and vacuum testing. The mechanical monitors for ETO include time, temperature, and pressure recorders that provide data via computer printouts, gauges, and/or displays.

Chemical indicators are convenient and indicate that the item has been exposed to the sterilization process. Chemical indicators should be used in conjunction with biological indicators, as they indicate sterilization at marginal sterilization time and because only a biological indicator consisting of resistant spores can measure the microbial killing power of the sterilization process. If the internal and/or external indicator suggests inadequate processing, the item should not be used. An air-removal test (Bowie–Dick Test) must be performed daily in an empty dynamic-air-removal sterilizer (e.g. pre-vacuum steam sterilizer) to ensure air removal.

Biological indicators are recognized by most authorities as being closest to the ideal monitors of the sterilization process because they measure the sterilization process directly by using the most resistant microorganisms (i.e. *Bacillus* spores), and not by merely testing the physical and chemical conditions necessary for sterilization. Since the Bacillus spores used in biological indicators are more resistant and present in greater numbers than are the common microbial contaminants, the demonstration that the biological indicator has been inactivated strongly implies that other potential pathogens in the load have been killed.

If biological monitors are not used routinely to monitor the sterilization process, instead, release of sterilizer items is based on monitoring the physical conditions of the sterilization process that is termed "parametric release." Parametric release requires that there is a defined quality system in place at the facility performing the sterilization and that the sterilization process be validated for the items being sterilized. Periodic sterilization control rounds to areas using sterilizers to standardize the sterilizer's use may identify correctable variances in operator competence; documentation of sterilization records, including chemical and biological indicator test results, sterilizer maintenance and wrapping, and load numbering of packs. These rounds also may identify improvement activities to ensure that operators are adhering to established standards.

Chemical characteristics	**Hydrogen peroxide (7.5%)**	**Peracetic acid (0.2%)**	**Glutaraldehyde (≥2.0%)**	**OPA (0.55%)**	**Hydrogen peroxide/peracetic acid (7.35%/0.23%)**
High-level disinfectant claim	30 min @ 20 °C	Not applicable	20–90 min @ 20–25 °C	12 min @ 20 °C, 5 min @ 25 °C in AER	15 min @ 20 °C

OPA, *ortho*-phthalaldehyde (FDA cleared as a high-level disinfectant, included for comparison to other chemical agents used for high-level disinfection).

Sterilization method	Advantages	Disadvantages
Steam	• Nontoxic to patient, staff, environment • Cycle easy to control and monitor • Rapidly microbicidal • Least affected by organic/inorganic soils among sterilization processes listed • Rapid cycle time • Penetrates medical packing, device lumens	• Deleterious for heat-sensitive instruments • Microsurgical instruments damaged by repeated exposure • May leave instruments wet, causing them to rust • Caution: potential for burns
Hydrogen peroxide gas plasma	• Safe for the environment • Leaves no toxic residuals • Cycle time is 28–75 min (varies with model type) and no aeration necessary • Used for heat- and moisture-sensitive items since process temperature <50 °C • Simple to operate, install (208 V outlet), and monitor • Compatible with most medical devices • Only requires electrical outlet	• Cellulose (paper), linens, and liquids cannot be processed • Sterilization chamber size from 1.8 to 9.4 ft^3 total volume (varies with model type) • Some endoscopes or medical devices with long or narrow lumens cannot be processed at this time in the United States (see manufacturer's recommendations for internal diameter and length restrictions) • Requires synthetic packaging (polypropylene wraps, polyolefin pouches) and special container tray • Caution: hydrogen peroxide may be toxic at levels greater than 1-ppm TWA
100% Ethylene oxide (ETO)	• Penetrates packaging materials, device lumens	• Requires aeration time to remove ETO residue

Sterilization method	Advantages	Disadvantages
	• Single-dose cartridge and negative-pressure chamber minimizes the potential for gas leak and ETO exposure • Simple to operate and monitor • Compatible with most medical materials	• Sterilization chamber size from 4.0 to 7.9 ft^3 total volume (varies with model type) • ETO emission regulated by states but catalytic cell removes 99.9% of ETO and converts it to CO_2 and H_2O • ETO cartridges should be stored in flammable liquid storage cabinet • Lengthy cycle/aeration time • **Caution: ETO is toxic, a carcinogen, and flammable**
ETO mixtures 8.6% ETO/91.4% HCFC 10% ETO/90% HCFC 8.5% ETO/91.5% CO_2	• Penetrates medical packaging and many plastics • Compatible with most medical materials • Cycle easy to control and monitor	• Some states (e.g. CA, NY, MI) require ETO emission reduction of 90–99.9% • Chlorofluorocarbon (CFC) (inert gas that eliminates explosion hazard) banned in 1995 • Potential hazards to staff and patients HCFC = hydrochloro-fluorocarbon • Lengthy cycle/aeration time • Caution: ETO is toxic, a carcinogen, and flammable
Peracetic acid	• Rapid cycle time (30–45 min) • Low temperature (50–55 °C) liquid immersion sterilization • Environmentally friendly by-products • Sterilant flows through endoscope that facilitates salt, protein, and microbe removal	• Point-of-use system, no sterile storage • Biological indicator may not be suitable for routine monitoring • Used for immersible instruments only • Some material incompatibility (e.g. aluminum anodized coating becomes dull) • One scope or a small number of instruments processed in a cycle • Caution: potential for serious eye and skin damage (concentrated solution) with contact

3

Sterile Manufacturing Facilities

The focus of sterile manufacturing is on state-of-the-art advanced aseptic techniques in tubing set design to cover challenging areas of aseptic manufacturing and demonstrate expert abilities to manufacture complex products that require borderline limits of process parameters. Strategic business and master planning in advanced aseptic facility design employ barrier technologies such as restricted-access barrier systems (RABS), automated isolators that are equipped with clean-in-place (CIP) and steam-in-place (SIP) with full automation of product filling to mitigate risk of contamination and increase reliability of product integrity [9], and the use of single-use disposable technologies in view of managing global regulatory and compliance requirements.

The use of critical utilities in the manufacture of sterile products is fundamental to ensuring the sterility of that product. Water for injection (WFI), heating ventilation and air conditioning (HVAC), clean steam (CSM), clean compressed air (CCA), and clean nitrogen (N_2) are key utilities that a sterile manufacturing site will rely on in the component preparation, compounding, mixing, and filling operations of a sterile product.

Preservation of pharmaceutical product quality attributes is key in the selection of sterilization method. Although terminal sterilization of finished product is always preferable, selection of an aseptic fill/finish route might be the choice to preserve the efficacy of active pharmaceutical ingredient (API) and compounding excipients. Accordingly, the selection of equipment in sterile manufacturing is determined based on product requirements to ensure efficacy and safety of the product. Irrespective of process selection, trending data of WFI and limulus amebocyte lysate (LAL)-endotoxin is imperative to tracking sterility conditions and process reliability.

Due to potential contamination by personnel, mitigation of risk in sterile operations needs to focus on methods to minimize exposure and contact of sterile materials and components, and processing surfaces, including instrumentation

Sterile Processing of Pharmaceutical Products: Engineering Practice, Validation, and Compliance in Regulated Environments, First Edition. Sam A. Hout.

and equipment. In addition, recent developments in RABS and isolator technologies with advanced programmable logic controller (PLC) automated controls are adding value to quality by design (QbD) and risk management in sterile manufacturing. The cost of such implementations remains a high price tag and should constantly be scrutinized as it relates to pharmaceutical product market penetration and value, and product handling–specific requirements.

The design of clean rooms requires an integrated understanding of the cascading requirements from the aseptic core to the wrap around corridors [10]. Air filtration is channeled through full plenum high-efficiency particulate air (HEPA) system in ISO 5 zones to ensure unidirectional laminar flow over product filling and capping operations. Airflow uniformity control is verified through flow regimes and validated by smoke studies. Differential pressure cascading from high to low ensuring cleanliness of linked filling and lyophilization suites, compounding suites, component preparation suites, aseptic corridors, and transfer rooms or pass-throughs for material flow are all considerations in the design of sterile facilities to continuously establish conditions of effective biocontainment [11].

The focus of main engineering design is to streamline process flow in terms of personnel flow and material flow with the key criteria of mix prevention to avoid cross-contamination [12, 13]. Hence, line clearance, mix prevention, and cleaning validation are major aspects of process design in sterile manufacturing. The selection and construction of pharmaceutical facilities requires an understanding of materials of construction to ensure product life cycle. Therefore, QbD as described in International Society for Pharmaceutical Engineering (ISPE) guidelines and quality risk management based on International Council for Harmonization (ICH) Q8, Q9, and Q10 are important considerations in design criteria.

Sterile manufacturing requires strict control to mitigate the risks of potential contaminants such as particulates, microorganisms, and endotoxins. Cleanliness is key to ensuring that aseptic processing or terminal sterilization is effective in controlling endotoxins levels and minimizing bioburden [14]. There are specific challenges to the various types of APIs or compounded products in processing liquids, colloids, suspensions, emulsions, powders, or semisolids. Considerations regarding if the formulation supports microbial growth, high potency, toxic category, would require design provisions in terms of construction and installation requirements, and protection of product environment with procedure to safely handle and store materials that could potentially harm personnel during manufacturing or in testing laboratories.

4

Sanitary Process Piping and Equipment

Quality assurance (QA) in the fabrication and installation of sanitary process piping is a key core value that requires an organization to be totally committed to providing high-quality products and services. All original, repair, replacement, or modification work shall be performed in strict compliance with Quality Assurance program (QAP). These piping systems affect water for injection (WFI) system and solution transfer system (STS) that are core quality requirements for the formulation of pharmaceutical therapeutic drugs.

This Statement of Policy and Authority has been established to assist and ensure that these objectives are realized and to further deem that the QAP is to be followed by all team members. QA for organization is responsible for the preparation, implementation, revision, and issuance of the Quality Assurance Manual for all field locations. QA has the complete authority and organizational freedom to identify QA issues, initiate actions that result in solutions, and verify the implementation of the solutions to those issues [15]. The Quality Department personnel shall in no way be hindered in accomplishing their tasks or fulfilling their duties with respect to verifications of process piping systems.

Supplier executive management team supports and promotes the QAP in its entirety. All team members shall support the requirements of the QAP while conducting their assigned tasks in the manner that ensures compliance. In doing so, the organization can ensure the delivery of the highest quality products and services.

All levels of the organization are responsible for their assigned activities to ensure compliance with the QAP. Teams may delegate certain permissible tasks of their assigned duties to qualified individuals, but they shall retain the ultimate responsibility. In the event of disagreements, quality management will decide on the problem for resolution in which the resolution itself shall not negate the requirements of the QAP.

Sterile Processing of Pharmaceutical Products: Engineering Practice, Validation, and Compliance in Regulated Environments, First Edition. Sam A. Hout.

The QAP will be reviewed and revised as necessary, on an annual basis, to ensure that current technology, Code or regulatory changes, and improved industry practices are always incorporated into our program.

QA Procedures

QA procedures and guidelines must be in conformance with specified Codes and Standards and in compliance with the current Good Manufacturing Practice (cGMP). QAP addresses the documentation and control of activities performed during the fabrication, installation, and acceptance testing of systems to be used in any facility determined to fall under the requirements of cGMP, as interpreted and enforced by the Food and Drug Administration (FDA) of the United States.

Executive management shall appoint the National QA/QC leadership who shall be responsible for creating and distributing an approved QAP to be implemented by QA/QC personnel and by each project team.

Executive management shall assign operation managers for the project disciplines of operations, engineering, labor, detailing QA/QC project team who are knowledgeable, capable, and have the necessary resources to perform their assigned activities and responsibilities.

QA/QC leadership shall obtain information from the project team as needed to allow for the preparation of site-specific supplement to the standard operating procedure (SOP) manual (as required) which follows the client's specifications, codes, standards, and any other additional contract agreements.

This program is developed in response to the demands of the biotech and pharmaceutical industries for an increased control over manufacturing processes. The manufacturer cannot ensure control of their manufacturing process if the system handling process was not installed in a rigorously documented and controlled manner.

The universally recognized method of maintaining control over a process is through the development, practice, and enforcement of written quality procedures. These quality procedures must govern all aspects of the process system installation that may impact the performance of the manufacturing process. In effect, a compliance program is a rigorously implemented QAP based on GMP regulations and guidelines. These principles have as their goal the production of articles that are fit for their intended uses and may be stated as follows:

- Quality, safety, and effectiveness must be designed and built into the product [16].
- Quality cannot be inspected or tested only into the final product.
- Each step of the manufacturing process must be controlled to maximize the probability that the finished product meets all quality and design specifications.

Standard Operating Procedures – cGMP Installations

This covers the purchasing of material and equipment for critical systems in facilities producing products regulated under the Food, Drug, and Cosmetic Act. A statement on the request that all material must have Material Tests Reports (MTRs) and/or Certificates of Conformances (COCs) as applicable. The following is observed:

- Connection sizes and types
- Surface treatment of wetted parts
- Installation details
- Material composition of wetted parts
- Electrical or pneumatic requirements.

Heat Exchangers

Design and capacities of the exchanger shall be as scheduled on the drawings. Heat exchanger shall be designed to minimize space requirements while maximizing heat transfer efficiency. Heat exchangers for sanitary solutions shall be fabricated in accordance with requirements of the cGMP regulations of the FDA. Minimum allowable design conditions will be 150 psig at 350 °F for both shell side and tube side.

Heat exchangers shall be constructed in accordance with the requirements of the Tubular Exchanger Manufacturers Association (TEMA) in addition to American Society of Mechanical Engineers (ASME) requirements. Heat exchangers shall be of the double-tube sheet, removable U-tube type, and be self-draining in the horizontal or vertical position. Shell or tube side materials in contact with sanitary solutions shall be 316L stainless steel. Baffles, tie rods, spacers, and other internal components (away from sanitary solution contact) shall be of 304L stainless steel. Shell side material (away from sanitary solution contact) shall be 304L stainless steel. End connections shall be 150# ANSI flanged for non-sanitary service, unless noted otherwise on the drawings. End connections shall be 316L stainless steel sanitary clamp type for sanitary solution service. Process tube sheet shall be sealed using a single ethylene propylene diene monomer (EPDM) O-ring located as close to the diversion and drain ports as possible. Drain slots and "gasketless" seals on the sanitary tube sheet shall be designed and fabricated in a manner that will not compromise the performance of the heat exchanger. Process contact surfaces shall be cleanable with no pockets or crevices. For U-tube type exchangers, the bend surface shall not be considered effective when calculating the heat transfer area. Bolts and nuts shall be stainless steel or another noncorroding alloy. Welds in contact with sanitary solutions shall be passivated after fabrication. Suggested Manufacturers: Allegheny Bradford, ITT Standard.

Sanitary Pumps

Pump casing, impeller, shaft, and backplate are to be 316L stainless steel on all sanitary-solution contact surfaces. Pump impellers shall be secured to the drive shaft by retainer clips. Impellers shall be dynamically balanced; shaft shall be statically balanced. Pumps shall be equipped with double-mechanical seals with WFI flush. Suction and discharge shall be of the sanitary clamp type. Casing halves shall be secured with a quick-opening clamp. Casing shall be equipped with drain and 6D valve, or less. Internal weld seams are to be ground smooth and flush with the adjacent surface area. All welds are to be thoroughly blended into the surrounding areas and are to be completely free from pits, ripples undercutting, and other surface defects. Each piece of equipment shall have a nameplate constructed of corrosion-resistant material (e.g. stainless steel), attached and sealed, or attached with a corrosion-resistant wire loop. The nameplate shall contain all required descriptive data and the item identification number for the project. Pump and motor shall be close-coupled. Pump motor shall be TEFC ("Totally Enclosed, Fan Cooled"). Pump internals shall be totally self-cleaning and self-sanitizing when operated at 95 °C with WFI media. Suggested Manufacturers: Fristam pumps, Tri-Clover, Waukeshau.

Sanitary Tanks

Vessels shall be manufactured in accordance with the requirements of the ASME Boiler and Pressure Vessel Code (BPVC), Section VIII, Div., 1 and be stamped with the ASME "U" symbol and be registered with the National Board. Vessels for sanitary service shall also comply with the requirements of cGMPs described in 21 CFR Part 211 as issued by the FDA and also shall conform to ANSI F2.1, Food, Drug, and Beverage Equipment. The number of vessel penetrations is to be minimized. The bottom and top head of the vessel shall be of the ASME flanged and dished shape. The rated maximum operating volume shall not be greater than 90% of the total internal volume of the vessel. The minimum design conditions shall be 50 psig at 250 °F and full vacuum rating. Vessel shell and heads, all fittings used for nozzles, and all other metals in contact with sanitary solutions shall be constructed of 316L stainless steel. All structural steel and insulation jacketing shall be 304 stainless steel. This shall include support frame, trunnions, trunnion pads, crotch pads, cross arms, and "T" bolts.

Exterior surfaces of the tank shall be a #4 finish, with welds blended and polished. Any insulation materials must be certified to be free of chloride ions. Vendor shall submit cleaning and passivation procedures for review and approval. Suggested Manufacturers for stainless steel tanks: DCI, Cherry Burrell, Precision Stainless, Walker Stainless, Feldmeyer.

Instruments

Instruments shall be certified and calibrated. All wetted surfaces shall be of 316L stainless steel material with finish to match the requirements of the tubing system at the point of installation.

Pressure Gauges

(a) Gauges shall be pharmaceutical grade with diaphragm seal and meet 3A sanitary requirements.
(b) Gauges shall have a minimum 3½″ dial, ±2% accuracy, range suitable for the application, glycerin-filled case of 304 SS, tamper-proof re-zeroing for field adjustment, and sanitary clamp fitting.
(c) Suggested Manufacturers: Ashcroft or Anderson.

Pressure Transmitter (Various Sources)

Temperature Gauge

(a) Indicators shall be bimetallic dial type with sanitary thermowell and be suitable for sanitary service.
(b) Thermometer and thermowell shall be of polished 316L stainless steel.
(c) Dial size shall be 3″ minimum, ±1% accuracy with external adjustment, and range shall be suitable for the service location and read in both °F and °C.
(d) Adjustable tilt face option shall be used where the indicator cannot be read from the ground.
(e) Suggested Manufacturers: Tri-Clover, Anderson, Ashcroft.

Temperature Transmitter (Various Sources)

Flow Switch

(a) Suggested Manufacturers: Flowline
    Flow Meter (Rotameter)
    On/Off Valve (various sources)
    Control Valve (various sources)
    Conductivity/Resistivity Meter

(b) Suggested Manufacturers: Thornton Instruments
pH Transmitter (various sources)
Relief Valves (various sources)
Various Recorders (pressure, temperature, pH, etc.)
Transducers (various sources)

A Welder Performance Qualification (WPQ) qualifies a welder for ranges of wall thicknesses, diameters, base materials, and filler materials. Welding process to be used: Examples are GTAW Manual (Man. TIG); GTAW Automatic (Orbital TIG). Base material is to be welded in production. Base material shall be specified by its American Society for Testing and Materials (ASTM) or ASME specification number along with the material type: Examples are SA-240 type 316L, thicknesses of the base material to be welded in production, diameter of the base material to be welded in production, and nominal pipe size (NPS) or outside diameter (OD) shall be specified for diameter.

All welder or welding operator qualification tests on documentation are required to be sent to client QA/QC to generate a WPQ record. A copy of the WPQ is required to be included into the final turnover package.

The "Control Isometric" shall contain shop or field weld, weld number (from the front of the isometric), weld date (date the weld is completed), welder's identification number and initials, weld machine identification (serial number or identification assigned for the project), weld head identification (serial number), purge gas type (typically argon), purge gas number (bulk tank serial number), OD and wall thickness (Nominal), material type (316L, AL6XN, 304L C-22, etc.), joint type (components welded together i.e. 90/T (Tube), and inspection date, results, and inspector's initials and number.

The weld sample coupon log as a minimum shall contain coupon number, coupon size, date and time the coupon was made, Welder Identification who made the coupon, welding machine serial number, welding head serial number, and welder initials. Weld sample coupons are made to assure that the welding machine parameters as set are capable of producing an acceptable weld.

All welds are to be allowed to cool in ambient air. Quenching with water or a wet rag is not permitted. All production welding is to be logged on the back of the Control Isometric Drawing.

Note: The Control Isometric Drawing is the single most important fabrication document used in the Installation Qualification. As such, the project team should safeguard against loss or damage to this document. It is recommended that on large projects, this document be turned in after each shift for review and safekeeping.

Any purge gas used for automatic or manual welding on systems in facilities producing products regulated under the Food, Drug, and Cosmetic Act is only to be used for on-site or fabrication shop testing of gases. Vender certification of gas purity meets the intended use.

The following items are necessary for the use of performance gases: oxygen analyzer, moisture meter, vendor's gas analysis certificate (at time of gas delivery), and bulk gas distribution systems.

The inspection of tube, fittings, and valve body material is key to performance of welded process piping systems, therefore establishing two separate material control areas, which are clearly labeled as follows: One area shall be used exclusively for receiving and inspecting materials. This area shall be clearly labeled as the "Unreleased Material" storage area. Another area shall be used exclusively for holding inspected and accepted material, ready for incorporation into the facility. This area shall be clearly labeled as the "Released Material" storage area. The "Released Material" storage area shall be separated from the "Unreleased Material" storage area by a physical partition. Both storage areas shall protect components from physical damage, mixing, cross-contamination, or environmental contamination. Stainless steel tubing and components shall be stored separate from any carbon steel materials. Equipment required for performing the inspection includes Profilometer, Dial Calipers accurate to "0.001," video Borescope Unit.

Note: It is not recommended that borescope inspection be performed on the long seam of electropolished tubing due to the abrasions that may be caused during the borescope inspection process unless there is physical evidence found during the visual inspection or during fabrication that discontinuities (such as incomplete electropolishing, bursts or lack of penetration/incomplete fusion) may exist.

The combination of documents, including C Cs, MTRs, and/or CMTRs for all equipment and component product contact surfaces, shall include the following information, as a minimum:

(a) ASME Bioprocessing Equipment Standard (BPE) standard, including year date
(b) Material type
(c) Heat number
(d) Chemical composition
(e) American Welding Society (AWS) Classification of filler metal, if used
(f) Alloy designation and material specification of insert, if used
(g) Post weld heat treatment (PWHT) documentation, if applicable
(h) Mechanical properties must be accurate to the raw material specification.

Visually examine 100% of all inspected fittings for cleanliness and obvious defects resulting from poor manufacture or abuse. Twenty percent (20%) of each line item (by Heat Number) will be inspected for the following:

(1) MTR verification
(2) OD
(3) Wall thickness
(4) ID surface finish.

Guidelines for the documentation of slope in cGMP piping systems involve basic technique for the verification of slope in critical systems in facilities producing products regulated under FDA to obtain slope acceptance criteria on critical system lines. Measurements should follow the Control Isometric Drawings for direction of slope. Slope measurement shall be made with a digital level or a digital protractor. The instrument used should be capable of displaying slope in degrees, percent, and in./ft (mm/m).

Slope measurements shall only be made before insulation has been installed, after hangers/pipe supports have been installed, adjusted, and fixed in-place, before the introduction of any fluids, such as liquids or process gases (pure oxygen, nitrogen, steam, etc.), and when the system is at ambient pressure and temperature. For piping or tubing systems, slope measurements shall be made at the following locations: Between hangers/pipe supports, at each change in direction, at any other location deemed necessary by the inspector, such as between welds or any apparent change in slope. Slope should be measured every 10 ft on horizontal runs. Slope measurements may be made on either the top or bottom of the tubing/piping.

For slope measurements made on skids or modules, ensure that the base is level in all directions. Then, make sure that all slope measurements are made relative to the base. Slope shall be verified after the fabricator has completed, or corrected, the piping installation, and set the slope. Document acceptance of the slope is by initialing the section of line checked on the front of the Control Isometric Drawing, initialing in the location for Slope in the title block of the Control Isometric Drawing, completing the information in the "slope acceptance block stamped on the drawing to document the slope verification if applied."

The examination of welds in cGMP piping systems covers standard examination methods suitable for welded critical systems used in the manufacture of products regulated under the FDA such as the use of a borescope with the following features: Probe diameter: 4, 6, 8, or 10 mm, as applicable for desired flexibility versus Rigidity. Probe length: at least long enough to insert through one full length of tubing. Borescope tips that provide for the appropriate viewing angle. Video recording capabilities, and mirrors for direct visual examination of easily accessible welds (as required).

Verify spool/fabrication configuration against the isometric drawing. Verify that the tubing heat number and BPE internal surface finish classification are vibro-etched 4–6″ from the weld on both ends of fabricated tube and that the BPE internal surface finish classification meets the specification requirements for the system fabricated. Verify the weld ID is correct and has Isometric Drawing Number, welder identification, date the weld was made. Verify the BPE markings on the fittings and that the BPE internal surface finish classification meets the specification requirements for the system fabricated. Verify the angular degree of the fitting against the drawing. Verify the heat numbers on the isometric drawing match the heats numbers on the spool/fabrication.

The frequency of borescope examination of the weld interior surface shall be 100% of all shop welds. Video recording is not required for this examination unless specifically stipulated by the project contract documents or Client Specification. One hundred percent (100%) of all accessible field welds will be borescope examined. Alternate methods of weld examination and acceptance (radiography, ultrasonic, etc.) may be utilized with the prior approval.

To establish guidelines for identifying, documenting, and resolution of nonconformances, nonconformance report (NCR) will be initiated by the person that has identified a material or procedural violation that does not conform to the project specifications. Chronic problems or issues that cannot be readily resolved will be documented by an NCR. This is to ensure that the issue is not overlooked and will be resolved. Material NCRs will be initiated for the following violations of the project specifications:

(a) Materials that have been damaged or show evidence of inferior workmanship to the extent they are not suitable for installation in accordance with the project specifications.
(b) Materials that did not come with the required documentation form the vendor.
(c) Materials that do not conform to the required chemical, physical, or mechanical properties as detailed within the project specifications (i.e. ASTM specifications).
(d) Materials that do not have the required markings per the Client Specification or the BPE.

Procedural NCRs will be initiated for the following violations of the project specifications:

(a) Materials receipt, storage, and handling techniques
(b) Fabrication, installation, and testing techniques
(c) Documentation (consistent, accurate, and controlled)
(d) Protocol (communications, transmittals, and turnover information).

Standard Operating Procedures – cGMP Installations

To provide guidelines for the preparation of field installation drawings in conformance with the relevant contract documents and the requirements of cGMP, a list of all contract drawings with the latest revision is required. Review all contract drawings to verify extent of engineered field installation drawing requirements. Keep the pharmaceutical isometric tracking log updated as drawings are completed, submitted, or approved.

Prepare a coordinated piping orthographic drawing (if required) prior to starting detailed isometric drawings. The orthographic should include column line references, and any available background information, including walls, ductwork, mechanical piping, plumbing piping, sprinkler piping, columns, structural steel, pipe chases and floor openings, electrical conduit racks, cable tray, major supports, equipment location and connection details, line identification, working point elevations, direction of grade, and isometric sheet breaks. For critical systems such as WFI, a total system isometric should be prepared to facilitate review by owner and regulatory authorities.

The following information is included on isometric drawings: Specification, system, material list, insulation thickness, location within building, line number and size, orthographic reference, Piping and Instrumentation Diagram (P&ID) reference, continuation, weld bubbles, column line reference, instrumentation to be installed, equipment centerlines, elevations at the end of each line and at elevation changes, equipment ID, accurate bill of materials (BOMs), include part numbers on valves, show all dimensions and elevations from centerline, and show supports and anchor points.

The following critical systems include:

- Direction and degree of slope
- Valve angle and orientation to ensure drain ability
- Minimum distance requirements
- Route piping in the most direct manner
- Maintain a clearance between the piping and any other object a minimum of 3″
- Maximum distance requirements, if any (Maximum allowable dead-leg is 6D from main line centerline).

To provide guidelines for the revision of field installation drawings in conformance with the requirements of cGMP, this covers the revision of installation drawings for only those systems with direct product contact or with incidental product contact. Determine the drawings and/or systems affected by the changes to specifications, scope of work, or field conditions. Obtain the pharmaceutical isometric tracking log. Review and mark the affected drawings on these logs as "On Hold For Revision." Consult with proper authorities to arrive at the best solution to the revised conditions.

The sequence of field installation is to be determined after receiving isometric spools from the fabrication shop "cGMP Piping System Fabrication." Use the piping orthographic drawing and isometric drawing marked "Issued for Construction." Use permanent supports during installation of systems. Where it is not feasible to use permanent supports, and temporary supports are required, isolate the stainless steel tubing from contact with carbon steel. Beginning at a fixed point (major equipment or header), verify elevations and break dimensions against isometric. Continue placing spools in supports until a complete line is formed between fixed points. Verify slope of line.

Guidelines for purging cGMP piping systems cover purging both manual and automatic orbital welding of systems used in the manufacturing of FDA-regulated products. Assemble purge manifold using separate flow meters for each line being purged and for each welding machine. Test purge manifold for leaks using argon gas. Attach purge gas source to the fabrication or system to be welded. For OD less than 1″: Use S.S. "Swagelok" connections with nylon ferrules. For OD 1″ and larger: Use plastic cap with S.S. "Swagelok" bulk-head fitting, fastened to tube end with pressure-sensitive tape. For line sizes 1½″ and larger, the use of a diffusion cup or purge block is recommended if the joint to be welded or tacked is within 2 ft of the purge source. Force air out of the fabrication or system using the selected purge gas. Use a handheld O_2 meter to check the quality of the exiting purge gas. Begin welding when confident that gas purge is adequate. The heat number and BPE surface finish classification shall be situated so that it will not be consumed or obscured by subsequent welding, cutting, or facing operations.

Welds not accessible by the borescope probe on the jobsite shall be categorized as "blind" welds. Blind welds are those that cannot be visually/remotely inspected and shall be kept to a minimum through good planning and weld sequencing. Fabrication of non-stainless steel components should not be permitted in the vicinity of high-purity stainless steel welding to avoid cross-contamination by airborne particles from cutting, grinding, etc.

To provide guidelines for the pre-cleaning, passivation, post-rinsing, and neutralization of stainless steel cGMP piping systems after flushing and testing operations are complete and prior to activation of the piping system covers all stainless steel piping in facilities producing FDA-governed products. The piping system shall have been flushed clean. The piping system shall have successfully passed a pressure test.

Check the system integrity by filling with treated filtered water and circulating the fluid while inspecting the pipe, fittings, and equipment for leaks. Size the circulating pump to maintain the flowrate at a linear bulk velocity of at least 2 ft/s. Maintaining a flowrate of 5 ft/s is preferred. Flush this water from the system through low-point drains and strainers to check for particles exiting the system. Continue to drain from the low points until no additional particles are found in the strainers. Add additional water to maintain sufficient water volume as needed. Prepare the pre-cleaning solution using the appropriate mixture as specified by the manufacturer. The solution used may involve a detergent wash of one of the following types:

(1) Surfactant solution; anionic or nonionic
(2) Alkaline solution; caustic-soda-based cleaner or sodium gluconate (The gluconate formulations are quite versatile and can be used hot, cold, as soaks, sprays, ultrasonically or electrolytically.).

Exercise care while mixing chemicals and use proper safety gear such as, but not limited to, gloves and a full-face shield. Prepare the solution in a separate 300-series stainless steel tank and inject the mixture into the system being cautious not to exceed the maximum concentration of the designated specific weight percent within the system. Use only new "chemical reagent" grade material purchased for this particular application. Recycled material from previous operations is not acceptable. Hydrotest water may be used in the precleaning process if mixing can be done without exceeding the specified maximum concentrations. File purchase order and the inspection report for the reagent used with the passivation documentation.

Circulate the cleaning solution throughout the system for a minimum of one hour (or as per manufacturer's recommendations) after the specified temperature has been reached (usually $160 \pm 20\,°F$). Cycle flush all valves that do not discharge to the atmosphere during the cleaning cycle. Cycle flush all valves that discharge to the atmosphere during the cleaning cycle. Record all information pertaining to the passivation process. Record the circulating start time and finish time.

(a) Record the solution temperature
(b) Record the estimated linear bulk velocity through all tubing
(c) Document the time and position of all valves as they are being cycle-flushed
(d) Record the diaphragm valves removed for cleaning and note that they have been replaced
(e) File these data sheets with the passivation documentation form upon their completion.

Flush the system with filtered treated water at $120 \pm 20\,°F$

(a) Cycle all valves during the flushing process
(b) Exercise care to assure that the precleaning solution is thoroughly rinsed from the system prior to the start of passivation.

Remove all diaphragm valves from the system and clean any imbedded metal from the diaphragms. The second step to the process is the acid wash treatment for passivation. The passivation solution is one of two types:

(a) Mineral acid: Nitric acid diluted with water to 10–30% used at 120–160 °F for a typical time of four hours
(b) Organic acid: A chelating agent, ammonium citrate acid, diluted with water to 5–10% at 120–175 °F for a typical time of four to six hours.

Prepare the passivation acid solution to the required mixture as specified by the chemical manufacturer. Circulate the passivation solution throughout the system for the specified time at temperature and at the same required flowrates as specified in precleaning. The next step is to post-rinse and neutralize the piping system by removing the acid wash solution. Flush the system with filtered treated water at $120 \pm 20\,°F$. Check the pH with a pH meter or with litmus paper to assure that the pH of the remaining solution is approximately 7. The system may be blown dry using a supply of clean dry air or an inert gas. The system may be steam sanitized and started into service.

To provide guidelines for thermal insulation installation procedures for equipment, piping, and tubing in cGMP systems. Do NOT insulate exposed chrome-plated plumbing fixture piping, most drain lines, buried piping, or fire protection piping. Insulation shall be installed after slope and pressure tests acceptance. DO NOT insulate over equipment handholes, manways, cleanouts, nameplates, or identification stamping unless noted on the drawings to provide removable insulation sections.

To provide standard weld acceptance criteria in accordance with ASME BPE for welds on tube that are contained in cGMP piping systems. Misalignment of a maximum of 15% of the nominal wall thickness except for 4″ Tube may have a maximum of 0.015″ misalignment on the OD surface, and 6″ Tube may have a maximum of 0.030″ misalignment on the OD surface. Concavity allowance is maximum of 10% of the nominal wall thickness. OD and ID concavity shall be such that the wall thickness is not reduced below the minimum thickness required. Cracks, fusion, undercut, arc strikes, or incomplete penetration are none allowed.

To provide standard acceptance criteria for purge gases used in the welding of cGMP piping systems, standard purge acceptance criteria for high-purity argon gas total overall argon purity: 99.998% minimum. Maximum Impurities by volume:

(a) Oxygen: Less than 3.0 ppm
(b) Moisture: Less than 2.0 ppm
(c) Nitrogen: Less than 5.0 ppm
(d) Total hydrocarbons: Less than 1.0 ppm

Fittings and process components, including but not limited to tubing, valves, pumps, filter housings, and instrumentation, shall be visually examined for manufacturer's name, logo, or trademark, material type, description, including size and configuration, heat number/code, product contact surface finish symbol, reference to ASME BPE, and pressure ratings for valves.

ASME BPE criterion shown next is for fittings and process components. For tubing tolerances, use ASTM A270 Supplement 2:

OD Tolerances

Size	OD variation
¼″–1″	(+).005″
1½″–2″	(+).008″
2½″–3″	(+).010″
4″ OD	(+).015″
6″ OD	(+).030″

Wall Thickness Tolerances

Size	Wall TK (mech. polish)		Wall TK (electropolished)	
1/4″ and 3/8″ OD	(+).003	(−).004″	(+).003	(−).006″
1/2″–3″ OD	(+).005	(−).008″	(+).005	(−).010″
4″ OD	(+).008	(−).010″	(+).008	(−).012″
6″ OD″	(+).015	(−).015″	(+).015	(−).017″

Size	Nominal wall thickness
1/4″ and 3/8″ OD	0.035″
1/2″–3″ OD	0.065″
4″ OD	0.083″
6″ OD	0.109″

Surface Finish

Acceptance criteria for stainless steel and higher alloy mechanically polished product contact surfaces shall be in accordance with ASME BPE-2009 Table SF1. Acceptance criteria for stainless steel and higher alloy mechanically polished and electropolished product contact surface finish shall be in accordance with ASME BPE 2009 Table SF2. Product contact surface finishes (as required by the Client/Owner), shall meet the Ra criteria as specified.

Ra Readings for Product Contact Surfaces

Designation	µin.	µm
Mechanically Polished Ra Max Surface		
SF0	Finish requirement	Finish requirement
SF1	20	0.51
SF2	25	0.64
SF3	30	0.76
Mechanically Polished and Electropolished Ra Max		
SF4	15	0.38
SF5	20	0.51
SF6	25	0.64

Note: No single Ra reading shall exceed the Ra max. value in this table.

MTRs are required for all tube, weld fittings, components, and component parts. Separate test line item required for each weld-end component of fitting. Any material that is not properly traceable to its ASTM specification, grade, and heat number shall not be used.

Material 316L	%
Carbon, max. (C)	0.035
Chromium (Cr)	16.00–18.00
Manganese, max. (Mn)	2.0
Molybdenum (Mo)	2.00–3.00
Nickel (Ni)	10.00–15.00
Phosphorus, max. (P)	0.045
Silicon, max. (Si)	1.00
Sulfur (S)	0.005–0.017

To provide standard acceptance criteria for slopes required in the installation of cGMP piping systems. Standard slope of horizontal lines shall be 1/8″ per foot. Isometric drawings shall note slope direction. There are exceptions to the 1/8″ per foot slope criteria. Examples of acceptable exceptions where the piping is to be installed flat include:

- Spray ball piping in between spray ball POC's, i.e. from tank nozzle to tank nozzle.
- Transfer panel nozzles and piping immediately adjacent to panel connections.
- In general, in tight sections of pipe where fitting-to-fitting components are installed horizontally to equipment, flat piping is acceptable.

Standard slope is not to be considered a maximum. Increased slope is acceptable when necessary for fabrication routing.

To provide standard acceptance criteria for test pressures, it holds duration required in the installation of cGMP piping systems. This covers lines being tested to assure safety and reliability by either hydrostatic or pneumatic means in accordance with Pressure Testing Procedures. For hydrostatic testing, pressure shall be at least one and one-half times (1.50) the maximum operating pressure or 50 psig, whichever is greater. The pneumatic test pressure shall be at one and one-tenth times (1.10) the maximum operating pressure. Ensure that the system test pressure is limited to a lower value if any nonremovable components in the system would be damaged by exposure to the above-described test pressure and cannot be isolated from the test pressure. In no instance, will a test pressure lower than system operating pressure be acceptable? Minimum length of time a system shall be held at pressure before acceptance of the system integrity shall be 10 minutes (ASME Code B31.3, para. 345.2.2).

To provide standard acceptance criteria for defining "minimum" and "dead leg" dimensions as specified on the installation drawings of cGMP piping systems. This section is applicable to all high-purity piping installed in cGMP high-purity piping distribution systems. "Dead-End Branch" is a piping segment through which no fluid flows during a 24-hour period or during a process step, whichever is shorter. This situation is present in applications of steam in place (SIP), clean-in-place (CIP), Clean Steam Distribution, USP Recirculating Water, and process lines. "Six-*d*" is measured from the axis of the main run to the end point of the dead-end branch. The referenced diameter "*d*" is the nominal diameter of the smaller branch. Applications for "Six-*d*" branch dimensions are seen in steam sanitization of process, Clean Steam Distribution, USP Recirculating Water, and Process Lines. "One-and-a-half-*d*" is measured from the nearside edge of the main run to the end point of the dead-end branch. The referenced diameter "*d*" is the nominal diameter of the smaller branch line. "1.5-*d*" branch dimensions are most often seen on process lines that will not be cleaned through during the routine cleaning cycle. "Minimum" notations on a P&ID, standard sheet, or detail sheet define locations where "6*d*" or "1.5*d*" dimensions specifically apply. This situation is present in applications of SIP, CIP, Clean Steam Distribution, USP Recirculating Water, and Process Lines. "Dead leg" is an area of entrapment in a vessel or piping run that could lead to contamination of the product.

(a) Dead legs are measured by the term "L/D", where L is the leg extension from the ID wall normal to the flow pattern or direction, and D is the ID of the extension or leg of a tubing fitting or the nominal dimension of a valve or instrument.
(b) Tables SD-1 and SD-2 of ASME BPE indicate L/D values based on the BPE definition for various tubing geometrics and configurations.

Welding

Implementing welding design, fabrication, and inspection requirements in a site facility to ensure that welded joints:

- Do not alter the safety, identity, strength, quality, or purity of product solution by:
  - promoting biological growth
  - being reactive, additive, or absorptive
- Comply with industry safety standards for welding.

Applicability

The procedure applies to plant personnel designing, procuring/contracting, fabricating, and inspecting hygienic (DW/WFI, STS, or process clean air [PCA] contact) and/or safety-significant (e.g. structures, piping designed to operate at pressures at or above 15 psig) weldments on new systems or substantial renovations or repairs to existing system.

This establishes design requirements for welded joints and implements the site welding program, which is structured to comply with AWS standards for welding and inspection.

- Client at site can implement AWS code requirements using plant personnel, where ASME requirements may be used as supplemental guidance by plant personnel in meeting AWS design requirements.
- The only entities that can weld, repair welds, or inspect welds on pressure vessels (PVs), boilers, or structural members thereof (e.g. supports) are outside contractors or suppliers holding an ASME "R" certificate (*certified for ASME Section I or VIII, as applicable, per ASME Section IX*).

When there is a conflict between this SOP and a design code or specification to which the equipment or item is being constructed, the conflict shall be documented in the weld record(s) and the applicable design specification or code shall govern.

American Society of Mechanical Engineers (ASME) Codes:

- o BPE:2016, Bioprocessing Equipment (industry guidance)
- o Boiler and Pressure Vessel Code (BPVC) Section VIII:2017, Rules for Construction of Pressure Vessels
- o BPVC Section I:2017, Rules for Construction of Power Boilers
- o BPVC Section IX:2017, Welding and Brazing Qualifications

Note: Section IX of the BPVC relates to the qualification of welders and welding operators as well as welding procedures employed in accordance with the BPVC and B31.3, Process Piping.

- o B31.3:2016, Process Piping (pipe welding)
- o B46.1:2009, Surface Texture
- o ASME Codes in Tech Street: https://subscriptions.techstreet.com/subscriptions/index

American National Standard (ANSI):

- o Z49.1:2012, Safety in Welding, Cutting, and Allied Processes: https://www.aws.org/standards/page/ansi-z491

Definitions

Term	Definition
Arc strike	A discontinuity resulting from an arc, consisting of any localized remelted metal, heat-affected metal, or change in the surface profile of any metal object
Burn-through	A hole or depression in the root bead of a single-groove weld due to excess penetration
Crack	A fracture-type discontinuity characterized by a sharp tip and high ratio of length and width to opening displacement. Reference Appendices A–E
Crater crack	A crack initiated and localized within a crater
Clean/pure steam	Steam used for sanitization, sterilization, or process in which steam comes in contact with product or product contact surfaces
Compendial water	Pharmaceutical water that meets the requirements and recommendations stated in applicable pharmacopoeia documents
Contractor	Includes manufacturers, contractors, subcontractors, fabricators, erectors, or other organizations that fabricate weldments that are intended to meet the requirements of this specification
Crevice	Narrow gap between a piece of metal and another piece of metal
Critical systems	A facility system that has the potential to directly impact the quality of the product produced such as, but not limited to Process Water, Clean Steam, Compressed Air, Compressed Gas, and Pest Control systems
Defect	A discontinuity or discontinuities that by nature or accumulated effect render a part or product unable to meet minimum applicable acceptance standards or specifications. The term designates reject ability
Discoloration	Any change in surface color from that of the base metal. Usually associated with oxidation occurring on the weld and heat-affected zone on the outside and inside diameter of the weld joint as a result of heating the metal during welding. Colors may range from bluish-gray to deep blue and from pale straw color to a black crusty coating
Discontinuity	Interruption of the typical structure of a weldment, such as a lack of homogeneity in the mechanical, metallurgical, or physical characteristics of the material or weldment. A discontinuity is not necessarily a defect
Essential variable	A variable in which a change beyond that described in "welding variables" section of subject code is considered to affect the mechanical properties of the weldment and shall require requalification of the welding procedure specification (WPS) if changed

Term	Definition
Fabricator	Includes contractors, subcontractors, erectors, or any organizations that produce weldments that are intended to meet the requirements of this specification
Fissure	Crack in weld
Fusion	The melting together of filler metal and base metal, or of base metal only that results in coalescence. Reference Appendix E for details
Gas tungsten arc welding (GTAW) aka tungsten inert gas (TIG) welding	An arc welding process that uses a non-consumable tungsten electrode to produce the weld. Reference illustration at right. 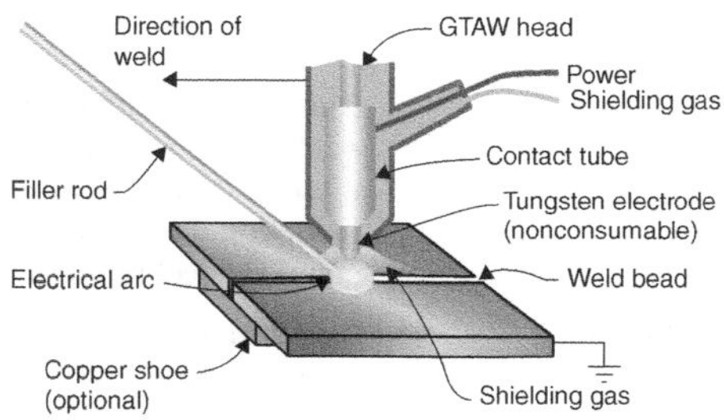
Hygienic	Pertaining to equipment and piping systems that by design, materials of construction, and operation provide for the maintenance of cleanliness so that products produced by these systems will not adversely affect human health
Hygienic design	A design employing materials and construction practices that provides surfaces, both product contact and non-product contact, which are non-particulate generating and non-deteriorating under conditions of intended use and which are smooth, readily accessible, and easily cleanable utilizing standard cleaning practices. May also be referred to as sanitary design
Hygienic service (*at Site*)	Material/surfaces in contact with fluids that could end up in an IV Bag: production use distilled water (DW)/water for injection (WFI), solution (STS), or process air (PCA)
Icicles	Localized regions of excessive penetration, which usually appear as long, narrow portions of weld metal on the weld under-bead
In operation	Condition where the cleanroom or classified area installation is functioning, and a specified number of personnel are present and working as required
Inclusion	Entrapped foreign solid material, such as slag, flux, tungsten, or oxide
Indication	A condition or an anomaly, of a localized area, that has not been classified as being accepted or rejected
In-house fabrication	Welding and weld inspection performed by client in-house personnel (not by supplier or contactor) in accordance with the Site welding program as defined in this SOP

Term	Definition
Joint penetration	The depth that a weld extends from its face to a joint, exclusive of reinforcement. Reference Appendix B for details
Non-compendial water	Water that meets the requirements of drinking water, but with additional treatment to meet specified process requirements
Orbital welding	Automatic or mechanized welding of tube or pipe in which a welding arc rotates (orbits) around a stationary weld joint
Overlap (fusion welding)	The protrusion of weld metal beyond the weld toe or weld root
Oxidation	The formation of an oxide layer on a metal surface. When oxidation occurs as a result of welding, it is visible as a discoloration. The discoloration or a heat tint produced by oxidation has been associated with the onset of corrosion in stainless steel piping systems
Passivation	Removal of exogenous iron or iron compounds from the surface of stainless steels and higher alloys by means of a chemical dissolution, most typically by a treatment with an acid solution that will remove the surface contamination and enhance the formation of the passive layer
Pipe	A pressure-tight cylinder is used to convey a fluid or to transmit a fluid pressure, ordinarily designated "pipe" in applicable material specifications. Materials designated "tube" or "tubing" in the specifications are treated as pipe when intended for pressure service
Pipe support	Pipe-supporting elements consist of fixtures and structural attachments as follows: (a) Fixtures: Fixtures include elements that transfer the load from the pipe or structural attachment to the supporting structure or equipment. They include hanging type fixtures, such as hanger rods, spring hangers, sway braces, counterweights, turnbuckles, struts, chains, guides, and anchors; and bearing type fixtures, such as saddles, bases, rollers, brackets, and sliding supports (b) Structural attachments: Structural attachments include elements that are welded, bolted, or clamped to the pipe, such as clips, lugs, rings, clamps, clevises, straps, and skirts
Piping	Assemblies of piping components used to convey, distribute, mix, separate, discharge, meter, control, or snub fluid flows. Piping also includes pipe-supporting elements but does not include support structures, such as building frames, bents, foundations, or equipment

Term	Definition
Pit	A surface discontinuity the depth of which exceeds 25% of the smallest surface dimension, and the length of which does not exceed three times its width
Plant air	Compressed air that does not come in direct or indirect contact with product or components and has no impact on product quality
Porosity	Cavity-type discontinuities formed by gas entrapment during solidification or in a thermal spray deposit
Prequalified welding procedure specification (PWPS)	A welding procedure specification in compliance with the stipulated conditions of a particular welding code or specification and therefore acceptable for use under that code or specification without a requirement for qualification testing
Pressure boundary (*for pressure vessels and pressure service piping*)	• The welding end of the first circumferential joint for welded connections • The first threaded joint for screwed connections • The face of the first flange for bolted, flanged connections • The first sealing surface for proprietary connections or fittings • Nonpressure parts welded directly to a pressure-retaining surface • Pressure retaining covers and pressure relief devices
Pressure service	Items designed for peak operation >15 psig. Reference "pressure boundary" definition
Pressure vessel (PV)	Vessel (*not a pipe*) designed for peak operation >15 psig and inside diameter to be ≥6″. Reference "pressure boundary" definition
Process air (PCA)	Compressed air that is used in a manufacturing process that may have direct product contact or may affect product quality
Process water	Water used in manufacturing that may have product quality impact
Product	Any material in any form (liquid, powder, solid, slurry, etc.) for which the tank, vessel, or other pieces of sanitary equipment is designed to store, contain, or process
Product contact surface	Any surface that is exposed to the product or any surface from which materials may drain, drop, or be drawn into the product
Roughness average (Ra)	The numerical value derived from the measurement of surface roughness on a given material. (*Log of the arithmetic mean of the surface profile*)

Term	Definition
Safety-significant service	System or item whose failure could result in significant personal injury (e.g. loss of life) and fluid systems designed for peak operation >15 psig: • Steam/heating boilers, PVs, and pressure service piping • Structures, including supports, relied upon for human safety
Sanitary weld	A weld made by purge gas welding to preclude carbide precipitation and oxide formation. The weld must be free of pits and fissures as viewed with the naked eye
Sanitization	Process to reduce the number of all forms of microbial life, including fungi, viruses, and all forms of bacteria, including their spores
Sanitary design	A system of design that meets standards, specification, codes, regulatory and industrial guidelines, and acceptable engineering design methods to reach a degree of sanitation required by pharmaceutical processing. May also be referred to as hygienic design
Sanitary system	System must be designed to minimize microbial proliferation and prevent chemical contamination. Systems must contain smooth surfaces, be drainable, contain crevice-free connections, and prevent backflow from systems of lesser quality. Design must be approved for use by and/or meet the requirements of governing regulatory agency and meet the requirements of a recognized industry standards
Slag	A nonmetallic product resulting from the mutual dissolution of flux and nonmetallic impurities in some welding processes
Supplied item	Item provided in final fabricated state. Fabrication conducted by supplier in their facility in accordance with the supplier's welding program
Surface inclusion	Particles of foreign material in a metallic matrix. The particles are compounds such as oxides, sulfides, or silicates but may be a substance foreign to and essentially insoluble in the matrix
Surface residuals	A foreign substance that adheres to a surface by chemical reaction, adhesion, absorption, or ionic bonding (for example, corrosion, rouging, and staining)
Tank	Vessel (*not a pipe*) designed for peak operation ≤15 psi
Tube/tubing	See definition for "pipe"
Tungsten inert gas (TIG) welding	See definition for Gas Tungsten Arc Welding (GTAW)

Term	Definition
Undercut	A groove melted into the base metal adjacent to the weld toe or weld root and left unfilled by weld metal
Underfill	A groove weld condition in which the weld face or root surface is below the adjacent surface of the base metal
Undersized	Weld that is smaller than specified size for more than 10% of the length unless otherwise allowed or required by design. Reference Appendix C for details
Weld map	An illustration that assigns information to a weld joint (Weld ID) to enable easy identification of its design (WPS), production (welders, date welded), quality (inspection, date inspected), and traceability (heat #s of materials joined)
Welding procedure qualification record (WPQR)	A record of welding variables used to produce an acceptable test weldment and the results of tests conducted on the weldment to qualify a welding procedure specification
Welding procedure specification (WPS)	A document providing the required welding variables for a specific application to assure repeatability by properly trained welders and welding operators

Process/Procedure

Hygienic and safety-significant welding shall be performed and inspected by qualified welders and inspectors per requirements established by engineering and documented in weld records in accordance with the client corporate specifications and local SOPs in addition to corporate standards for typical cases.

(a) Determine if welding applications meet the criteria for "hygienic" and/or "safety-significant" service so that applicable requirements can be applied and communicated to fabricator.

(b) Establish technical (design) and quality (inspection) requirements for welding/weld repair in accordance with Appendix A and applicable portions of this SOP regarding design, welding, repair, finishing, cleaning, inspection, records, and additional required documentation.

   i. Material of construction (MOC) selection shall be based on compatibility with operating environment and cleaning methods and requirements for maintenance and repair.

   ii. Welding symbols in design drawings shall be in accordance with AWS A2.4 or other industry standards such as ASME.

iii. *Hygienic design*:
- All surfaces shall not be reactive, additive, or absorptive so as to alter the safety, identity, strength, quality, or purity of the product solution quality beyond the established specifications.
- Materials and design should facilitate room environmental conditions by minimizing degradation and particulate generation and inhibit microbial growth. The following summarizes requirements and standards for materials and surface finish (*reference only, verify details within governing spec, SOP, and standard and apply more stringent if multiple/conflicting*):

Quality attribute	Application(s)	Requirement
Metal type and grade	DW/WFI	316L stainless steel (*CMTR required for piping and valves*)
	STS	316L stainless steel
	PCA piping and components, including fittings in mixing, filling, and over-pouch	316L stainless steel
	PCA piping outside wet clean room	304(L) or 316(L) unless otherwise approved as noted on drawing PAS-0300-01 (PCA header system)
Surface finish	DW/WFI (*contact materials, weld surface excluded*)	Interior surface finish of ≤30 µin. Ra (0.8 µm Ra)
	STS (*longitudinal pipe welds*)	
	STS tanks (*internal welds*)	

DW/WFI and STS piping and process connections shall be per industry standard and either sanitary ferrule or welded.

Welding in Controlled Environments

(a) For welding, grinding, or cutting (*potential for particulate generation/release*) in a controlled environment, apply requirements in CritSys5A, Appendix B, as applicable.

(b) If emergency welding/grinding is required during production hours on STS piping, tanks, or equipment, preapproval must be obtained prior to beginning work in accordance with CRIT SYS 5E, Solution Transmission System.

Pressure Vessels

(a) Vessel repairs:

i. Repair of vessel welds and vessel structural member weldments affect the certification of a PV, as all domestic PVs are ASME-coded and -stamped PVs.

ii. Repairs must be performed by an ASME Code welder from a company with a National Board "R" stamp and so evidenced.

(b) Vessel modifications: Modifications affecting vessel structural integrity, pressure-bearing components, or safety devices must not be performed without appropriate prior approval. Modifications or repairs to a vessel must be done by a certified welder, must require inspection by an approved insurance inspector for PVs, and must conform to applicable ASME and/or local authority regulations. Inspections required or done by inspectors must be performed BEFORE returning an affected vessel to service.

(c) Hydrostatic tests are made at 1.3 times their indicated working pressure, unless otherwise required by applicable SOP or design requirements in accordance with ASME Boiler and Pressure Vessel Code (BPVC) VIII (Division 1, Part UG-99).

Boilers

(a) Welding on the boilers must be performed by certified welder for boilers (ASME BPVC Section I, Rules for Construction of Power Boilers) in accordance with ASME BPVC Section IX, Welding and Brazing Qualifications.

(b) Welding permitted by plant maintenance personnel starts at the first flange of the main header post the boiler stopcock.

Records (General)

Weld record is required to document that technical and quality requirements were fulfilled with welder(s) and inspector(s) who are qualified for applicable welding processes/procedures and inspection criteria in accordance with requirements.

Welding

(i) Equipment (e.g. press brake) and tools (e.g. clamp, brush) used in the welding and preparation of stainless steel shall not have been used on carbon steel.

(ii) Stainless steel items and welding work areas shall be segregated from the carbon steel welding area unless passivated after welding.

(a) *Pipe/tube welding (DW/WFI and STS only)*:

i. Pipe welding should be performed using an automatic welding machine (orbital welder), when practical.

   Note: Ensure any plugs are removed after welding and prior to sanitization.

ii. Manual welds on piping/tubing are allowed where machine orbital welding is not feasible and prior approval is obtained by a Client Quality Representative (documented on the weld record).

(b) *Purge gas*: Components being welded must be purged with argon gas to remove oxygen in the weld area. The argon gas must be at least 99.99% pure and should contain no more than 2-ppm oxygen and 1-ppm moisture.

(c) *Sample welds/coupons:*
   i. Sample weld(s)/coupon(s) shall be produced in the following cases and visual inspection in accordance with applicable requirements prior to proceeding (to minimize potential for rework):
      - Start of welding for each day a welder is welding (e.g. beginning of each work shift).
      - Change in purge bottle, WPS, or welding machine (orbital welder) setup: pipe size, setting, or head.
   ii. Inspection results of sample welds/coupons are to be documented in the weld record.
   iii. Weld samples/coupons and printouts may be discarded after of inspection results are documented in the weld record.

Post-Weld Cleaning and Finishing

(a) Welds must be cleaned of particulates, residue, and unacceptable weld discoloration. Color chart in Appendix I should be used as a color guide, if applicable, where Examples 4–10 in Appendix I are unacceptable. The weld bead shall be free of discoloration on the inside. Slight discoloration (light straw or light blue color, or hueing) in the heat-affected zone and discoloration that is tight to the surface and unable to be removed during normal operation is acceptable.
(b) Surfaces should be finished suitably smooth, nonporous, and free from cracks and cavities. Welds must be continuous (*no profile interruption*) without cracks or crevices.
(c) Tank welds shall be ground smooth and flush without visual pits or defects.
(d) Additional Requirements for STS:
   i. Longitudinal piping weld (seam) surface finish shall be ≤30 Ra (0.8 μm).
   ii. STS Tank welds shall be ground smooth and flush in accordance with GQP-08-10.

Inspection

(a) *Inspector*: Qualified/certified client personnel or third party industry certified inspectors shall inspect welds.
(b) *Inspection plan*: Visually inspect using the following inspection plan criteria:
   i. 100% Inspection: The following types of welds shall be inspected at 100% per acceptance criteria:
      - Manual welds conducted on piping/tubing.
        Note: Preapproval is required by a Quality Rep for manual welds on DW/WFI and STS piping/tubing per Section 6.4.1.b.ii.
        Note: Surfaces of welds that are not easily accessible should be inspected/viewed using an appropriate borescope, if possible.
      - Exterior of welds (external portions of welds).

- Interior of pipe/tubing welds that are easily assessable for visual inspection (e.g. elbows, ferrules, flanges) using mirror and/or flashlight.
- Interior welds on storage tanks.

ii. Sampling inspection: A minimum of 20% of machine/orbital welds associated with the same weld record must be inspected internally and documented.
- Machine/orbital weld internal surfaces that are inaccessible to a borescope or visual inspection are not required to be inspected if the specific weld process and setup is documented as prior qualified with production of an acceptable sample weld/coupon in accordance with Section 6.4.1.d (*existing requirement provides reasonable assurance of required quality*).
- If a failure is noted upon internal inspection of 20% of the sampled welds, an additional 20% (or more based on deviation severity) of welds associated with the weld record must be inspected per acceptance criteria. If a failure is noted in the second sampling of welds, then all remaining welds associated with the weld record shall be internally inspected per acceptance criteria.

(c) *Acceptance criteria*: Welds are deemed acceptable if they meet the following requirements.

i. Welds meet established/specified design requirements (e.g. type, size), reference Section 6.1.

ii. Welds are cleaned and finished per Section 6.4.2, as applicable.

iii. Welds are free of defects such as the following (reference Section 4 for definitions):
- Arc strikes.
- Burn-through.
- Cracks (fissures) or Crater Cracks.
- Crevices (gap between metal pieces).
- Embedded or protruding material.
- Icicles (excessive penetration).
- Incomplete fusion (reference Appendix E).
- Overlap, undercut, or underfill.
- Pits or porosity (rounded indications) on wetted surfaces.
  Note: For nonwetted surfaces: porosity under 1/32″ (diameter) and less than three rounded indications separated by 1/16″ or less edge to edge in any direction is permitted.
- Undersized (reference Appendix C).

iv. Weld penetration: Butt welds shall be full penetration to the inside of the wall thickness. In the case of two different wall thicknesses, the weld shall penetrate to the inside of the thickest wall. Tack welds must be fully consumed by the process.

v. Joints: Joints shall be square facing and properly aligned. The maximum misalignment/offset allowed shall be less than both of the following criteria:
   - 15% of nominal wall thickness.
   - 25% of the weld bead width, measured as a deviation from the weld centerline.

vi. Weld bead:
   - The weld bead shall be straight and uniform around the entire weld circumference. The minimum weld bead width shall not be less than 50% of the maximum weld bead width.
   - Convexity and concavity are ≤0.012 in.

vii. Slag: Slag-like surface impurities greater than 1/16 in. diameter are unacceptable. No more than four slag-like surface impurities shall be present in any four linear inches of weld.
   Note: Slag-like surface impurities less than 1/64 in. shall be disregarded.

viii. Rejected welds: Rejected welds shall be cut out and replaced with new welds or repaired in accordance with Section 6.4.1.e.

Records

Weld records (including weld maps as needed) and contractor/supplier turnover packages shall be generated to document the following:

(a) Welding procedure/WPS used (*attach if not referenceable in client system*).
(b) Unique weld numbers (unless sample), weld location (*e.g. welds traceable to weld map*), and weld repairs.
(c) Weld-type identification: Automatic/machine (orbital) or manual (Quality Rep preapproval required for manual DW/WFI and STS pipe welds).
(d) Machine/Orbital Welding Machine # (*Automatic Welding Operator*) and Weld Head used (if applicable).
(e) Description of materials being joined, size of material, and material heat numbers (*include Client Stock Room part # if applicable and readily available*).
(f) Surface roughness test results for internal surfaces associated with STS longitudinal pipe (seam) welds and STS Tank welds, unless otherwise specified.
(g) Welder identification and weld date.
(h) Identification of Purge Gas used and verification of compliance
(i) Inspection plan applied and results of sample weld inspection(s) if applicable.

(j) Inspection requirements per technical requirements, as applicable, and inspection results (pass/fail).
(k) Inspector identification and inspector's signature and date.
(l) Reference to passivation record # for spot passivated welds, if applicable.
(m) The following documentation is to be attached, as applicable:
   i. CMTRs for materials (including weld rod/wire) for all materials not issued by Client Stock Room (*materials issued by the Client Stock Room are previously received and verified for conformance to material and surface finish requirements specified with certification documentation/CMTRs filed by heat/lot #*).
   ii. WPS for weld type(s) unless internal WPS(s) used and is referenceable in client system (e.g. SOP Attachment).
(n) Additional Attachments Requirements for Welding by Contractor/Supplier (include in weld record or turnover package):
   i. Certificate of Compliance or other documentation (e.g. Supplier program evaluation) demonstrating contractor/supplier compliance to applicable portions of this SOP established and communicated.
   ii. Welder and Inspector Certifications unless internal welder/inspector.
   iii. Purge gas certification or other purity verification for purge gas to document compliance.
   iv. Weld Procedure Qualification Records (WPQ/WPQR) for applicable WPS(s).

Passivation (*DW/WFI and STS only*)

(a) Individual welds, including welds resulting from system repair or modification, shall be passivated (*typically spot passivated*) either before or after in spection unless the exception defined applies. Passivation shall also cover Weld #s etched on surfaces.
(b) *Exception*: When repairs result in the addition of inaccessible welds, cleaning and passivation of the interior of the weld are not required if the following conditions are met:
   i. Any small piping segments or small pieces of equipment added to the system as a result of the repair were cleaned and passivated prior to being welded into the system.
   ii. The repair and/or modification to the loop does not require more than two new inaccessible welds.
   Note: Requiring more than two welds would necessitate the cleaning and passivation of the affected piping.

Applicable Codes and Standards

Category	Case, item, weld types	Applicable code requirement(s)[a]
Hygienic service and safety-significant	Pressure vessels (PVs) and pumps designed for operation above 15 psig – hygienic air and water systems (*including pressure boundary and structural members*)	ASME BPVC Section VIII, Rules for Construction of Pressure Vessels[b]
	Piping (*designed for operation above 15 psig*)	AWS D18.1, Specification for Welding of Austenitic Stainless Steel Tube and Pipe Systems in Sanitary (*Hygienic*) Applications, as supplement to AWS D1.6, Structural Welding (*stainless steel*) OR
		ASME B31.3, Process Piping (*pipe welding*), Chapter X (*high purity*) [*comparable code*][a]
	Structures	AWS D1.6, Structural welding (*stainless steel*)
	ASME BPE, Bioprocessing Equipment, should be used as guidance when needed for hygienic applications	
Hygienic service and non-safety-significant	Tanks, chillers, and pumps operating designed for operation up to 15 psig – hygienic air and water systems that are not safety-significant	AWS D18.3, Specification for Welding of Tanks, Vessels, and Other Equipment in Sanitary (*Hygienic*) Applications, as supplement to AWS D1.6:2017, Structural welding (*stainless steel*)
	Piping (*designed for operation up to 15 psig*)	AWS D18.1, Specification for Welding of Austenitic Stainless Steel Tube and Pipe Systems in Sanitary (*Hygienic*) Applications, as supplement to AWS D1.6, Structural Welding (*stainless steel*)
	ASME BPE, Bioprocessing Equipment, should be used as guidance when needed for hygienic applications	
Non-hygienic and safety-significant	Pressure vessels (PVs) and pumps designed for operation above 15 psig – non-hygienic air and water systems (*including pressure boundary and structural members*)	ASME BPVC Section VIII, Rules for Construction of Pressure Vessels[b]

Applicable Codes and Standards

Category	Case, item, weld types	Applicable code requirement(s)[a]
	Wood Boiler (*up to first flange of the main header past the boiler stopcock*)	ASME BPVC Section I, Rules for Construction of Power Boilers[b]
	Piping (*designed for operation above 15 psig*)	AWS D10.12, Guide for Welding Mild Steel Pipe (mild/carbon steel)
		AWS D1.6, Structural welding (*stainless steel*) OR
		ASME B31.3, Process Piping (*pipe welding*) [*comparable code*][a]
	Structures	AWS D1.1, Structural Welding Code (steel)
		AWS D1.6, Structural Welding (*stainless steel*)
Non-hygienic, non-safety-significant	Everything else	cGMP and code requirements as required by the design such as AWS D9.1, Sheet Metal Welding

a) Comparable/more stringent ASME Codes may be applied to satisfy AWS code requirements in all categories, but AWS codes shall not be applied to satisfy ASME Code requirements in any category.

b) The only entities that can weld or repair welds and inspect welds on pressure vessels (PV), boilers, or structural members thereof (e.g. supports) are outside contractors holding an ASME "R" certificate (not by plant personnel).

Automatic Orbital Welding

Weld Acceptance Criteria for Automatic Orbital Weld Qualifications

Visual acceptance criteria for automatic orbital welding process are intended to be used as a uniform guide of weld acceptance quality control examiners and fabrication. Standard weld acceptance criteria shall be as follows:

i. Angular misalignment shall be held to a maximum of one degree.
ii. Burn through will not permitted.
iii. Concavity and convexity shall not exceed 10% of the wall thickness welded.
iv. Cracks shall not be permitted.
v. Crater pit or weld crater pit in an orbital weld is unacceptable.
vi. Lack of fusion shall not be permitted.
vii. Lack of penetration shall not be permitted.

viii. Meandering shall be limited to a maximum of one-third of the weld bead width.
ix. OD misalignment shall be within 10% of wall thickness Welded.
x. Oxidation: The interior surface or heat-affected zone of a stainless steel weld may exhibit a slight straw to light bluing. The heat-affected zone shall be free of a dark blue to heavy black appearance. A granular texture indicates serious oxidation and shall be rejected.
xi. Porosity that is visually detected on the surface of the weldment or adjoining base material shall not be permitted.
xii. Scratches on the interior surface are of concern in the high purity and biopharmaceutical industries. The weld operator must be able to prep, fit, and weld the coupons without scratching the interior or marring the exterior surfaces. The proctor will check the coupons before they are issued to the weld operator to ensure they are free of scratches. Scratching the interior surface during the weld test will be a cause for rejection.
xiii. Tacks shall be surface only and shall be completely consumed by weld.
xiv. Tungsten inclusions visually detected in the surface of the weld or adjoining base material shall not be permitted.
xv. Undercut shall not exceed 10% of the wall thickness welded.

5

Passivation

To establish the process for performing, documenting, and controlling periodic deactivation of stainless steel surfaces, stainless steel surfaces do not alter the safety, quality, or purity of product solution by being corroded, reactive, or otherwise additive. Passivation is required for solution and distilled water (DW)/water for injection (WFI) contact surfaces prior to initial release for production use [17]. Re-passivation is required if surfaces required to be passivated are compromised (e.g. grinding), welded, or as required by preventive maintenance program. In addition, passivation of stainless steel external surfaces is required for certain items in clean rooms subject to wash down as specified in local standard operating procedures (SOPs). Preexisting items are to be adequately cleaned, if necessary, prior to beginning the passivation process described herein.

Passivation: Passivation is defined as "Removal of exogenous iron or iron compounds from the surface of stainless steels and higher alloys by means of a chemical dissolution, most typically by a treatment with an acid solution that will remove the surface contamination and enhance the formation of the passive layer."

Passivation shall be accomplished via processes specified herein. Electropolished parts are not required to be passivated. Items shall be adequately cleaned prior to start of passivation process. Chemicals used in cleaning, passivation, and neutralization processes must be water soluble and managed if planned for routine use and on-site storage. Citric acid–based chemicals are preferred for standard passivation processes. Listing of chemicals routinely used for cleaning, passivation, neutralization, and concentration testing must be kept on file.

Note: Nitric acid–based passivation solutions are not permitted for passivation of sanitary system components because nitric acid can be a precursor for the formation of nitrosamines–nitrosodimethylamine (NDMA) impurities.

EHS considerations require those performing passivation processes that are responsible to implement adequate safety measures as specified in the passivation process used. Passivation performed shall implement EHS precautions as a

Sterile Processing of Pharmaceutical Products: Engineering Practice, Validation, and Compliance in Regulated Environments, First Edition. Sam A. Hout.

minimum in addition to precautions identified on passivation chemical Safety Data Sheets (SDSs) and manufacturer's recommendations:

(1) Personal protective equipment (PPE) requirements: Employees must wear the following articles of PPE when preparing cleaning and passivation chemicals (e.g. citric acid solution) or performing passivation to prevent chemical contact with eyes and skin:
   - Safety glasses and face shield
   - Chemical-resistant gloves and apron with sleeves.
(2) Disposal of waste solution: Used passivation chemicals shall not be disposed of to plant drains or via neutralization process due to heavy metals in solution. Used passivation chemicals (considered hazardous waste) must be disposed of off-site unless contained within a paper towel or rag (e.g. wiped-off spot passivation gel), which may be disposed of in a typical waste container.

   Note: Used cleaning and neutralization chemicals (not expected to include heavy metals) may be disposed of through the neutralization tank process. Rinse water (water with trace amounts of cleaning, passivation, and/or neutralization chemicals) may be disposed of via plant drains.
(3) Passivation Records shall be numbered, for example, PN-20-008. Passivated parts shall be identified with passivation identification number as described next. Systems passivated via circuit passivation are not required to be identified with passivation identification number, but the associated Passivation Record is required to be documented in associated paperwork.

   For items required to be passivated per requirements (e.g. DW/WFI and Solution Transfer System [STS]): the Passivation Record # shall be referenced in installation paperwork associated with the item (e.g. EMR, CW, Trackwise).
(4) The Passivation Identification Number will be used as needed for identification of items in Passivation Record and consist of the "Passivation Record #" and sequence # separated with a hyphen – for example, PN-20-0032-03 (Passivation Record # PN-20-0032, sequence # 03). Quantity of digits used for sequence #s should be based on the quantity of parts in the batch being passivated (i.e. under 10: use one digit, under 100: use two digits).

   Passivation in Vats:
   (a) Parts to be passivated shall be identified with passivation identification number as follows: unless the part is too small: engraving/etching (prior to passivation) on external surface, if practical.
   (b) Parts not practical for etching passivation identification number on external surface (e.g. too small) shall be associated with passivation identification number otherwise such as placed in an appropriate container (e.g. sealed bag) following passivation and drying (if closed container) with the

assigned passivation identification number labeled on/affixed to the container. Containerized passivated parts:

(1) should be associated with the same passivation identification number (since further traceability is not possible).

(2) shall remain in the identified container or transferred to another identified container until installed to maintain traceability between parts and passivation identification number.

(5) Items passivated by supplier (purchased items) shall be identified with the Passivation Report # via item or container labeling or tagging.

(6) Association of Passivation Report # with stock room parts:

(a) Passivated stock room parts shall be associated with a Passivation Record # prior to storage.

(b) Passivation Records # are assigned to stock room passivated parts as follows:

(1) In-house passivation (by maintenance): assigned by maintenance and identified with part

(2) Supplier passivation (pre-passivated purchased part): It is assigned by the engineering. Part is to be identified for passivation documentation verification via enhanced receipt inspection, and stock room personnel is to label or otherwise identify associated parts (e.g. packaging) with the Passivation Record #.

(c) Stock room personnel should provide or otherwise communicate Passivation Report # with issuance of passivated parts to recipient if part is planned for installation in DW/WFI, STS, or process compressed air (PCA) system.

Instrumentation used to achieve passivation process and verify adequate rinsing of chemicals shall be calibrated or otherwise verified in accordance with local SOPs. pH meters used shall be verified as accurate within ± 0.02 pH units against pH buffers. pH meter verifications for the purposes of this procedure (e.g. pH difference to verify chemical absence) are valid for up to 10 days.

In-house Passivation Using Vats

Test Reagents and Instrumentation: Verify that testing reagents (e.g. phenolphthalein, 0.2 N HCl) are not expired (*established at six months from date originally made*), verify pH meter accuracy using buffers identified in the Passivation Record, and verify that thermometer and conductivity meter are within calibration valid period.

Expiration (six months from being made) and concentrations are within target ranges (verified within last three days), and temperatures are within target range. Make cleaning and passivation chemicals, if needed, and label vat(s) with expiration date six months from date originally made.

- Sodium hydroxide (NaOH): 0.75 ± 0.25 wt% sodium hydroxide
- Citric acid ($C_6H_8O_7$): 8 ± 2 wt% citric acid.

Important: Always add acid to water to avoid splatter from the mixing reaction.

Identify parts to be passivated with passivation identification number. Items to be containerized are to be segregated for future containment in container identified with passivation identification number. Surface prep/clean using sodium hydroxide solution (high pH) [0.75 ± 0.25 NaOH] @ 140 ± 20 °F for 30+ minutes. The resulting pretreated surface shall be substantially free of oil, grease, rust, scale, and other foreign matter. Isolate or remove parts of system that cannot be exposed to the chemicals being used. Any openings created by the removal of components shall be filled to allow for recirculation of the chemical solutions.

Remove surface prep/cleaning chemical: Thoroughly rinse the cleaned items/surfaces with filtered[1] water with influent pH of 6–9 until the following "Rinse Completion Criteria" has been achieved. Cycle valves at least once during the rinsing period. All items in lot may be held within basket and rinsed at same time with effluent pH measured at discharge of rinse stream. Rinse Completion Criteria: effluent pH is within ±0.5 pH units of the influent and effluent conductivity is within ±0.2 μS/cm of the influent.

Verify clean: Visually verify cleanliness[2] assisted by a 250+ ft candles (lumens) light source (e.g. flashlight) and tools such as scope and mirrors necessary to see cleaned surfaces. Scrub any unclean area(s) with a 316L stainless steel wire brush or otherwise, as necessary.

Passivate with citric acid solution (low pH) [8 ± 2% Citric Acid] @ 130 ± 10 °F for 30+ minutes.

Note: Transmission lines that are to be tri-clamped into the main recirculating system and which are too long to be submerged in a vat, may be passivated by assuring that the pipeline is filled with passivating solution.

Remove passivation chemical: Thoroughly rinse the passivated items/surfaces with filtered water with influent pH of 6–9 until the following "Rinse Completion Criteria" has been achieved. Cycle valves at least once during the rinsing period. All items in lot may be held within basket and rinsed at same time with effluent pH measured at discharge of rinse stream. Rinse Completion Criteria: effluent

---

1 *Note*: Water filtered with standard water filter contains maximum TDS of 200 ppm.

2 *Clean*: free of oil, grease, dirt, rust, scale, slag, and other foreign matter.

pH is within ±0.5 pH units of the influent, and effluent conductivity is within ±0.2 μS/cm of the influent.

Verify passivity of sample representing the lot, passivity test for stainless steel. If the sample representing the lot is found to be ineffectively passivated, consult with maintenance for action to be taken (e.g. re-passivate lot).

Spot Passivation

Production equipment or sanitary piping that has been previously passivated may require passivation and cleaning of a single area (e.g. weld) due to repair or minor alteration. Spot passivation of surfaces (e.g. weld) shall be performed using gels (e.g. Astro Pak UltraPass Gel passivation kit), pastes, or electro-chemicals in accordance with manufacturer specified processes for surface preparation, passivation, and neutralization.

Note: When a repair results in an inaccessible weld, cleaning and passivation of welds is not required if the item installed (e.g. piping) has been passivated and the repair does not involve more than two inaccessible welds.

If an area (e.g. weld) has been polished, ensure that the entire area that has been polished is cleaned and passivated. If passivating in a clean room, rinse water shall be compatible with the room requirements based on conditions (e.g. use DW if the room is clean). The Spot Passivation Record # should be referenced in documentation that triggered the activity such as work order and/or weld record.

Astro Pak UltraPass Gel Passivation

Verify chemicals: 70% IPA (Isopropyl Alcohol), UltraPass Gel (six months life typically), and Neutralizer (Sodium Bicarbonate, three year life typically) are not expired. Also, identify the water source for use in neutralization mix and rinsing.

Instrumentation: Verify pH meter accuracy using buffers identified in the Passivation Record, and conductivity meter is within calibration (not expired).

Verify surfaces: Inspect the area to be passivated (e.g. weld) assisted by a 250+ ft candles (lumens) light source (e.g. flashlight) to ensure it is free of scale, slag, and other foreign matter. Implement corrections as necessary (e.g. polish).

Clean weld or surface area with 70% IPA by wiping with a clean wipe soaked in 70% IPA. Wipe off excess cleaner and then rinse with filtered[3] water with influent pH of 6–9 or DW by spraying, brushing, flushing, or wiping.

---

3 *Note*: Water filtered with standard water filter contains maximum TDS of 200 ppm If passivating in a clean room, rinse water shall be compatible with the room requirements based on conditions (e.g. use DW if the room is clean).

Apply passivation gel to spot/area surface with applicator brush. Reapply passivation gel every ~15 minutes for a minimum of 90 minutes. Wipe off passivation gel from surface area with dry wipe.

Mix and apply neutralization solution: Mix ~1 part sodium bicarbonate (neutralizer, essentially baking soda) to ~10 parts DW (approximate by volume) to make neutralization solution. Apply neutralization solution to passivation surface area until foaming or reaction ceases.

Rinse passivated and neutralized surface with filtered water (with influent pH of 6–9) or DW flowing to drain until passivated surfaces are visually free from chemical residue and effluent pH is within +/0.5 pH units of the influent and effluent conductivity is within ±0.2 μS/cm of the influent.

Verify passivity of spot(s) passivated, passivity test for stainless steel. If the spot(s) is found to be ineffectively passivated, consult maintenance for action to be taken (e.g. re-passivate).

Alternative Process

An alternative process may be used provided it complies with the applicable portions of this SOP and passivation requirements for Sanitary System Design; the procedure and record to be used have been preapproved by engineering (*compliance to applicable requirements*), technical services (*chemical absence verification criteria*), and EHS (*chemical use and disposal considerations)*; and the procedure/record document the following:

1. Chemicals and concentrations to be used and disposal method(s)
2. Safety considerations and chemical disposal methods.

Notes:

- Spots passivated and location on system drawing, if applicable
- Procedure used, contents of record, and basis (validation documentation) for chemical absence verification criteria
- Process steps with dates and times for critical durations
- Verification of passivation and chemical absence (following the passivation).

Subsequent uses of previously approved alternative processes/procedures may reference the initial approval instead of obtaining additional approvals unless there are changes to the procedure, record template, or application.

Circuit Passivation

(1) Record includes components passivated, location on system drawing, and circuit diagram.

(2) Procedure used meets the following criteria:
  (a) Complies with passivation requirements for Sanitary System Design in ASTM A967 or other industry standard; and applicable requirements detailed (e.g. no use of Nitric Acid).
  (b) Includes process steps with dates and times for critical durations, contents of record/record template.
  (c) Specifies chemicals, concentrations to be used, disposal method(s), and safety practices plan.

(3) Client Standard with the following clarifications/exceptions:
  (a) Flushing and cleaning methods shall be as specified in SOP.
  (b) Typical filtered water (maximum TDS of 200 ppm) with influent pH of 6–9 or DW shall be used for flushing or rinsing. If passivating in a clean room, rinse water shall be compatible with the room requirements based on conditions (e.g. use DW of the room is clean).
  (c) Hydrotest prior to cleaning or passivating at 125% of the maximum pressure expected during cleaning, passivating, and rinsing processes.
  (d) Passivation process and passivity verification methods shall be as specified in SOP.
  (e) Chemical removal methods, chemical absence verification methods and criteria (following passivation), and chemical absence of criteria basis (e.g. validation documentation) is included in procedure/documentation.

(4) The following steps shall be followed or otherwise considered for cleaning surfaces and verifying systems prior to passivation, as applicable:
  (a) Isolate or remove parts of system that cannot be exposed to the chemicals being used. Any openings created by the removal of components shall be filled to allow for recirculation of the chemical solutions.
  (b) Inspect water system components for signs of excessive rouging. The system shall be de-rouged if it is determined that the rouge layer is adversely affecting the performance of the water system or the final product. Recommended areas for inspection include pump casings, heat exchanger bonnets, use point valve bodies, and spray ball impingement areas. When de-rouging, it is acceptable to only clean those parts of the system that have been identified to show excessive amounts of rouging.
  (c) Connect the cleaning solution tank and pump to system. Closed stainless steel tanks should be equipped with spray balls to wet product contact surfaces with cleaning solutions.
  (d) Cleaning processes shall ensure surfaces are substantially free of oil, grease, rust, scale, and other foreign matters prior to passivation.
  (e) Processes are started with enough water or solution to maintain circulation in the system. Required pressure testing is completed before the cleaning process is started. The circuit and components (including heat

exchangers, fittings, and seals) are checked for leaks prior to introducing chemicals to the system. Leaks are repaired before continuing.
   (f) Verify that all items/materials (e.g. gaskets) in a loop to be passivated can withstand the process (chemicals, temperatures) and implement methods to mitigate damage prior to passivating.
(5) Checkpoints detailed and included in the passivation record unless the preapproved passivation procedure/record includes similar checks/verifications, as applicable.
(6) Procedures and records used shall comply with (or state compliance to) the above with any clarifications/exceptions being detailed in the preapproved client passivation record prior to passivation.
(7) Subsequent uses of previously approved processes/procedures (including preapproved clarifications and exceptions to requirements) may reference the initial approval instead of obtaining additional approvals unless there are changes to the procedure, record template, or application.

Supplier (Purchased Item) Passivation

(1) *DW/WFI or STS items*: Passivation of purchased items subject to passivation requirements or exemption thereto shall be documented unless re-passivated in-house. Electropolished parts are not required to be passivated.
    Passivation documentation shall include the following:
    Supplier passivation procedure to include/verify the following:
    - Chemicals used for cleaning and passivating are safe for use in drinking water systems and/or the pharmaceutical industry.
    - Components were cleaned prior to completing passivation.
    - Verification of passivation process compliance with industry standards such as ASTM A967 (typical passivation process standard).
    - Procedure used includes chemicals with concentrations and temperatures, and process steps including final verification chemical absence.

    Passivation of part documentation:
    - Identification of the item(s) passivated (e.g. model and serial number(s)).
    - Documentation that the subject item(s) were passivated per the above-evaluated procedure and date of passivation.

    Subsequent uses of previously approved processes/procedures (including preapproved exceptions to requirements) may reference the initial approval instead of obtaining additional approvals unless there are changes to the procedure, record template, or application.

(2) *PCA (clean air) items*: Items passivated to comply with requirements for items in clean rooms subject to wash down (e.g. process air component) may be received and declared passivated based on documentation of passivation of items via a process that complies with industry standards such as ASTM A967.

Control of Passivated Items

Items requiring passivation are to be maintained in a segregated, secured area until they are removed for passivation. This area must be clearly identified that only non-passivated parts are to be located in the area and that the parts are not to be used until passivated. Passivated parts are not to contact non-passivated metal surfaces – for example, be placed on metal table.

Chemicals Routinely Used for Cleaning, Passivation, Neutralization, and Concentration Testing

Chemical/ manufacturer	Purpose	Removal	Absence verification
50% Sodium hydroxide (NaOH), Brenntag	Mix to 0.75 ± 0.25 wt% and clean items prior to passivating in vats	Thoroughly rinse the cleaned/ passivated items/surfaces with filtered water with influent pH of 6–9 for at least 30 min	Effluent pH is within ±0.5 pH units of the influent and effluent conductivity is within ±0.2 μS/cm of the influent
Citric acid anhydrous ($C_6H_8O_7$), 50 lb bag, Manuf. PN 622965/JBL	Mix to 8 ± 2 wt% and passivate items in vats		
IPA 70%	Clean surfaces prior to spot passivating	Wipe, then rinse with distilled water	N/A
UltraPass Gel CN/Astro Pak	Gel for spot passivation	Wipe gel, apply diluted neutralizer, then rinse with distilled water	Foaming/reaction stops. Then, (after rinsing), passivated and neutralized surfaces are visibly free from chemical residue, effluent pH is within ±0.5 pH units of the influent, and effluent conductivity is within ±0.2 μS/cm of the influent
Sodium bicarbonate (SDS # 144-55-8-8, CAS # 144-55-8)/ Genesis Alkali (procured from Astro Pak)	Dilute ~10 : 1 in DW to neutralize Astro Pak UltraPass Gel used for spot passivation		
Phenolphthalein T.S. (CAS # 77-09-8) VWR Catalog # EM-PX0520-1, EMD Millipore Model PX0520-1)/EMD Millipore (procured from VWR)	Testing chemical concentration, reagent	N/A	N/A

(*Continued*)

Chemicals Routinely Used for Cleaning, Passivation, Neutralization, and Concentration Testing

Chemical/ manufacturer	Purpose	Removal	Absence verification
HCl 0.2 N Standard, JT5612-2 Hydrochloric Acid (CAS # 7647-01-0)/ Avantor Performance Materials, Inc. (procured from VWR)	Testing concentration of Sodium Hydroxide used for passivation	N/A	N/A
NaOH 0.2 N Standard, JT5665-2 Sodium Hydroxide (CAS # 1310-73-2)/Avantor Performance Materials, Inc. (procured from VWR)	Testing concentration of citric acid used for passivation	N/A	N/A

Preparing, Testing, and Adjusting Cleaning and Passivation Chemicals (Passivation in Vats)

A. Sodium Hydroxide Solution (NaOH)

(1) Preparation of sodium hydroxide solution:

(a) The NaOH solution may be prepared using tank car caustic from the boiler room. Water used in the solution must at least be filtered.

(b) The dimensions (area) of the sodium hydroxide passivation vat are 2.4 ft × 24 ft. Calculate the volume of solution desired by multiplying the length by the width by the depth of solution desired (*in feet*).

(c) The final concentration of the sodium hydroxide solution must be 0.75% ± 0.025% w/v.

(d) The amount of caustic solution needed (in gallons) is equal to:

$$\frac{XY}{(Z)(1.52)}$$

where

X = volume of solution desired (in gallons)

Y = concentration of solution desired (0.75% w/v)

Z = concentration of caustic solution

1.52 = approximate specific gravity of caustic solution.

EXAMPLE using 2.4 ft × 24 ft vat:

$$X = (2.4\ \text{ft})(24\ \text{ft})(2\ \text{ft})(7.48\ \text{gal/ft}^3) = 862\ \text{gal}$$

Width Length Depth

Y = concentration desired = 0.75% w/v

Z = 50.0% caustic solution

1.52 = approximate specific gravity of caustic solution.

$$\text{Gallons caustic needed} = \frac{(862\ \text{gal})(0.75\%)}{(50.0\%)(1.52)} = 8.5\ \text{gal}$$

(2) Testing of sodium hydroxide solution:

A member of the analytical lab will analyze a representative sample of the sodium hydroxide solution and record raw data and applicable calculations on the Passivation Record.

Concentration test of solution within three days prior to each passivation. Pipet 10 ml of the NaOH solution into a flask. Two drops of phenolphthalein T.S. will be added. Titrate the solution with 0.2 N HCl (purchased standard) to a colorless endpoint.

$$\%\text{NaOH}\ \ (\text{w/v}) = \frac{(\text{ml HCl})(0.2\text{N HCl})(4.00)}{10\ \text{ml NaOH sample}}$$

If the concentration is *outside the limits* (0.50%–1.00% w/v), perform the solution adjustments per applicable steps that follow.

(3) Adjustment of the sodium hydroxide solution

(a) Solution is too weak (less than 0.50% w/v)

Calculation:

$$X = \frac{(Z)(C - Y)}{(W)(1.52)}$$

where

X = gallons of caustic needed to adjust solution

Y = current concentration of NaOH solution

Z = current volume at NaOH solution

W = concentration of caustic solution to be added

C = target concentration of NaOH solution

1.52 = approximate specific gravity of caustic solution

EXAMPLE:

Y = 0.450% w/v = current concentration of NaOH solution

Z = 820 gal = current volume of NaOH solution

$W = 50.0\% =$ concentration of caustic solution to be added
$C = 0.75\%$ w/v $=$ target concentration of NaOH solution
$X =$ gallons caustic solution required for adjustment (*round up to nearest measurable fraction*).

$$X = \frac{(820\text{ gal})(0.75\%\text{ w/v} - 0.450\%\text{ w/v})}{(50.0\%)(1.52)} = 3 - 1/4\text{ gal}$$

(b) Solution is too strong (greater than 1.00% w/v)
Calculation:

$$X = \frac{(Z)(Y-C)}{C}$$

where

X = gallons of water required to adjust solution
Y = current concentration of NaOH solution
Z = current volume of NaOH solution
C = target concentration of NaOH solution

EXAMPLE:
$X =$ gallons of water required for adjustment
$Y = 1.10\%$ current concentration of NaOH solution
$Z =$ current volume of NaOH solution
$C = 0.75\% =$ target concentration.

$$X = \frac{(785\text{ gal})(1.10\% - 0.75\%)}{0.75\%} = 366 - 1/3\text{ gal}$$

B. Citric Acid Solution ($C_6H_8O_7$)

(1) Preparation of citric acid solution
  (a) This solution is prepared using citric acid which is stored in the production supply cage. Water used in the solution must at least be filtered.
  (b) The dimensions (area) of the citric acid passivation vat are 2.4 ft by 24 ft. Calculate the volume of solution desired by multiplying the length by the width by the depth (in feet) of solution desired.
  (c) The final concentration of the citric acid solution must be 8% ± 2% w/v in accordance with ASTM A967 (tighter range).

The amount of citric acid needed (in pounds) is equal to $(0.0835)(X)(Y)$, where

X = volume of solution desired (in gallons)
Y = % w/v solution target
0.0835 = conversion factor (g/ml to lb/gal)

EXAMPLE using 2.4 ft × 24 ft vat:

$X = (2.4\ \text{ft})(24\ \text{ft})(2\ \text{ft})(7.48\ \text{gal/ft}^3) = 862\ \text{gal}$

$Y = \text{concentration target} = 8.0\%\ \text{w/v}$

$$\text{Pounds citric acid required} = (862\ \text{gal})(8.0\%)(0.0835) = 575.8\ \text{lb}\ (\sim 11.5\ \text{bags})$$

Note: Each bag of citric acid weighs ~50 lb.

(2) Testing of the citric acid solution

A member of the analytical lab will analyze a representative sample of the citric acid solution and record raw data and applicable calculations on the Passivation Record.

Concentration test of solution within three days prior to each passivation. Pipet 5.0 ml of the citric acid solution into a flask. Add two drops of phenolphthalein T.S. (purchased reagent) and titrate to a pale pink endpoint with 0.2 N NaOH (purchased standard).

$$\%\ \text{Citric acid (w/v)} = \frac{(\text{ml NaOH})(0.2\text{N NaOH})(64.04)}{(5.0\ \text{ml citric acid sample})(10)}$$

If the concentration is *outside the limits* (6.0%–10.0% w/v), perform the solution adjustments per applicable steps that follow.

(3) Adjustment of citric acid solution

(a) Solution is too weak (less than 6.0% w/v)

Calculation

$$X = (0.0835)(V)(C2 - C1)$$

where

X = pounds of citric acid required to adjust solution to desired concentration

V = current volume of solution (area × depth), in gallons

$C1$ = current concentration of solution

$C2$ = target concentration of solution

EXAMPLE:

$V = 824\ \text{gal}$

$C1 = 5.5\%\ \text{w/v}$

$C2 = 8.0\%\ \text{w/v}$

$$X = (0.0835)(824)(8.0\% - 5.50\%) = 172\ \text{lb}\ (\textit{approx. } 3^1/_2\ \textit{bags})$$

(b) Solution is too strong (greater than 10.0% w/v)
Calculation:

$$X = \frac{V(C1 - C2)}{C2}$$

where

X = additional water required to bring solution to target concentration
V = current volume, in gallons
$C1$ = current concentration
$C2$ = target concentration

EXAMPLE:
$V = 836$ gal
$C1 = 11.0\%$ w/v
$C2 = 8.0\%$ w/v

$$X(\text{gallons } H_2O \text{ needed}) = \frac{(836\text{ gal})(11.0\% - 8.0\%)}{8.0\%}$$
$$= 313.5\text{ gal}$$

6

Chilled Water System

This defines the user requirements specification (URS) for chiller system. These are user requirements for equipment that is used to support the production of finished pharmaceuticals in a current good manufacturing practice (cGMP) environment. An equipment validation plan is developed to outline the planned tasks and expectations for qualification of the equipment. The URS serves as the input to subsequent project risk assessment activities, design objectives, control strategy, and acceptance criteria for testing and qualification of the manufacturing system.

Term	Title
ANSI	American National Standards Institute
ASME	American Society of Mechanical Engineers
ASTM	American Society for Testing and Materials
cGMP	Current Good Manufacturing Practice
NEC	National Electrical Code
NEMA	Nat'l Electrical Manufacturers Association
NFPA	National Fire Protection Association
UL	Underwriter's Laboratories

Sterile Processing of Pharmaceutical Products: Engineering Practice, Validation, and Compliance in Regulated Environments, First Edition. Sam A. Hout.

Abbreviation	Definition
AHU	air handler unit
BAS	building automation system
CC	change control
CDA	clean dry air
CPU	central processing unit
CR	change request
CS	carbon steel
DS	design specification
FAT	factory acceptance test (at supplier site)
FCU	fan coil unit
FDA	Food and Drug Administration
FDS	functional design specification
FS	functional specification
GMP	good manufacturing practices
HDS	hardware design specification
HMI	human–machine interface
IPC	industrial PC (personal computer)
IQ	installation qualification
ISO	International Organization for Standardization
OIT	operator interface terminal
OQ	operational qualification
P&ID	process and instrumentation diagram or piping and instrumentation diagram
PAC	programmable automation controller
PAS	plant automation system
PLC	programmable logic controller
PQ	performance qualification
RA	risk analysis
SCADA	supervisory control and data acquisition
SAT	site acceptance testing
SOP	standard operating procedure
TCU	temperature control unit
URS	user requirements specification

Chilled water system provides chilled water to AHU, FCU, and TCU for the entire plant. Chilled water system consists of the following:

- A central chiller building which houses chillers and distribution pumps, water treatment skid, and controls
- Cooling towers near the new central chiller building, including stream filtration units for each tower, and cooling tower pumps and controls
- CHW distribution system
  - o Primary pumps provided for chillers
  - o Variable speed secondary pumps, pumping into a common secondary distribution loop
  - o Variable speed tertiary pumps, pumping into tertiary loops to end users

Process Description

The chillers, towers, and distribution pumps will provide chilled water at a nominal 40 °F leaving set point and a 10 °F system temperature differential initially. In addition, the system set point might be raised to 44 °F with a 14 °F temperature differential. Chillers, towers, and pumps will be staged on and off as needed to meet the cooling loads of the plant with consideration to having the most energy-efficient combination of equipment at all times.

Chilled water is pumped from the chillers into the secondary loop of the plant. Tertiary pumps draw chilled water from the secondary loop and provide chilled water to the end users on each of the tertiary distribution loops. All CHW supply and return piping will be a nominal schedule 40 carbon steel configuration. The piping is mounted on a combination of fixed steel and roller supports. The piping is insulated and heat traced (piping below 12″ diameter) to prevent heating of the chilled water or freezing in ambient temperatures below 32 °F.

- The operation and control of the cooling towers, chillers, and distribution pumps is executed by plant control system. Flowmeters, temperature, and pressure/pressure differential sensors are utilized to control the operation of the chillers, towers, and pumps to match the generation of the chilled water to the actual chilled water loads in the most energy-efficient manner possible.
- Piping connections for future connections to Expanded Chilled Water system are provided in the form of tees, isolation valves, and blind flanges.
- Consideration for pipe and electrical power sizing for future expansion of the chilled water system in chiller building and cooling towers are provided. Each requirement has a unique identifier and a designation for the level of importance. Refer to the "Requirement Key" for the designation definitions. Each of the requirements detailed will be tested at the relevant qualification stage as defined in the applicable commissioning plan. Process descriptions, specifications, flowcharts, diagrams, etc. are added as attachments where appropriate.

Requirement Key

M	Mandatory	This requirement has to be met. It impacts critical parameters; product quality, safety, or efficacy, or personnel safety. The equipment/system is not acceptable unless this requirement is provided in an agreed manner
I	Important	The equipment/system will provide significantly greater, product quality, safety, or efficacy, personnel safety, or business benefit if this requirement is met, but at client's discretion may not make the system unacceptable if it is not delivered
D	Desirable	These requirements might be worthwhile, but it will not make the system unacceptable if not delivered
IN	Information	Information provided for the supplier which may be useful for equipment design or calculation of the offer

Requirement statement	Requirement				Responsibility client/vendor	URS to be tested during this activity			
	M	I	D	IN		FAT	IQ	OQ	PQ
General requirements	☒	☐	☐	☐	Vendor	☐	☒	☐	☐
All system components shall be designed to meet or exceed the International Building Code, state, and local code requirements	☒	☐	☐	☐	Vendor	☐	☒	☐	☐
All system components shall be designed as per the Energy Efficiency and Critical Systems checklist for new construction	☒	☐	☐	☐	Vendor	☐	☒	☐	☐
All system components shall be designed to be compliant with guidelines	☒	☐	☐	☐	Vendor	☐	☒	☐	☐
New chiller building architectural requirements	☒	☐	☐	☐	Vendor	☐	☒	☐	☐
The exterior of the building is to be designed with materials that are cost-effective and still maintain the aesthetic consistency of the site buildings	☒	☐	☐	☐	Vendor	☐	☒	☐	☐
Vehicular access provided on one end via roll-up door	☒	☐	☐	☐	Vendor	☐	☒	☐	☐
Brick façade to match existing building	☒	☐	☐	☐	Vendor	☐	☒	☐	☐

Requirement statement	Requirement				Responsibility client/vendor	URS to be tested during this activity			
	M	I	D	IN		FAT	IQ	OQ	PQ
No active components on roof, no louvers, minimal roof penetrations	☒	☐	☐	☐	Vendor	☐	☒	☐	☐
Personnel access doors, including emergency egress doors provided	☒	☐	☐	☐	Vendor	☐	☒	☐	☐
Access to roof via ladder/safety cage	☒	☐	☐	☐	Vendor	☐	☒	☐	☐
New chiller building mechanical requirements	☒	☐	☐	☐	Vendor	☐	☒	☐	☐
Chiller room ventilated and heated	☒	☐	☐	☐	Vendor	☐	☒	☐	☐
Electrical room ventilated, heated, and cooled	☒	☐	☐	☐	Vendor	☐	☒	☐	☐
Floor drain system provided in foundation	☒	☐	☐	☐	Vendor	☐	☒	☐	☐
Instrument air provided for controls	☒	☐	☐	☐	Vendor	☐	☒	☐	☐
Makeup water provided for chilled water and tower systems	☒	☐	☐	☐	Vendor	☐	☒	☐	☐
Multiple primary and secondary pump	☒	☐	☐	☐	Vendor	☐	☒	☐	☐
Vendor provided chemical skid (for chilled water and tower systems)	☒	☐	☐	☐	Vendor	☐	☒	☐	☐
Chilled water and tower water piping	☒	☐	☐	☐	Vendor	☐	☒	☐	☐
Chiller refrigerant detection and ventilation system	☒	☐	☐	☐	Vendor	☐	☒	☐	☐
New chiller building to be sprinkled	☒	☐	☐	☐	Vendor	☐	☒	☐	☐
New chiller building electrical requirements	☒	☐	☐	☐	Vendor	☐	☒	☐	☐
Electrical power components located in dedicated electrical room	☒	☐	☐	☐	Vendor	☐	☒	☐	☐

Requirement statement	Requirement				Responsibility client/vendor	URS to be tested during this activity			
	M	I	D	IN		FAT	IQ	OQ	PQ
Electrical power components located outside wall of chiller building	☒	☐	☐	☐	Vendor	☐	☒	☐	☐
Site electrical power should be designed in a way to eliminate a single point of failure for the generation and distribution equipment. This should at a minimum include more than one feed from the utility substation and multiple transformers	☒	☐	☐	☐	Vendor	☐	☒	☐	☐
Power can be medium voltage or low voltage as required by the equipment installed	☒	☐	☐	☐	Vendor	☐	☒	☐	☐
Interior lights, fire alarm system, house power, etc.	☒	☐	☐	☐	Vendor	☐	☒	☐	☐
Lighting shall be LED based	☒	☐	☐	☐	Vendor	☐	☒	☐	☐
No standby power is required for the chilled water system building or equipment	☒	☐	☐	☐	Vendor	☐	☒	☐	☐
New chiller building structural requirements	☒	☐	☐	☐	Vendor	☐	☒	☐	☐
Foundation and structural steel designed based upon geotechnical report	☒	☐	☐	☐	Vendor	☐	☒	☐	☐
Roof steel to provide support for roof, as well as interior CHW and tower piping	☒	☐	☐	☐	Vendor	☐	☒	☐	☐
Monorail beam systems to lift chiller and pump components during maintenance	☒	☐	☐	☐	Vendor	☐	☒	☐	☐
Maintenance ladders/platforms provided as needed for access to elevated equipment	☒	☐	☐	☐	Vendor	☐	☒	☐	☐
Handrails/fall protection provided on roof	☒	☐	☐	☐	Vendor	☐	☒	☐	☐

Requirement statement	Requirement				Responsibility client/vendor	URS to be tested during this activity			
	M	I	D	IN		FAT	IQ	OQ	PQ
New cooling towers	☒	☐	☐	☐	Vendor	☐	☒	☐	☐
Nominal ton cooling capacity	☒	☐	☐	☐	Vendor	☐	☒	☐	☐
Nominal tower entering temperature of 95 °F and leaving water temperature of 85 °F	☒	☐	☐	☐	Vendor	☐	☒	☐	☐
Worst-case wet bulb shall be 80 °F	☒	☐	☐	☐	Vendor	☐	☒	☐	☐
Towers shall be capable of resetting leaving tower water set point to track ambient wet bulb, with leaving tower water set point as low as 65 °F	☒	☐	☐	☐	Vendor	☐	☒	☐	☐
Towers shall be provided with full bypass to cold basin capability to avoid fill freezing during low ambient temperature conditions	☒	☐	☐	☐	Vendor	☐	☒	☐	☐
Tower fans shall be provided with VFD fan motors and controls	☒	☐	☐	☐	Vendor	☐	☒	☐	☐
Tower fans provided with vibration alert	☒	☐	☐	☐	Vendor	☐	☒	☐	☐
Towers capable of operation with tower fan motor speed from 100% to 50%	☒	☐	☐	☐	Vendor	☐	☒	☐	☐
Tower static lift shall not exceed 15 ft	☒	☐	☐	☐	Vendor	☐	☒	☐	☐
Cooling towers provided with side stream filtration system, including basin sweeping piping	☒	☐	☐	☐	Vendor	☐	☒	☐	☐
Towers provided electric basin heaters	☒	☐	☐	☐	Vendor	☐	☒	☐	☐
Towers shall provide 100% redundancy for one tower and pump set	☒	☐	☐	☐	Vendor	☐	☒	☐	☐
Tower system shall be capable of operation with one tower and any one tower pump not in service	☒	☐	☐	☐	Vendor	☐	☒	☐	☐

Requirement statement	Requirement				Responsibility client/vendor	URS to be tested during this activity			
	M	I	D	IN		FAT	IQ	OQ	PQ
Towers shall be provided with galvanized structural steel, and stainless steel hot and cold basins	☒	☐	☐	☐	Vendor	☐	☒	☐	☐
New cooling tower pumps	☒	☐	☐	☐	Vendor	☐	☒	☐	☐
Shall be constant speed pumps	☒	☐	☐	☐	Vendor	☐	☒	☐	☐
Base mounted	☒	☐	☐	☐	Vendor	☐	☒	☐	☐
Rated for outdoor service	☒	☐	☐	☐	Vendor	☐	☒	☐	☐
Tower pumping system shall be configured such that any one tower pump can be removed from service without impacting operation	☒	☐	☐	☐	Vendor	☐	☒	☐	☐
New chilled water pumps	☒	☐	☐	☐	Vendor	☐	☒	☐	☐
The primary pumps shall be constant speed	☒	☐	☐	☐	Vendor	☐	☒	☐	☐
The secondary and tertiary pumps shall utilize VFDs	☒	☐	☐	☐	Vendor	☐	☒	☐	☐
Base mounted	☒	☐	☐	☐	Vendor	☐	☒	☐	☐
Secondary pumps rated for indoor service	☒	☐	☐	☐	Vendor	☐	☒	☐	☐
Tertiary pumps rated for outdoor service	☒	☐	☐	☐	Vendor	☐	☒	☐	☐
Chiller and cooling tower auxiliary support systems	☒	☐	☐	☐	Vendor	☐	☒	☐	☐
Chilled water system provided with expansion tank, air, and dirt separator	☒	☐	☐	☐	Vendor	☐	☒	☐	☐
Chilled water system provided with automatic makeup water system	☒	☐	☐	☐	Vendor	☐	☒	☐	☐
Chilled water system provided with interface to vendor-provided chemical injection skid	☒	☐	☐	☐	Vendor	☐	☒	☐	☐

Requirement statement	Requirement				Responsibility client/vendor	URS to be tested during this activity			
	M	I	D	IN		FAT	IQ	OQ	PQ
Cooling tower system provided with interface to vendor-provided chemical injection skid	☒	☐	☐	☐	Vendor	☐	☒	☐	☐
Cooling tower provided with automatic makeup water system and drains	☒	☐	☐	☐	Vendor	☐	☒	☐	☐
System controls	☒	☐	☐	☐	Vendor	☐	☒	☐	☐
Shall interface with existing plant control system (Rockwell Automation Plant BAS)	☒	☐	☐	☐	Vendor	☐	☒	☐	☐
Operation of the system components, including all chillers, towers, pumps, valves, etc. shall be fully automatic, but capable of manual override	☒	☐	☐	☐	Vendor	☐	☒	☐	☐
BMS system capable of trending historical data	☒	☐	☐	☐	Vendor	☐	☒	☐	☐
Changes to control system made through the BMS, and not through local control panels or HMIs	☒	☐	☐	☐	Vendor	☐	☒	☐	☐
Control hardware shall be Allen Bradley ControlLogix PLCs and I/O modules	☒	☐	☐	☐	Vendor	☐	☒	☐	☐
Upon power loss, equipment will default to fail safe condition	☒	☐	☐	☐	Vendor	☐	☒	☐	☐
Controls shall be provided with uninterruptible power supply (UPS)	☒	☐	☐	☐	Vendor	☐	☒	☐	☐
System will be PLC controlled with graphic user interface; Allen Bradley Panelview, minimum 15″ is preferred terminal	☒	☐	☐	☐	Vendor	☐	☒	☐	☐
Control system access via multiple levels of security with individual user log on/password, or other options by vendor	☒	☐	☐	☐	Vendor	☐	☒	☐	☐

Requirement statement	Requirement				Responsibility client/vendor	URS to be tested during this activity			
	M	I	D	IN		FAT	IQ	OQ	PQ
Graphic user interface shall provide the following system process control functions: o Process overview screen (P&ID) with current states/process data o Menu-driven screen selection o Popup controller graphics o The ability to manually control devices/instruments with appropriate security privileges o Alarm management screen	☒	☐	☐	☐	Vendor	☐	☒	☐	☐
Miscellaneous system components	☒	☐	☐	☐	Vendor	☐	☒	☐	☐
Meters and gages provided shall comply with the requirements, meters and gages for HVAC piping	☒	☐	☐	☐	Vendor	☐	☒	☐	☐
General duty valves provided shall comply with the requirements	☒	☐	☐	☐	Vendor	☐	☒	☐	☐
Butterfly valves provided shall comply with the requirements	☒	☐	☐	☐	Vendor	☐	☒	☐	☐
Piping hangers and supports provided shall comply with the requirements	☒	☐	☐	☐	Vendor	☐	☒	☐	☐
Heat tracing provided shall comply with the requirements	☒	☐	☐	☐	Vendor	☐	☒	☐	☐
Identification of piping and components provided shall comply with the requirements	☒	☐	☐	☐	Vendor	☐	☒	☐	☐
Piping insulation provided shall comply with the requirements	☒	☐	☐	☐	Vendor	☐	☒	☐	☐
Commissioning of the system shall comply with the requirements	☒	☐	☐	☐	Vendor	☐	☒	☐	☐
Flow instruments provided shall comply with the requirements	☒	☐	☐	☐	Vendor	☐	☒	☐	☐

Requirement statement	Requirement				Responsibility client/vendor	URS to be tested during this activity			
	M	I	D	IN		FAT	IQ	OQ	PQ
The system sequence of operations shall be implemented	☒	☐	☐	☐	Vendor	☐	☒	☐	☐
Hydronic piping provided shall comply with the requirements	☒	☐	☐	☐	Vendor	☐	☒	☐	☐
Hydronic piping specialties provided shall comply with the requirements	☒	☐	☐	☐	Vendor	☐	☒	☐	☐
Water treatment for the chilled water loop provided shall comply with the requirements	☒	☐	☐	☐	Vendor	☐	☒	☐	☐
System design loads • Existing chilled water peak load in tons • Future load in tons	☒	☐	☐	☐	Vendor	☐	☒	☐	☐
Chillers	☒	☐	☐	☐	Vendor	☐	☒	☐	☐
Proper service and operational clearances shall be provided, including tube pulling/cleaning activities	☒	☐	☐	☐	Vendor	☐	☒	☐	☐
Chillers shall be water cooled with centrifugal compressors as manufactured by Trane, Carrier, York (JCI), Daikin (McQuay), or Smardt	☒	☐	☐	☐	Vendor	☐	☒	☐	☐
Chiller evaporators shall be 2 pass with marine water box and hinged cover	☒	☐	☐	☐	Vendor	☐	☒	☐	☐
Chiller condenser shall be 2 pass with marine water box and hinged cover	☒	☐	☐	☐	Vendor	☐	☒	☐	☐
All chillers should be designed to utilize a unit mounted VFD drive with 480 V input power	☒	☐	☐	☐	Vendor	☐	☒	☐	☐
Chillers shall be factory tested prior to shipment	☒	☐	☐	☐	Vendor	☐	☒	☐	☐
Chillers shall provide chilled water at 40 °F, with a 10 °F delta	☒	☐	☐	☐	Vendor	☐	☒	☐	☐

Requirement statement	Requirement				Responsibility client/vendor	URS to be tested during this activity			
	M	I	D	IN		FAT	IQ	OQ	PQ
Chiller manufacturers shall provide selections showing the lowest possible chilled water supply temperature achievable with chilled water design flowrate and design condenser water flow and temperatures. At a minimum, each chiller selection shall be capable providing a supply water temperature of 1 °F below the design chilled water supply temperature of 40 °F	☒	☐	☐	☐	Vendor	☐	☒	☐	☐
Chillers and chilled water pumping system shall be designed to operate at full capacity with chilled water system supply and return water temperature differentials from 8 to 15 °F	☒	☐	☐	☐	Vendor	☐	☒	☐	☐
Chillers shall be designed for entering and leaving condenser water temperatures of 85 and 95 °F, respectively, and entering condenser water temperature as low as 65 °F	☒	☐	☐	☐	Vendor	☐	☒	☐	☐
Chiller manufacturers shall provide selections showing the highest possible entering condenser water temperature achievable with design chilled water flowrate and temperatures and design condenser water flow. At a minimum, each chiller selection shall be capable providing design supply water with an entering condenser water temperature of 1 °F above the design entering condenser water temperature of 85 °F	☒	☐	☐	☐	Vendor	☐	☒	☐	☐
Chiller shall be capable of operation at 20% of full capacity with 85 °F entering condenser water without compressor surge or use of hot gas bypass	☒	☐	☐	☐	Vendor	☐	☒	☐	☐
The chiller system shall be designed in a configuration that will allow any chiller to be taken out of service and still support the facility	☒	☐	☐	☐	Vendor	☐	☒	☐	☐

Commissioning of HVAC

The purpose of commission is to ensure that work has been completed as specified and that systems are functioning in the manner as described in specified system operating criteria. It will assist operating staff training and familiarization with new systems. It will serve as a tool to reduce post-occupancy critical system's operational difficulty or failure. It will, also, be used to develop test protocol and record the associated test data in an effort to advance the building systems from a state of substantial completion to a full dynamic operation. Commission will commence after preliminary punch list items are completed by subcontractors. The steps associated with commissioning are outlined next:

1. Installation verification
2. System start-up
3. Functional performance testing.

Operational staff training is essential to the commission process and will run concurrently with steps one through three. The commissioning team will include representatives of the owner, construction and installing subcontractors, test and balance subcontractor, and engineering acting as commissioning agent.

Commissioning plan as prepared by engineering shall provide functional performance tests (FPTs) procedures for the heating, ventilation, and air conditioning (HVAC) systems. The commissioning plan shall outline the organization, scheduling, team members, and documentation pertaining to the overall commissioning process. A narrative description of the design intents of the systems and their intended modes of sequences of operation is included. Procedures at the minimum shall consist of Testing Prerequisites, which contains verification that primary mechanical, electrical, and controls systems that support or interact with the system are completed, tested, and operational.

Installation verification contains verification that the system installation is completed and is ready for commissioning. The commencement of functional performance testing records the date and time of the start of system commissioning. System condition prior to staring performance testing section records the current set points and parameters of the system at the start of commissioning. FPT shall provide the following:

(1) Sequential steps required to set parameters and conditions required to test component and functions throughout intended ranges of operation.
(2) Full range of checks and tests carried out to determine if electric and pneumatic connections, components, subsystems, systems, and interfaces between systems function in accordance with the contract documents and design intents.
(3) All modes and sequences of control operations, interlocks, and conditional control responses and specified responses to abnormal emergency conditions.

At the end of FPT, the date and time of the end of system commissioning is recorded.

Field notes on record notes or remarks during system commissioning are collected. This shall include systems modifications, list problems discovered during commissioning that were corrected. List problems discovered during commissioning that were not corrected. List recommendations for further actions

The subcontractors shall be responsible for performing procedures presented in specification and contract drawings as detailed in the FPTs. Members of the designated commissioning team shall witness various portions of the commissioning process. Commissioning team members shall sign-off on appropriate sections after verifying installation, operation, or documentation. Final sign-off shall be by the owner and commissioning agent.

Any test ports, gauges, test equipment, etc., needed to accomplish the FPTs shall be provided by subcontractors. Subcontractors shall provide to the commissioning team documentation of calibration of controls. Documentation shall include dates, set points, calibration coefficients, control loop verification, and other data required to verify system checkout. Documentation shall be dated and initialed by field engineer or technician performing the work.

System narrative descriptions will be prepared by the commission agent and supported by flow diagrams, one-line diagrams, and appropriate specification sections for major systems to be commissioned. The commissioning agent will coordinate meetings with members of facility management and maintenance department groups to review system description documentation. The meetings will provide an overview of major system features, components, and arrangements as part of training and to provide for system acceptance and transfer to owner. The subcontractor and associated manufacturer's representatives shall provide required training to operational staff after the system description meetings have occurred. The subcontractor training sessions shall provide a more detailed analogy of systems operation and maintenance.

Instrumentation will be provided by the subcontractor. Instruments used for measurements shall be accurate. Calibration histories for each instrument shall be available for examination. Calibration and maintenance of instruments shall be in accordance with the requirements of applicable standards.

The installing subcontractor shall be responsible for collection of pertinent data during system start-up and functional performance testing. The subcontractor shall submit to the commissioning agent documentation of tests performed prior to and after system start-up. Documentation shall also include start-up procedures as approved by commissioning team.

Before system start-up begins, the commission team shall conduct a final installation verification audit. The subcontractor shall be responsible for completion of work, including change orders and punch list items to the owner's satisfaction. The audit shall include, checking of:

(a) Piping specialties, including balance, control, and isolation valves.
(b) Control sensor types and location.
(c) Identification of piping, valves, equipment, controls, etc.
(d) Major equipment, pumps, valves, starters, gauges, thermometers, etc.
(e) Documentation of prestart-up tests performed, including manufacturer's factory tests.

If work is found to be incomplete, incorrect, or nonfunctional, the subcontractor shall correct the deficiency before system start-up work proceeds.

A start-up plan shall be developed and submitted by the installing subcontractor. Start-up plan to include the following:

(a) Flushing and cleaning of pipe
(b) Filters, strainers, and screens
(c) Valve/damper positions
(d) Electrical tests
(e) Pressure tests
(f) Safety features testing
(g) Chemical treatment applications testing
(h) Manufacturer's tests.

The start-up plan will be reviewed and a prestart-up inspection performed by designated members of the commissioning team. The installing subcontractor shall commence with system start-up after approval has been given to start up plan and the prestart-up inspection is completed. Designated members of the commissioning team shall witness system start-up and list system and equipment deficiencies noted during start-up. The subcontractor shall take corrective action on system deficiencies noted and demonstrate to the commissioning team members' suitable system operation.

Designated systems requiring test and balance work shall have this activity commence after systems have successfully completed start-up. System and equipment deficiencies observed during this activity are to be noted and corrected.

Functional performance testing begins after operational testing, adjusting, and balancing of the systems have been completed by the subcontractors; and the system description and hands-on training sessions have been completed. The objective of the functional performance testing is to advance the building systems from a state of substantial completion to full dynamic operation in accordance with the specified design requirements and design intent. Attaining this objective will be accomplished by developing individual systems testing protocols which, when implemented by the subcontractor, will allow the commissioning team to observe, evaluate, identify deficiencies, recommend modifications, tune, and document the systems and systems equipment performance over a range of load and functional levels. FPTs for the systems to be commissioned are defined in the commissioning plan. These tests are intended to be conclusive but may require minor modifications as system operation dictates.

Installation Verification – X Ton Chiller System

Item	Observations/ comments	Acceptable		
		Yes	No	N/A
Date of delivery of X ton chiller				
Inspection of chiller				
Document condition of chiller system as received; note any damage or other issues:				
Verify chiller manufacturer, model number, nominal capacity (tons), and design, maximum and minimum flowrates (gpm):	Manufacturer: Model: Tons: Design flow, gpm: Maximum flow, gpm: Minimum flow, gpm:			
Verify pump manufacturer, model number and design, maximum and minimum flowrate (gpm), and head (feet):	Manufacturer: Model: Design flow: Maximum flow: Minimum flow: Head:			
Verify connection size of pump inlet and outlet (inches)	Chiller inlet/outlet connection size:			
Verify electrical power (voltage and amperage) and electrical connection:	Skid supply voltage: Skid supply amperage:			
Verify chiller skid is installed level and stable on ground:				
Verify carbon steel chilled water piping is correct size, supported properly, insulated and heat traced:				
Verify that all piping connections, valves, instrumentation, etc. are installed:				
Verify that chiller has been connected to carbon steel piping				

Item	Observations/ comments	Acceptable		
		Yes	No	N/A
Verify that the system has been filled with water and air removed from system				
Verify maintenance access space is adequate per manufacturers requirements				
Verify electrical connections are tight, grounded, etc.				
Verify chiller safeties energized and tested				
Verify initial operating set points set and high and low limit safety set points set				
Verify written certification from chiller manufacturer that all specified features, control, and safeties have been installed and are functioning properly and that the installation and application comply with the mfgr's recommendations				
Installation and startup manual provided				
All control devices and wiring complete				
Chiller stamped with ASME label from successful factory test				
Chemical treatment system installed				
Power to chiller/pump installed per contract documents and manufacturer's installation requirements				
Local safety disconnects installed in accessible location per NEC and labeled correctly				
Pump motor rotation in the proper direction				
Field-adjustable switches and circuit-breaker trip ranges set as indicated				
Verify equipment installed as per approved drawings: isolation valves, control valves, strainers, check valves, etc.				
Piping is cleaned per approved flushing and cleaning plan				
Makeup water supply provided				
List of problems found and resolved				

Item	Observations/ comments	Acceptable		
		Yes	No	N/A
List of problems found, and not resolved; identify lead to resolve problems				
Verify carbon steel chilled water piping is correct size, supported properly, insulated and heat traced:				
Verify that all piping connections, valves, instrumentation, etc. are installed:				
Verify that chiller has been connected to carbon steel piping				
Verify that the system has been filled with water and air removed from system				
Verify maintenance access space is adequate per manufacturers requirements				
Verify electrical connections are tight, grounded, etc.				
Verify chiller safeties energized and tested				
Verify initial operating set points set and high and low limit safety set points set				
Verify written certification from chiller manufacturer that all specified features, control, and safeties have been installed and are functioning properly and that the installation and application comply with the mfgr's recommendations				
Installation and startup manual provided				
All control devices and wiring complete				
Chiller stamped with ASME label from successful factory test				
Chemical treatment system installed				
Power to chiller/pump installed per contract documents and manufacturer's installation requirements				
Local safety disconnects installed in accessible location per NEC and labeled correctly				
Pump motor rotation in the proper direction				
Field-adjustable switches and circuit-breaker trip ranges set as indicated				

Item	Observations/ comments	Acceptable		
		Yes	No	N/A
Verify equipment installed as per approved drawings: isolation valves, control valves, strainers, check valves, etc.				
Piping is cleaned per approved flushing and cleaning plan				
Makeup water supply provided				
List of problems found and resolved				
List of problems found, and not resolved; identify lead to resolve problems				

System Startup – X Ton Chiller

Item	Observations/comments	Acceptable		
		Yes	No	N/A
Date of system start up test:				
Startup of chiller and pump				
Verify that power is available for chiller and pump, as well as system instrumentation and controls:				
Motor starter overload elements correctly sized and installed.				
Unit checked, aligned, and certified prior to start up and report submitted.				
Unit and motor lubricated before startup				
Pump shaft rotates easily with power turned off				
System starts and runs without any unusual noise or vibrations				
Start-up strainer removed (after 24 h)				
Unit is free of unusual noise or vibration				
Motor voltage and amps verified – each phase of each motor				

Item	Observations/comments	Acceptable Yes	No	N/A
Using plant control system, open the flow control valves for 100% open. Set flowrate for AHU to 50% of design				
Start chiller pump. Document any issues or problems with pump flow. Document flow to AHU, and bypass flowrate, as well as total chiller/pump flow (gpm):	Design flowrate for chiller (gpm): Test flowrate for chiller (gpm): 50% Design flowrate for AHU (gpm): Test flowrate for AHU (gpm): Bypass design flowrate (gpm): Bypass test flowrate for bypass (gpm):			
Start chiller. Document any issues or problems with operation of chiller. Document supply and return temperatures at the chiller. Document pump flow (gpm)	Problems or issues with chiller or pump noted: Chiller supply temp: Chiller return temp: Pump flow, gpm:			
List of problems found and resolved				
List of problems found and not resolved; identify lead to resolve problems				

Functional Testing – X Ton Chiller

Item	Observations/comments	Acceptable Yes	No	N/A
Date of functional testing:				
Functional testing documented by:				
Function test 1: Normal system operation				
Start rental chiller and pump; allow to run for a minimum of 5 min. Verify no issues with operation (flow or supply or return temperatures):				
Using the plant control system, place AHU in automatic control mode.				

Item	Observations/comments	Acceptable Yes	No	N/A
Verify total pump flow (gpm):	Pump flow, gpm:			
Verify flow for AHU	AHU flow, gpm:			
Verify total CHW system flow calculate bypass flowrate through	Total CHW system flowrate, gpm: Bypass flowrate:			
Verify chilled water supply and return temperatures, and chiller operating amps, % load:	CHW supply temperature: CHW return temperature: Chiller operating amps: Chiller % load:			
Verify chiller and pump flowrates are not above maximum or minimum flowrates for chiller and pump:				
Function test 2: No AHU flow; minimum flowrate for chiller and pump				
Start chiller and pump; allow to run for a minimum of 5 min. Verify no issues with operation (flow or supply or return temperatures):				
Using plant control system, index the flow control valves for AHU fully closed				
Verify total CHW system flow (gpm)	Total CHW pump flow, gpm:			
Verify chilled water supply and return temperatures, and chiller operating amps/% load:	CHW supply temperature: CHW return temperature: Chiller operating amps: Chiller % load:			
Verify that flowrate is not below minimum flowrate for chiller or pump:				
List of problems found and resolved:				
List of problems found and not resolved; identify lead to resolve problems				

7

Clean-In-Place (CIP) Systems

Clean-in-place (CIP) systems, a normal range of equipment operation, are designed specifying equipment that is used to support the production of finished pharmaceuticals in a current good manufacturing practice (cGMP) environment. An equipment validation plan is developed to outline the planned tasks and expectations for qualification of the equipment. The user requirements specification (URS) serves as the input to subsequent project risk assessment activities, design objectives, control strategy, and acceptance criteria for testing and qualification of the manufacturing system. The qualification activities [18] are planned as follows:

- Stage 1: Design qualification
- Stage 2: Equipment qualification (IQ/OQ/PQ)
- Stage 2: Process performance qualification (PPQ)
- Stage 3: Continued process verification (CPV)
- Cleaning validation (CV)

Each requirement below is assigned an Item ID, description, and category. The category will be one of the following:

Good manufacturing practice (*GMP*): Any requirement for a function or process that may have an impact on product safety, quality, purity, identity, or efficacy. Requirements categorized as GMP will be qualified.

Safety: Requirement dictated by site, corporate, or other safety-related agency or body, such as Occupational Safety and Health Administration (OSHA) or National Fire Protection Association (NFPA). Requirements are categorized as safety and impact the control system or any GMP aspect will be qualified.

Sterile Processing of Pharmaceutical Products: Engineering Practice, Validation, and Compliance in Regulated Environments, First Edition. Sam A. Hout.

Abbreviation/Term	Definition
FDS	functional design specification
CIP	clean-in-place
CNC	controlled not classified
EBR	electronic batch record
HDS	hardware design specification
CONTROL SYSTEM	mix automation
P&ID	piping and instrumentation diagram
Process air	compressed air that directly contacts product
PTS	powder transfer system
SDS	software design specification
STS	solution transmission system
WFI	water system tested to USP WFI criteria (water for injection)

The CIP system is used to clean and sanitize all solution manufacturing equipment, including PTS, mix tanks, and STS piping. The CIP system will consist of permanently installed CIP skids, associated piping, and chemical supply containers. Sanitization will be performed by recirculating hot water throughout the system. An ambient-once through WFI rinse will be performed after sanitization. The equipment and process will be controlled by the controls automation system. Data collection will be handled by the control systems and communicated to the manufacturing execution system (MES) system and EBR.

The PTS and mix tanks will be cleaned through the tank's discharge piping. The STS circuits will be cleaned independently. The general process steps will include an ambient water rinse, chemical cycle, ambient water rinse, hot water sanitization, cool down. Cleaning recipes will be developed for each of the cleaning processes.

PTS

Spray devices will be installed. The control system for the PTS will control the cleaning process and will request water or chemical solution from the CIP skids depending on which step is active in the process.

Mix Tank and Discharge Piping

Spray devices will be installed in the top and bottom of the mix tanks. Control systems will control the cleaning process and will request water or chemical solution from the CIP skids depending on which step is active in the process.

STS

The control system for the STS will control the cleaning process and will request water or chemical solution from the CIP skids depending on which step is active in the process.

Life Cycle Requirements

Vendor is expected to provide the following during the design and installation cycle of the project:

- Written communication of any deviations from client specifications.
- Single point of contact for communication.
- Provide an FDS, SDS, and HDS as required.
- Written communication of changes to the design specification.
- Manufacturing/completion project schedule.
- Written communication of delays in the schedule within one day of the change.

Product and Process User Requirements

Cleaning Solutions

Name	Description	pH	Concentration
CIP-100	Alkaline detergent	12.4	Up to 1%
CIP-200	Acid detergent	2.0	Up to 1%

Process Quality Requirements

Critical Quality Attributes (CQAs)

Process step	CQA	Specification	Acceptance
Cleaning	Chemical carryover	12.5 ppb	Cleaning validation
Cleaning	Endotoxin removal	≤0.125 EU/ml	Cleaning validation
Cleaning/ sanitization	Bioburden removal	≤ 10 cfu/100 ml	Cleaning validation
Cleaning/ sanitization	Product carryover	Residual material removed	Cleaning validation

Process Parameter Requirements

Critical Process Parameters (CPPs)

Process step	CPP	Operating range	Associated CQA
Cleaning	Rinse time	As developed	Chemical carryover
Cleaning	Supply flow	≥ 5 ft/s	Bioburden/endotoxin
Sanitization	Temperature	≥80 °C	Bioburden removal
Sanitization	Time	As developed	Bioburden removal
Cleaning	Hot water rinse	As developed	Product carryover
Cleaning	CIP-100 concentration	0.1–1%	Bioburden removal
Cleaning	CIP-200 concentration	0.1–1%	Bioburden removal
Cleaning	CIP-100 conductivity	1.227–11.14 μS/cm	Bioburden removal
Cleaning	CIP-200 conductivity	1.547–7.614 μS/cm	Bioburden removal

Installation User Requirements

Material of Construction Requirements (MOC)

Description	Category
All product contact steel parts are to be made from 316L stainless steel. Non-product contact may be 304 SS	GMP
Painted materials are not acceptable. Stainless steel shrouded parts may be allowed if approved by Client Engineering	GMP
All welds must be continuous and of the same parent material	GMP
Welding certifications are required for all product contact welds	GMP
Acceptable nonmetallic product contact materials are Viton, polytetrafluoroethylene (PTFE), Buna-N, ethylene propylene diene monomer (EPDM), silicone, and polyurethane	GMP
Direct product contact parts must have a ≤25-μ in risk analysis (RA) finish	GMP
Porous, fiber-producing, or absorbing materials are not allowed	GMP
Material certifications are required for all product contact parts	GMP
Any exceptions to these material requirements must be approved by Client Engineering	GMP
All product contact surfaces are required to be passivated. Electropolishing can be done in lieu of chemical passivation	GMP

Construction Requirements

Description	Category
Distribution system piping shall be appropriately supported and sloped at a minimum of 1/16″ per foot. Slope measurements shall be made between pipe hangers/supports and at each change of direction	GMP
Product piping and valves shall be constructed using ASME BPE sanitary standards	GMP
All system low points shall have a low point drain	GMP
Orbital welded connections are required for product contact piping. All welds will meet local governing standard operating procedure (SOP) requirements	GMP
All piping shall be clearly labeled with environmentally suitable placards indicating system and direction of flow per GES-UTIL-3020	GMP
Distribution pumps shall be mounted on base plates to minimize the effect of transient shock waves (water hammer)	GMP
The CIP skid shall be designed as a two-tank (wash and rinse) system	GMP
The CIP skid rinses will be directed to drain to the process waste system. Rinse to the process waste system will return to the CIP skid and then be directed to drain	GMP
CIP supply and return pumps will be centrifugal type and will include automated casing drains	GMP
The CIP skid shall be capable of delivering WFI, up to two different recirculated wash detergent solutions, and process air for air blow	GMP
The wash vessel will have the following features: Insulation Vortex breaker Vent filter with heat trace and insulation blanket	GMP
The rinse vessel will have the following features: Insulation Vortex breaker Vent filter with heat trace and insulation blanket	GMP
The heat exchanger will be capable of heating incoming WFI and maintaining the recirculating cleaning solution within the skid boundary at temperatures ≥176 °F (80 °C) prior to leaving the skid boundary and entering the distribution piping	GMP

Description	Category
CIP chemicals will be supplied from drums or totes using a suction lance and pumped by an air operated diaphragm pump	GMP
The flow meter will have the following features: Minimum flow span: 0 gal/min Maximum flow span: 180 gal/min Measurement accuracy: ±0.15% of rate from 10% of span to 100% of span Field calibrated span: 0–300 gpm NEMA 4× housing 0.5″ conduit connection Continuously adjustable zero and span 4-line backlit LCD display with touch control	GMP
The air-blow assembly will have the following features: Sanitary 316L ball check valve with automated air-blow system on the discharge side of the CIP heat exchanger The air-blow supply filter shall be a sanitary code-7 style with a 316L SS housing and 0.2-μm PTFE filter cartridge Vendor shall include a filter/regulator and pressure gauge for the air-blow compressed air supply	GMP
All piping will be fully drainable	GMP
The system will be designed to prevent cross contamination of unclean, clean in-process chemicals, and cleaned equipment sections	GMP

Lubrication Requirements

Description	Category
All lubricants used must be detailed in the design documentation. These Include lubricants used in the mechanical aspects of the equipment as well as the lubricants used in the product assembly	GMP
Where direct lubrication is required, the design and construction must be such that the lubrication cannot leak or drain into the product or into the product contact areas	GMP
Where pneumatic machine functions require the use of lubricated compressed air, only USP- and FDA-approved lubricants are acceptable: e.g. white mineral oil, USP	GMP
Lubricants with USDA and FDA approval and "AA" classification shall be used where incidental contact may occur	GMP

Pneumatic Requirements

Description	Category
A single compressed air supply must be utilized for both oiled and filtered air. Process air must not be lubricated	GMP
Process air piping must be made from 316L stainless steel. Tubing must be a non-sloughing plastic material	GMP
Airlines are to be color coded as follows: Yellow: Pressurized airlines for non-product contact air Clear: Process airlines Blue: Vacuum lines Black: Exhaust lines	GMP
All pneumatic lines must be accurately labeled at both termination points	General
All pneumatic manifold assemblies must have quick disconnects	General
All pneumatic components must be mounted within the machine base or enclosure	GMP
Use venturis for vacuum where possible. Vacuum pumps must be approved by Client Engineering	General
Air blow-offs are not acceptable in clean room installations	GMP
A 0.22-μm "point of use" filter will be installed by client, where practical, between the last air control device and product contact air supplies	GMP
Maximum compressed air pressure available is 90 psi	GMP

Electrical Requirements

Description	Category
Electrical systems must be designed to the provided Site Electrical and Control Specifications	GMP

Safety Requirements

Description	Category
All pinch points and hazardous areas must be guarded per OSHA standards using required materials of construction	General
E-stop buttons must be located near operator work areas and must release all potential energy immediately upon engagement	GMP
All access doors must have interlocks as described in the Site Electrical and Control Specifications	GMP

Description	Category
The design must permit the electrical, pneumatic, and hydraulic devices to be locked out during maintenance and repair activities. Lockout points must be easily accessible	General
The control system must be designed to operate in a fail-safe manner and meet OSHA safety requirements. The safety circuit must be category 3 or higher	General
Jog capability is acceptable only with a single momentary switch	General
Sharp corners and edges must be eliminated	General
Safety circuits must not be able to be bypassed	GMP
Access doors must be interlocked with magnetic mechanisms to not allow entry to the machine while it is running	GMP
Electrical panels with >50 V are to be designed with incident energy below 1.2 cal/cm^2	Safety
Fixed-mount human machine interfaces (HMIs) are to be set at a height of 58″ from the floor to the center of the screen. Adjustable mounts are to cover the range of 54″–68″ to center of the screen	Safety
When an emergency stop button is pushed, all sequences shall stop and valves shall return to their safe position. All stored energy must be immediately dissipated. Operator reset is required to resume operation	Safety
Noise level as measured at a 3-ft perimeter of the equipment shall not exceed 80 dBA	Safety
Enclosed spaces must be illuminated to a minimum 100 lx	Safety
Hot surfaces must be insulated or labeled if insulation is not possible to help protect personnel from harm	Safety
Chemical containers must be labeled with contents and SDS information	Safety

Operational Requirements

Process Constraints and Limitations

Description	Category
STS piping diameters will range from 1 to 3 in. depending on the line	General
Available utilities include: WFI: 150 gpm Steam: 55 psig Power: 480 V, 3 ph, 60 Hz Compressed air: 90 psig	General

Functions – FCT

Description	Category
The control system will include recipes with cleaning cycle parameters	GMP

Process Control System – PCS

Description	Category
The CIP skid shall be controlled by the Client North Mix Automation system and be capable of providing solution as requested by the system	General

Equipment Alarms and Warnings – EAW

Description	Category
Critical alarms shall automatically stop the equipment and notify the operator of the condition(s)	GMP
The operator will be required to acknowledge the alarm before the alarm can be reset and the system restarted	GMP
The condition leading to the alarm must be resolved before the alarm or warning can be reset	GMP
Warnings (Informational) will notify the operator of a condition but not stop the equipment or require further action	GMP
Discrete device valves, actuator position must be monitored in both directions open and closed. The system must fault if a position error is detected	GMP
Any CQA that is out of specification limits must trigger an alarm	GMP
System compressed air pressure outside of limits must trigger an alarm	GMP

Data – DAT

Description	Category
Data collection will be defined in the control systems project	General

Power Loss and Recovery – PLR

Description	Category
On power restoration, the system shall not restart without operator interface or communication-link input	GMP
No damage to machine will occur as a result of going to the safe state	General
The equipment manufacturer will supply instructions for recovery from catastrophic control system failures. It is generally accepted that the system shall protect in the following priority; personnel, equipment, and then product	GMP
The process may be able to continue operation upon loss of power communication with the server. The ability to continue will depend on where the process is stopped. In some cases, the process will be put into a hold state and resume once the power is restored or communication is reestablished. In some cases, the process may have to be aborted. Data stored during the outage will be uploaded once communication is restored	GMP

Cleaning and Sanitizing – CLE

Description	Category
The CIP skid will be cleaned and sanitized on a regular basis	GMP
Cleaning validation will be performed on the CIP skid	GMP

Material/Waste Movement Requirements (MWM)

Description	Category
Diluted CIP chemicals may be discharged to drain if concentration is $\leq$1%	General

Maintenance Requirements (MNT)

Description	Category
Preventive Maintenance procedures will be available at the factory acceptance test (FAT)	GMP
Spare parts lists will be available 1 mo prior to FAT	General
Original equipment manufacturer (OEM) part numbers are required for all spare parts	General
AutoCAD drawings will be supplied for all machined parts	General
The operations manual will include a section on troubleshooting and repair	General

Training and Documentation Requirements (DOC)

Description	Category
Material and weld certifications for direct product contact parts are required	GMP
Material cut sheets or specs are required for all process air connections	GMP
Training documents will be supplied for operation and maintenance procedures	General
Operator training will be available during commissioning, potentially on multiple shifts	General
Maintenance-specific training will be available during commissioning, potentially on multiple shifts	General
A draft operations manual will be available 1 mo prior to the FAT and the final version at the FAT	General
Manuals will be provided in both electronic and hard-copy versions	General

8

Computerized Automated Systems

Functional Requirement Specification

Process Automation System

This defines the process automation system (PAS) environmental monitoring system (EMS) area operations and functions. The EMS operations are monitored by Programmable Logic Controller (PLC)-based control system. The EMS is implemented on the PAS that is detailed in the PAS Core Functional Specification. The EMS PAS includes these major components.

This system will monitor environmental conditions for the following areas:

1. Chemical solvent storage
2. Warehouse
3. Mezzanine
4. Extrusion process (component manufacturing)
5. Quality control labs
6. Filling operations
7. Mixing operations
8. Component preparations
9. Sampling

Acronym or term	Definition
AUTO	Proportional–integral–derivative (PID) automatic, automatic
BAS	Building automation system
CAS	PID cascade, cascade
CIP	Clean in place
CM	Control module

Sterile Processing of Pharmaceutical Products: Engineering Practice, Validation, and Compliance in Regulated Environments, First Edition. Sam A. Hout.

Acronym or term	Definition
EMS	Environmental monitoring system
Ethernet	Broadcast networking system that carries digital data packets to local nodes. A 10-Mbps base and local area network (LAN), Ethernet evolved from the IEEE 802.3 standard. Ethernet is the transport vehicle for many upper level protocols, including transmission control protocol and internet protocol (TCP/IP), and Xerox network systems (XNS)
EM	Equipment module
HMI	Human machine interface (operator terminal)
I/O	Input/output signals to/from the control system and the field devices
MAN	PID manual, manual
OEM	Original equipment manufacturer
PAS	Process automation system
PD	Power distribution
PID	PID control, proportional–integral–derivative control
PIO	Point I/O – Allen Bradley distributed I/O system
P&ID	Piping and instrumentation diagram
PLC	Programmable logic controller
POU	Point of use
PW	Purified water
RIO	Remote I/O
SCADA	Supervisory control and data acquisition
SIP	Steam in place
SMDS	Software module design specification
SOP	Standard operating procedure
UPS	Uninterruptible power supply
URS	User requirement specification

EMS User Requirement Specification

Automation Overview

The PAS consists of a plant PAS system, which will provide monitoring, visualization, and control over areas contained within the PAS. The control system comprises an Allen Bradley ControlLogix controller (PLC) and color-thin client operator interfaces running FactoryTalk View SE client software.

The PAS EMS hardware and software components contain a summary of the system components.

Software

EMS control logic is integrated with the PAS control system. The EMS control system SCADA platform is Rockwell Automation's FactoryTalk View SE. Functional specification for the PAS system is documented in PAS Core Automation Functional Specification. The software used to program the PLC is Rockwell RS Logix 5000. The software used to program the graphics application is FactoryTalk View SE, network distributed.

Hardware

SCADA

All visualization software and configuration are installed on the PAS control system. The PAS control system is hosted on VMWare VSphere virtual servers located and maintained by IT group.

Thin client monitors are mounted in logical areas for operator interface. For example, these reside in the following locations:

- Manufacturing area between filling and component preparations
- Manufacturing area main hallway outside filling
- Preparations area outside gowning room
- Boiler control room
- Heating ventilation and air conditioning (HVAC) control room

Control Panels

The EMS system consists of a single PLC enclosure, PIO rack one is installed in this same enclosure, other PIO remotely network enclosures, and an instrument enclosure installed in each area of the EMS. Each PLC/PIO enclosure has a 120-V AC to 24-V DC power distribution enclosure that serves as the rack lockout point. By segregating the 120-V AC, it ensures all voltages inside the PLC enclosures are 24-V DC or less.

Main Control Panel

This panel houses low-voltage (24-V DC) control components. The panel contains the PLC and RIO rack one as well as power supplies for each, a network switch, terminals, circuit breakers, and fuses.

The PLC rack contains an A–B ControlLogix PLC, two Ethernet/IP modules, and one historian module. The RIO rack one contains an A–B PIO Ethernet/IP communication base and standard PIO I/O cards with a minimum of 20% spare. I/O are wired to terminal blocks, as required. These panels provide termination points for PIO rack one environmental monitoring of field I/O. An A–B Ethernet network switch will interface the RIO rack to the PLC via an Ethernet/IP network. I/O modules and rack layout are defined in the PAS EMS Area Design Specification.

Remote I/O Panels

Each RIO panel houses low-voltage (24-V DC) control components. Each panel contains a remote rack and required power supply, a network switch, terminals, circuit breakers, and fuses. These panels provide termination points for room environmental monitoring field I/O.

The RIO racks each contain an A–B PIO Ethernet/IP communication base, input, and output cards with a minimum of 20% spare. An A–B Ethernet network switch will interface the RIO racks to the main PLC control panel via an Ethernet/IP network. I/O modules and rack layout are defined in the PAS EMS area design specification.

Power Distribution Panels

Each of the PLC and RIO panels has an associated PD panel installed nearby. Each PD includes an external flanged non-fusible disconnect that serves as the area panel lockout. These panels house a mix of low-voltage (120-V AC and 24-V DC) control components. Each panel contains a 480-W 24-V DC power supply, a DIN-rail-mounted UPS, terminals, and circuit breakers. These panels will provide termination points for incoming plant 120-V AC power and 24-V DC supply power to the PLC/RIO panels.

Instrumentation Panels

Each Instrument panel houses low-voltage (24-V DC) control components. Each panel contains a pressure transmitter, alarm beacon, and terminals. A combination temperature–humidity transmitter is located external to the instrument panel. These panels will provide in-room termination points for environmental monitoring field I/O.

The instrument panels each have a single stainless steel "instrument tag" hanging in proximity of the panel. This tag lists the individual instrument enclosure number, client instrument numbers, manufacturer part numbers, associated drawing package, and voltage level.

System Functions

System Overview

The client PAS EMS scope is limited to monitoring and historization of critical environment parameters. The PAS will provide visualization and alarm functionality for this data.

Control Modules

Control modules implemented as a part of the PAS will be used to execute monitoring and control functions. For design details and a complete list of module classes refer to FRS.

Interfaces

User Interface (SCADA)

The PAS SCADA is designed so the system can accommodate multiple process areas. Interfaces specific to the client, EMS are described later. PAS system-level interfaces are described in the PAS Core Design Specification.

PAS System Overview

The PAS system overview depicts and allows navigation to the process areas in the client facility.

Graphical Screen Navigation

Graphical screen navigation within the client process area follows the navigation convention for the PAS. A general overview screen will allow navigation to graphical control and alarm screens related to the client systems.

Client EMS Area Graphical Screens

Graphical Control Screens

Display screens include tabular representations of data from the monitored equipment.

Chemical Solvent Storage Environmental Monitoring System

This contains graphical representations of the components pertinent to environmental monitoring in the chemical solvent storage area. All related monitoring is located on these screens and allows view from the faceplates of these modules.

Warehouse Monitoring System

This contains graphical representations of the components pertinent to environmental monitoring in the warehouse area. This area includes the preparations and subassembly areas. All related monitoring is located on these screens and will allow view from the faceplates of these modules.

Mezzanine Environmental Monitoring System

This will be split into graphics, filling, and mixing. These screens contain graphical representations of the components pertinent to environmental monitoring of mezzanine area. All related monitoring will be located on these screens and will allow view from the faceplates of these modules.

Extrusion Processing Environmental Monitoring System

This will contain graphical representations of the components pertinent to environmental monitoring in the extrusion area. All related monitoring will be located on these screens and will allow view from the faceplates of these modules.

Quality Control Labs Environmental Monitoring System

This will contain graphical representations of the components pertinent to environmental monitoring in the quality control labs area. All related monitoring will be located on these screens and will allow view from the faceplates of these modules.

Filling Environmental Monitoring System

This will contain graphical representations of the components pertinent to environmental monitoring in the filling area. All related monitoring will be located on these screens and will allow view from the faceplates of these modules.

Mixing Environmental Monitoring System

This will contain graphical representations of the components pertinent to environmental monitoring in the mixing area. All related monitoring will be located on these screens and will allow view from the faceplates of these modules.

Component Assembly Environmental Monitoring System

This will contain graphical representations of the components pertinent to environmental monitoring in the component assembly area. All related monitoring will be located on these screens and will allow view from the faceplates of these modules.

Sampling Environmental Monitoring System

This will contain graphical representations of the components pertinent to environmental monitoring in the sampling area. All related monitoring will be located on these screens and will allow view from the faceplates of these modules.

Warehouse Environmental Monitoring System

This will contain graphical representations of the components pertinent to environmental monitoring in the warehouse area. All related monitoring will be located on these screens and will allow view from the faceplates of these modules.

Interfaces to Equipment

The PAS will interface with EMS area field instrumentation and devices through Ethernet/IP protocols in addition to traditional hardwired I/O. Where applicable, the PAS will provide an EMS area user interface where operators can monitor system conditions and receive alarm notifications. The PAS EMS user interface will consist of multiple thin client interfaces. Each thin client interface will provide the ability for a user to monitor all EMS operations.

Interfaces to Other Systems

Ethernet Network

Wireless Connections

The EMS will retrieve data from several gateway units. These connections will be made in the chemical solvent storage and warehouse areas.

Nonfunctional Attributes

The control system will be available for continuous operation 24 hours a day, 7 days a week.

Power Failure Recovery

In the event of a power failure, the area UPS will maintain PLC and RIO operation for a limited time. If power outage persists, the PLC and RIO panels will turn off and field devices will default to their respective fail-safe state. Once power is restored, the operations staff will resume operations as required by client SOPs.

PLC Only Failure

If there is a failure in the PLC only, field devices will default to their respective fail-safe state. Since the HMI will not receive any response from the PLC, bad data quality will be indicated on the display for all affected points.

HMI Only Failure

If there is a failure of the HMI only, it will not affect process operations and the PLC will continue to operate as normal. If the HMI fails, there will be no way to monitor inputs to the system.

Network Only Failure

If there is a failure at the client network level, it will not affect process operations and the PLC will continue to operate as normal. There is a local historian module installed in the main PLC rack. This module will continue to monitor the devices located in the mezzanine control system area. Other areas will not collect data until the network is operational.

Maintainability

The control system will be maintained in accordance with site administrative and maintenance procedures and manufacturer's guidelines.

Continuous control hardware will utilize nonvolatile memory or battery backup to preserve programming, configuration, and setpoints, through a loss of power.

Backup and Recovery

System components will be backed up via client SOP. The system will be fully recoverable utilizing vendor-provided software in conjunction with back-up media.

Functional Requirement Specification

GLC Environmental Monitoring System

This defines the global logistics center (GLC) EMS area operations and functions at client facility. The EMS operations will be monitored by the supplied PLC-based control system.

The GLC EMS includes these major components:

This system will monitor environmental conditions for the following manufacturing areas:

- Refrigeration rooms
- Zone x
- Zone y
- Zone z

Definitions and Acronyms

Acronym or term	Definition
AUTO	PID automatic, automatic
BAS	Building automation system
CAS	PID cascade, cascade
CIP	Clean in place

Definitions and Acronyms

Acronym or term	Definition
CM	Control module
EMS	Environmental monitoring system
Ethernet	Broadcast networking system that carries digital data packets to local nodes. A 10-Mbps base and local area network (LAN), Ethernet evolved from the IEEE 802.3 standard. Ethernet is the transport vehicle for many upper level protocols, including TCP/IP and Xerox network systems (XNS)
EM	Equipment module
EMS	Environmental monitoring system
HMI	Human machine interface (operator terminal)
I/O	Input/output signals to/from the control system and the field devices
MAN	PID manual, manual
MKT	Mean kinetic temperature
OEM	Original equipment manufacturer
PAS	Process automation system
PD	Power distribution
PID	PID control, proportional-integral-derivative control
PIO	Point I/O – Allen Bradley distributed I/O system
P&ID	Piping and instrumentation diagram
PLC	Programmable logic controller
POU	Point of use
PW	Purified water
RIO	Remote I/O
SCADA	Supervisory control and data acquisition
SIP	Steam in place
SMDS	Software module design specification
SOP	Standard operating procedure
UPS	Uninterruptible power supply
URS	User requirement specification

Automation Overview

The GLC EMS will provide monitoring and visualization over Zone x, Zone y, and Zone z of the GLC at Client facility. The control system comprises an Allen Bradley

ControlLogix controller (PLC) and a color-thin client operator interface running FactoryTalk View SE client software.

System Components

The GLC EMS hardware and software components contain:

Software

The GLC EMS control system SCADA platform is Rockwell Automation's FactoryTalk View SE. The software used to program the PLC is Rockwell RS Logix 5000. The software used to program the graphics application is FactoryTalk View SE, network distributed.

Hardware

SCADA

All visualization software and configuration will be installed on the PAS control system. The PAS control system is hosted on VMWare VSphere virtual servers located and maintained by IT group.

Thin client monitor will be mounted for operator interface.

Control Panels

The GLC EMS system consists of a PLC enclosure and a 120-V AC to 24-V DC PD enclosure that serves as the rack lockout point. Segregating the 120-V AC ensures all voltages inside the PLC enclosures are 24-V DC or less.

Power Distribution Panel

The PD panel includes an external flanged non-fusible disconnect that serves as the area panel lockout. The PD panel houses a mix of low-voltage (120-V AC and 24-V DC) control components. It contains a 480-W 24-V DC power supply, a DIN rail mounted UPS, terminals, and circuit breakers. The PD panel will provide termination points for incoming plant 120-V AC power and 24-V DC supply power to the PLC panel.

Control Panel

Located in the GLC near the PD panel, the Control panel houses low-voltage (24-V DC) control components. The panel contains the PLC and a gateway as well as a 24 V chassis for the PLC, a network switch, a Prosoft module, terminals, circuit breakers, and fuses.

The PLC rack contains an A–B ControlLogix PLC, two Ethernet/IP modules, and one Historian module. An A–B Ethernet network switch will interface the gateway, Prosoft, and PLC via an Ethernet/IP network.

Gateways

The GLC EMS system includes two gateways, located in the warehouse and high bay areas. Temperature and humidity sensors wirelessly transmit to the Gateways, which communicate via Modbus to a Prosoft module in the Control panel.

System Functions

System Overview

The GLC EMS scope is limited to monitoring and historization of critical environment parameters. The PAS will provide visualization and alarm functionality for this data.

Control Modules

Control modules implemented as a part of the PAS will be used to execute monitoring and control functions. For design details and a complete list of module classes, see the PAS software module design specification.

Interfaces

User Interface (SCADA)

Interfaces specific to the GLC EMS are described later. PAS system-level interfaces are described in the PAS Core Design Specification.

GLC EMS Area Graphical Screen

The display screen will include tabular representations of data from the monitored equipment.

Global Logistics Center (GLC) Environmental Monitoring System

This graphic will contain graphical representations of the components pertinent to environmental monitoring in the GLC area. All related monitoring will be located on these screens and will allow view from the faceplates of these modules.

Interfaces to Equipment

The PAS will interface with GLC EMS area field instrumentation and devices through Ethernet/IP protocols in addition to wireless I/O. The PAS will provide the GLC EMS area user interface where operators can monitor system conditions and receive alarm notifications. The PAS GLC EMS user interface will consist of one thin client interface.

Interfaces to Other Systems

Ethernet Network

The GLC PLC will communicate over the Ethernet network to the EMS PLC. The EMS PLC will pass sensor data from the refrigeration room area, so that the data may be viewed in both areas.

Wireless Connections

The EMS will retrieve data from gateway units. These connections will be made in the GLC areas.

Nonfunctional Attributes

The control system will be available for continuous operation 24 hours a day, 7 days a week.

Power Failure Recovery

In the event of a power failure, the area UPS will maintain PLC and RIO operation for a limited time. The local historian module will continue to collect data until the UPS is exhausted. If power outage persists, the PLC will turn off and field devices will not have the data saved. Once power is restored, the data communication and recording will resume as required by Client SOPs.

PLC Only Failure

If there is a failure in the PLC only, field devices will default to their respective fail-safe state. Since the HMI will not receive any response from the PLC, bad data quality will be indicated on the display for all affected points.

HMI Only Failure

If there is a failure of the HMI only, it will not affect process operations and the PLC will continue to operate as normal. If the HMI fails, there will be no way to monitor inputs to the system, but the historian will continue to collect data.

Network Only Failure

If there is a failure at the client network level, it will not affect process operations and the PLC will continue to operate as normal. There is a local Historian ME module installed in the main PLC rack. This module will continue to monitor the devices located in the GLC control system area and will upload to the primary historian when the network is recovered.

Maintainability

The control system will be maintained in accordance with site administrative and maintenance procedures and manufacturer's guidelines. Continuous control hardware will utilize nonvolatile memory or battery backup to preserve programming, configuration, and setpoints, through a loss of power.

Backup and Recovery

System components will be backed up via Client SOP. The system will be fully recoverable utilizing vendor-provided software in conjunction with back-up media.

Software Module Design Specification

The purpose of this SMDS is to specify the functionality and design details for the module classes used in the software development of the process control system (PCS) implemented.

PCS is built on Rockwell PLCs and a FactoryTalk View SE PlantPAS platform. These products support the use of library templates that facilitate standardization and code reuse.

SMDS will detail items associated with the SCADA and PLC templates comprising the module class. The scope includes but is not limited to the following:

1. Template names and versions
2. Functional description
3. Parameters
4. Alarms
5. Historization
6. Scripts

Definitions

Term	Definition
DST	Device signal tag. A PLC structure that contains a data value and a data quality
PLC	Programmable logic controller
SCADA	Supervisory control and data acquisition
SMDS	Software module design specification

Appendix – Example – Functional Design Specifications

Functional Requirement Specification

Process Automation System

This document defines the PAS EMS area operations and functions. The EMS operations will be monitored by supplied PLC-based control system. The EMS will be implemented on the PAS that is detailed in the PAS Core Functional Specification.

The EMS PAS includes these major components:

This system will monitor environmental conditions for the following areas:

- Chemical solvent storage
- Warehouse
- Mezzanine
- Extrusion
- Quality control labs
- Filling
- Mixing
- Assembly
- Sampling
- Warehouse

Definitions and Acronyms

Acronym or term	Definition
AUTO	PID automatic, automatic
BAS	Building automation system
CAS	PID cascade, cascade
CIP	Clean in place
CM	Control module
EMS	Environmental monitoring system
Ethernet	Broadcast networking system that carries digital data packets to local nodes. A 10-Mbps base and local area network (LAN), Ethernet evolved from the IEEE 802.3 standard. Ethernet is the transport vehicle for many upper level protocols, including TCP/IP and Xerox network systems (XNS)
EM	Equipment module
EMS	Environmental monitoring system
HMI	Human machine interface (operator terminal)
I/O	Input/Output signals to/from the control system and the field devices
MAN	PID manual, manual

Definitions and Acronyms

Acronym or term	Definition
OEM	Original equipment manufacturer
PAS	Process automation system
PD	Power distribution
PID	PID control, proportional–integral–derivative control
PIO	Point I/O – Allen Bradley distributed I/O system
P&ID	Piping and instrumentation diagram
PLC	Programmable logic controller
POU	Point of use
PW	Purified water
RIO	Remote I/O
SCADA	Supervisory control and data acquisition
SIP	Steam in place
SMDS	Software module design specification
SOP	Standard operating procedure
UPS	Uninterruptible power supply
URS	User requirement specification

Automation Overview

The PAS will consist of a PlantPAS system, which will provide monitoring, visualization, and control over areas contained within the PAS. The control system comprises an Allen Bradley ControlLogix controller (PLC) and color-thin client operator interfaces running FactoryTalk View SE client software.

System Components

Software

EMS control logic will be integrated with the PAS control system. The EMS control system SCADA platform is Rockwell Automation's FactoryTalk View SE. Functional specifications for the PAS system are documented in PAS Core Automation Functional Specification. The software used to program the PLC is Rockwell RS Logix 5000. The software used to program the graphics application is FactoryTalk View SE, network distributed.

Hardware

SCADA

All visualization software and configuration will be installed on the PAS control system. The PAS control system is hosted on VMWare VSphere virtual servers located and maintained by the Client IT group.

Thin client monitors will be mounted in logical areas for operator interface. These will reside in the following locations:

- Manufacturing area hallway between filling and assembly.
- Manufacturing area main hallway outside filling.
- Assembly area outside garb room.
- Boiler control room
- HVAC control room

Control Panels

The EMS system consists of a single PLC enclosure, PIO rack one is installed in this same enclosure, nine PIO remotely network enclosures, and an instrument enclosure installed in each area of the EMS. Each PLC/PIO enclosure has a 120-V AC to 24-V DC PD enclosure that serves as the rack lockout point. By segregating the 120-V AC, it ensures all voltages inside the PLC enclosures are 24-V DC or less.

Main Control Panel

This panel houses low-voltage (24-V DC) control components. The panel contains the PLC and RIO rack one as well as power supplies for each, a network switch, terminals, circuit breakers, and fuses.

The PLC rack contains an A–B ControlLogix PLC, two Ethernet/IP modules, and one Historian module. The RIO rack one contains an A–B PIO Ethernet/IP communication base and standard PIO I/O cards with a minimum of 20% spare. I/O are wired to terminal blocks, as required. These panels will provide termination points for PIO rack one environmental monitoring of field I/O. An A–B Ethernet network switch will interface the RIO rack to the PLC via an Ethernet/IP network. I/O modules and rack layout shall be defined in the PAS EMS Area Design Specification.

Each RIO panel houses low-voltage (24-V DC) control components. Each panel contains a remote rack and required power supply, a network switch, terminals, circuit breakers, and fuses. These panels will provide termination points for room environmental monitoring field I/O.

The RIO racks each contain an A–B PIO Ethernet/IP communication base, input, and output cards with a minimum of 20% spare. I/O are wired to TB's, as required. An A–B Ethernet network switch will interface the RIO racks to the main PLC control panel, via an Ethernet/IP network. I/O modules and rack layout shall be defined in the PAS EMS area design specification.

Power Distribution Panels

Each of the PLC and RIO panels has an associated PD panel installed nearby. Each PD includes an external flanged non-fusible disconnect that serves as the area panel lockout. These panels house a mix of low-voltage (120-V AC and 24-V DC) control components. Each panel contains a 480-W 24-V DC power supply, a DIN rail mounted UPS, terminals, and circuit breakers. These panels will provide termination points for incoming plant 120-V AC power and 24-V DC supply power to the PLC/RIO panels.

Instrumentation Panels

Each Instrument panel houses low-voltage (24-V DC) control components. Each panel contains a pressure transmitter, alarm beacon, and terminals. A combination temperature–humidity transmitter is located external to the instrument panel. These panels will provide in-room termination points for environmental monitoring field I/O.

The instrument panels each have a single stainless steel "instrument tag" hanging in proximity of the panel. This tag lists the individual instrument enclosure number, client instrument numbers, manufacturer part numbers, associated drawing package, and voltage level.

System Functions

System Overview

PAS EMS scope is limited to monitoring of critical environment parameters. The PAS will provide visualization and alarm functionality for this data.

Control Modules

Control modules implemented as a part of the PAS will be used to execute monitoring and control functions.

Interfaces

User Interface (SCADA)

The PAS SCADA will be designed so the system can accommodate multiple process areas. PAS system-level interfaces are described in the PAS Core Design Specification.

PAS System Overview Screen

The PAS System Overview will depict and allow navigation to the process areas.

Graphical Screen Navigation

Graphical screen navigation within process area will follow the navigation convention for the PAS. A general overview screen will allow navigation to graphical control and alarm screens.

Graphical Control Screens

Display screens will include tabular representations of data from the monitored equipment.

Chemical Solvent Storage Environmental Monitoring System

This will contain graphical representations of the components pertinent to environmental monitoring in the chemical solvent storage area. All related monitoring will be located on these screens and will allow the view from the faceplates of these modules.

Warehouse Monitoring System

This will contain graphical representations of the components pertinent to environmental monitoring in warehouse area. This area includes assembly and subassembly areas. All related monitoring will be located on these screens and will allow the view from the faceplates of these modules.

Mezzanine Environmental Monitoring System

This will be split into three graphics: north filling east, north filling west, and north mixing. These screens contain graphical representations of the components pertinent to environmental monitoring of the north mezzanine area. All related monitoring will be located on these screens and will allow the view from the faceplates of these modules.

Extrusion Processing Environmental Monitoring System

This will contain graphical representations of the components pertinent to environmental monitoring in the north extrusion area. All related monitoring will be located on these screens and will allow the view from the faceplates of these modules.

Quality Control Labs Environmental Monitoring System

This will contain graphical representations of the components pertinent to environmental monitoring in the quality control labs area. All related monitoring will be located on these screens and will allow view from the faceplates of these modules.

Filling Environmental Monitoring System

This will contain graphical representations of the components pertinent to environmental monitoring in the south-filling area. All related monitoring will be located on these screens and will allow view from the faceplates of these modules.

Mixing Environmental Monitoring System

This will contain graphical representations of the components pertinent to environmental monitoring in the south mixing area. All related monitoring will be located on these screens and will allow view from the faceplates of these modules.

Assembly Environmental Monitoring System

This will contain graphical representations of the components pertinent to environmental monitoring in assembly area. All related monitoring will be located on these screens and will allow view from the faceplates of these modules.

Sampling Environmental Monitoring System

This will contain graphical representations of the components pertinent to environmental monitoring in sampling area. All related monitoring will be located on these screens and will allow view from the faceplates of these modules.

Warehouse Environmental Monitoring System

This will contain graphical representations of the components pertinent to environmental monitoring in the warehouse area. All related monitoring will be located on these screens and will allow view from the faceplates of these modules.

Interfaces to Equipment

The PAS will interface with EMS area field instrumentation and devices through Ethernet/IP protocols in addition to traditional hardwired I/O.

Where applicable, the PAS will provide an EMS area user interface where operators can monitor system conditions and receive alarm notifications. The PAS EMS user interface will consist of multiple thin client interfaces. Each thin client interface will provide the ability for a user to monitor all EMS operations.

Interfaces to Other Systems

Ethernet Network

Wireless Connections

The EMS will retrieve data from several gateway units. These connections will be made in the chemical solvent storage and warehouse areas.

Nonfunctional Attributes

The control system will be available for continuous operation 24 hours a day, 7 days a week.

Power Failure Recovery

In the event of a power failure, the area UPS will maintain PLC and RIO operation for a limited time. If power outage persists, the PLC and RIO panels will turn off and field devices will default to their respective fail-safe state. Once power is restored, the operations staff will resume operations as required by client SOPs.

PLC Only Failure

If there is a failure in the PLC only, field devices will default to their respective fail-safe state. Since the HMI will not receive any response from the PLC, bad data quality will be indicated on the display for all affected points.

HMI Only Failure

If there is a failure of the HMI only, it will not affect process operations and the PLC will continue to operate as normal. If the HMI fails, there will be no way to monitor inputs to the system.

Network Only Failure

If there is a failure at the client network level, it will not affect process operations, and the PLC will continue to operate as normal. There is a local Historian ME module installed in the main PLC rack. This module will continue to monitor the devices located in the north mezzanine control system area. Other areas will not collect data until the network is operational.

Maintainability

The control system will be maintained in accordance with site administrative and maintenance procedures and manufacturer's guidelines. Continuous control hardware will utilize nonvolatile memory or battery backup to preserve programming, configuration, and setpoints, through a loss of power.

Backup and Recovery

System components will be backed up via client SOP. The system will be fully recoverable utilizing vendor-provided software in conjunction with back-up media.

Software Design Specification

Prosoft Configuration

The following graphic depicts communication flow through the system. Sensors communicate wirelessly to the gateways. Gateways check into the Enterprise Manager whenever available. Gateways push sensor data to the Prosoft card, which pushes that data into PLC tags. The Prosoft client commands are configured as follows:

Prosoft has been configured to give an error code when applicable. An error code will be given in the PLC code per Modbus line when applicable. For example, if all of the sensors were unable to communicate with the server, the Prosoft Error array will show −33 in the first 15 lines of the array. If it were only the sensors connected to Gateway1, −33 would appear in the first five lines of the array. Error codes and other related information can be found in the PLX3_user manual. If this error occurs, ensure that the Enterprise VM is running and then power cycle the gateways.

The gateways and humidity/temperature sensors can be checked for connectivity, calibrated, removed, and properties manipulated using the Enterprise Manager VM. To edit any sensors or gateways via the Enterprise Manger VM, the "Default Server Interface" option must be enabled via an offline configuration mode accessed on power up of the gateways. Once the gateway has the default server interface active, open a web browser in the VM and then go to http://localhost/ to access the login page. After logging in with the user credentials for the specific network, navigate to the manage tab at the top of the screen to view all sensors and gateways associated with the selected network.

Enabling and Disabling of the Gateway Default Server Interface

Once finished with the Enterprise Manager utilities, follow the steps again to disable the default server interface to continue operating the sensors in offline mode.

1. Verify the LAN settings are correct to communicate to the gateway's static IP address. Connect an Ethernet cable from the Enterprise Manager VM computer to the Ethernet port on the back of the gateway.
2. While the device is booting, press and hold the reset button on the back until all three lights toggle red, green. Release the button and the device is in local configuration mode.
3. Use a web browser and go to the static IP address of the device. The lights will all turn solid green once this connection has been established.
4. Edit the configuration pages as needed and save. Once settings have been saved, you will need to reboot the device for them to take effect via the reboot button or power cycling the device.

Temperature Sensor Settings

The settings seen later are for every temperature sensor. After pressing save, the sensors will update on their next heartbeat interval check-in. If an extended time is used, it may be best to power cycle the sensors by removing the cover. Once one sensor has received the correct settings, the hyperlink on the left bottom corner ("Use these settings for other sensors") can be used to apply the settings to all desired sensors of that same type.

Note: All fields other than Sensor Name, Heartbeat Interval, and Aware State Interval have been left as default values. The Aware State Interval is not used but changes automatically if the Heartbeat Interval is a smaller interval. The question marks by each field give an explanation by hovering over them.

Humidity Sensor Settings

The settings seen later are for every humidity sensor. After pressing save, the sensors will update on their next heartbeat interval check-in. If an extended time is used, it may be best to power cycle the sensors by removing the cover. Once one sensor has received the correct settings, the hyperlink on the left bottom corner ("Use these settings for other sensors") can be used to apply the settings to all desired sensors of that same type.

Note: All fields other than Sensor Name, Heartbeat Interval, and Aware State Interval have been left as default values. The Aware State Interval is not used but changes automatically if the Heartbeat Interval is a smaller interval. The question marks by each field give an explanation by hovering over them.

Software Design Specification

Continuous Logic

Module Classes and Control Modules

Module classes are the foundation for system operation. Module classes are implemented in PLC PlantPAS library Add-On instructions and SCADA faceplates to promote ease of configuration and consistency of implementation. Module functionality is documented in PAS SMDS.

Room pressures are monitored by PAS. Where necessary, the room to room differential pressure is calculated by the EMS controller. The resulting value is referred to as the virtual differential pressure (VDP). The VDP is displayed on the appropriate graphic and has full alarming capability. VDP values are calculated. The calculation is Room A – Room B = VDP.

Virtual Differential Pressure Alarm Disable

During differential pressure calibration, there may be the need to disable room-specific VDP alarming. Each "Room A" will have a calibration button to disable all VDP alarms that affect that room. These calibration buttons will be located on top of the room on each graphic and will be made visible with a global visibility button that can only be accessed with proper security credentials.

Room Condition Indication

Each EMS instrument enclosure will have a light/horn combination beacon mounted on the bottom. The lights and horn will provide notification if the room conditions are out of specification. The amber "Alert" light will be activated when the alert set point is exceeded. The red "Shutdown" light will be activated when the shutdown set point is exceeded. Once a shutdown set point is exceeded for three consecutive minutes, the horn will chirp every 10 seconds for 2 minutes. Once this time has elapsed, the alarm will be triggered, and the horn will continuously sound until the alarm is acknowledged.

Navigation Links

There are navigational links contained within the body of the main EMS screen. Pressing any of the navigation links will navigate to the individual area graphic.

This describes the reporting generated for the EMS process for the PAS. This specification defines the software design for the reporting in the EMS process area. It does not address reporting in other process areas. This specification addresses the following:

1. Alarm Daily Report
2. Alarm Summary Report
3. Device Out of Service Report

Definitions and Acronyms

Acronym or term	**Description**
EMS	Environmental monitoring system
FRS	Functional requirement specification
FT	FactoryTalk (Generally a Rockwell software component)
HMI	Human machine interface
I/O	Input/Output
IP	Internet protocol
PAS	Process automation system
PLC	Programmable logic controller
RIO	Remote input output
URS	User required specification

Definitions and Acronyms

Acronym or term	Description
SCADA	Supervisory control and data acquisition system
SE	Site edition
SMDS	Software module design specification
SQL	Structured Query Language
SSRS	Structured Query Language Server Reporting Services

Alarm Daily Report

The Alarm Daily Report is used to return alarms and events and acknowledgments that took place in the 24 hours period beginning midnight of the previous date and ending midnight of the generation date.

The report Header lists the title of the report and a summary of the applied filters such as time range and lookup value. The report Body lists the result of the Alarm Summary Report in one table. The Device Out of Service Report returns devices that have been manually taken out of service using the HMI.

Example 8.1 ***Functional Requirement Specification***

WFI Still Equipment Module Class

The purpose is to define the functional requirements for the WFI Still equipment module class. This type of equipment module is used to manage the WFI Stills for the WFI system. This module controls a WFI Still's header supply valves and a drain valve and places the valves into different configurations based on the operational mode of the Still. The module may be accessed by the operator directly from the HMI.

Acronym or term	Definition
AUTO	PID automatic, automatic
CAS	PID cascade, cascade
CM	Control module
Ethernet	Broadcast networking system that carries digital data packets to local nodes. A 10-Mbps base and local area network (LAN), Ethernet evolved from the IEEE 802.3 standard. Ethernet is the transport vehicle for many upper level protocols, including TCP/IP and Xerox network systems (XNS)

Acronym or term	Definition
EM	Equipment module
HMI	Human machine interface (operator terminal)
I/O	Input/Output signals to/from the control system and the field devices
MAN	PID manual, manual
OEM	Original equipment manufacturer
PAS	Process automation system
PID	PID control, proportional–integral–derivative control
PIO	Point I/O – Allen Bradley distributed I/O system
P&ID	Piping and instrumentation diagram
PLC	Programmable logic controller
POU	Point of use
RIO	Remote I/O
SMDS	Software module design specification
SOP	Standard operating procedure
URS	User requirement specification
WFI	Water for injection
WFICS	Water for injection control system

This type of equipment module is used to manage the WFI Stills for the WFI system. This equipment module class is used to (i) select the operational mode (stop, maintenance, hold, advising [drain/hot/ambient], sanitize), (ii) manage the sanitization timers, including placing the equipment in the required state on expiry, and (iii) monitor and handle equipment faults as required. Details of the operational modes and fault conditions are described in the WFI Still Equipment Module Design Specification. This module only operates in AUTO mode, and the operational mode is selected by the operator from the HMI.

The Module controls the output valves of the still based on commands from the operator, status of the WFI, Still water conductivity, differential pressure, and each of the Headers' temperature command statuses. There are sanitization timers which are monitored by this equipment module. The OFF_SAN_TIMER and OFF_SAN_WARN_TIMER monitor the time that the skid is offline ([WFI_ON] is off). When the OFF_SAN_WARN_TIMER timer expires, a message is displayed to the operator to notify of the pending sanitization expiry. When the OFF_SAN_TIMER expires, the module will be placed in drain mode, and commands for Header A and Header B will not be honored until a sanitization is completed.

The SAN_TIMER and SAN_WARN_TIMER monitor the time from the last sanitization. When the SAN_WARN_TIMER timer expires, a message is displayed to the operator to notify of the pending sanitization expiry. When the SAN_TIMER expires, the module will be placed in drain mode, and commands for Header A and Header B will not be honored until a sanitization is completed and will remain in Drain mode.

Alias name	Description
HDR_A_VLV	Header A supply valve
HDR_B_VLV	Header B supply valve
DRAIN_VLV	Drain valve
WFI_ON	Input on
INSTR_AIR	Instrument air pressure switch
STILL_COND_HL	High conductivity switch
STILL_DP_LO	Low DP switch
STILL_COND	Conductivity transmitter
STILL_TEMP	Temperature transmitter
WFI_TEMP_CMD	Hot/ambient command
HDR_A_TEMP	Header A hot/ambient status
HDR_B_TEMP	Header B hot/ambient status
OFF_SAN_TIMER	Offline sanitization timer
OFF_SAN_WARN_TIMER	Offline sanitization warning timer
SAN_TIMER	Sanitization timer
SAN_WARN_TIMER	Sanitization warning timer
SAN_DUR_TIMER	Sanitization duration timer
STARTUP_TIMER	Startup timer

The command source determines which commands and settings will be used for the equipment module. The command sources used for equipment modules have the same look and functionality as the command sources used for the control modules. The operator controls the device. Operator commands and settings from the HMI will be accepted. If priority logic controls the device, operator commands will be superseded. For the WFI Still Equipment Module, activation of any hold condition will trigger the override command and transition the equipment module to the Hold Operational Mode.

Maintenance owns control of the device. Maintenance commands will supersede the override commands and operator commands and settings are accepted from the HMI. Out of service is initiated by maintenance personnel placing the equipment into maintenance mode and then out of service. When the Still is placed Out of Service, all underlying control modules are released so that they can be manipulated as required for the service being performed on individual pieces of equipment (valves/pumps/etc.). The Equipment Module is placed into an idle state. When a Maintenance Out of Service has been commanded, a sanitization is required before the equipment can be placed in service again. Before the sanitization can be initiated, the following must be filled out by Maintenance Personnel:

- Extraordinary Sanitization Number
- Extraordinary Maintenance Number
- Maintenance Sanitization Temperature (°F)
- Maintenance Sanitization Time (minutes)

Note that for all equipment modules, underlying control modules are locked in cascade mode before running a mode sequencing. Stop is initiated by the lack of an Input On [WFI_ON] signal from the WFI Still, by loss of air pressure, and by operator. While in Stop mode, a [WFI_ON] signal is received the WFI still will transition into a startup sequence. Hold mode is initiated exclusively by a hold condition and is not accessible to the operator. Hold mode overrides all other operator selectable modes. While in Hold mode, all controlled equipment is placed into a safe state and remains there until hold conditions have cleared. The still will automatically return to the previous state following completion of the delay timer. The safe states for equipment controlled by this EM are as follows:

Equipment	State
DRAIN_VLV	Open
HDR_A_VLV	Closed
HDR_B_VLV	Closed

OCMD_HEADER	STS_WFI_TEMP	HDR_A_TEMP	HDR_B_TEMP	Opened valve
Header A	Ambient	Hot	Hot	DRAIN_VLV
Header A	Ambient	Hot	Ambient	HDR_B_VLV
Header A	Ambient	Ambient	Hot	HDR_A_VLV
Header A	Ambient	Ambient	Ambient	HDR_A_VLV
Header A	Hot	Hot	Hot	HDR_A_VLV
Header A	Hot	Hot	Ambient	HDR_A_VLV
Header A	Hot	Ambient	Hot	HDR_B_VLV
Header A	Hot	Ambient	Ambient	HDR_A_VLV

OCMD_HEADER	STS_WFI_TEMP	HDR_A_TEMP	HDR_B_TEMP	Opened valve
Header B	Ambient	Hot	Hot	DRAIN_VLV
Header B	Ambient	Hot	Ambient	HDR_B_VLV
Header B	Ambient	Ambient	Hot	HDR_A_VLV
Header B	Ambient	Ambient	Ambient	HDR_B_VLV
Header B	Hot	Hot	Hot	HDR_B_VLV
Header B	Hot	Hot	Ambient	HDR_A_VLV
Header B	Hot	Ambient	Hot	HDR_B_VLV
Header B	Hot	Ambient	Ambient	HDR_B_VLV

Advancing mode is initiated either by the completion of the sanitization sequencing. In advancing mode, the header supply vales are controlled based on the following inputs from the operator:

1. Start Startup Duration timer [STARTUP_TIMER] for [STARTUP_DUR]
   a. If high conductivity [STILL_COND_HL] or high differential pressure [STILL_DP_HL] occurs, the timer will restart.
2. Upon timer completion, the operator selection of "Advancing" mode is enabled.
   - OCMD_HEADER (Header A/Header B)
   - OCMD_TEMP_MODE (Ambient/Hot)
     - WFI_TEMP_CMD is directly controlled by OCMD_TEMP_MODE while supplying a header.
     - Hot mode is only initiated by starting a sanitization mode or by completion of a sanitization.
     - If the attached Header is ambient then the still will automatically be set to ambient following the sanitization.

Based on these inputs and the temperature status of Header A and Header B, the listed "Opened Valve" will open and the other valves will close.

From advancing mode, they still can transition to any mode. The sanitization sequencing initiated by the operator from the HMI or by sanitization commanded from the selected header following a 5-minute timer. The sanitization sequencing is as follows:

1. Set [OCMD_TEMP_MODE] to Hot.
   a. Hot command is sent to WFI [WFI_TEMP_CMD].
2. Heat to [SAN_TEMP]
   a. Set [SAN_OK] sanitization permissive as long as this condition is true
3. Start Sanitization Duration timer [SAN_DUR_TIMER] for either [SAN_DUR] or [MAINT_SAN_DUR]

   a. [MAINT_SAN_DUR] is used if the previous mode was Out of Service
   b. If [STIL_TEMP] is less than [SAN_TEMP], the timer will restart
4. Upon timer completion, set Sanitization Complete Alarm and reset OFF_SAN_TIMER, OFF_SAN_WARN_TIMER, SAN_TIMER, SAN_WARN_TIMER.
5. Wait for operator acknowledgment of Sanitization Complete Alarm.
6. Set [OCMD_TEMP_MODE] to Ambient.
   a. Ambient command is sent to WFI [WFI_TEMP_CMD].

If the [SAN_OK] permissive is set and Sanitization is not required (not STS_SAN_EXPIRED), the still will open supply valves based on the following inputs from the operator:

- OCMD_HEADER (Header A/Header B)

Based on these inputs and the temperature status of Header A and Header B, the listed "Opened Valve" will open and the other valves will close.

OCMD_HEADER	STS_WFI_TEMP	HDR_A_TEMP	HDR_B_TEMP	Opened valve
Header A	Hot	Hot	Hot	HDR_A_VLV
Header A	Hot	Hot	Ambient	HDR_A_VLV
Header A	Hot	Ambient	Hot	HDR_B_VLV
Header A	Hot	Ambient	Ambient	HDR_A_VLV
Header B	Hot	Hot	Hot	HDR_B_VLV
Header B	Hot	Hot	Ambient	HDR_A_VLV
Header B	Hot	Ambient	Hot	HDR_B_VLV
Header B	Hot	Ambient	Ambient	HDR_B_VLV

At the end of the sanitization sequencing the mode is set to Advancing. The command temperature will automatically follow the attached header's commanded temperature.

Parameter description	Parameter tag name	Parameter type	Engineering units
Command Source (Operator, Override, Maintenance, Out of Service)	CMD_SRC	PlantPAx Command Source Module	N/A
Operator Operating Mode (Stop, Advancing, Sanitize)	OCMD_MODE	Integer	N/A
Operator Header Selection (Header A, Header B)	OCMD_HEADER	Integer	N/A

Parameter description	Parameter tag name	Parameter type	Engineering units
Offline Sanitization Time	OFF_SAN_TIME	Real	Hours
Offline Sanitization Warning	OFF_SAN_TIME_WARN	Real	Hours
Sanitization Time	SAN_TIME	Real	Hours
Sanitization Warning	SAN_TIME_WARN	Real	Hours
Sanitization Temperature	SAN_TEMP	Real	°F
Sanitization Duration	SAN_DUR	Real	Minutes
Post-Maintenance Sanitization Duration	MAINT_SAN_DUR	Real	Minutes
Startup Duration	STARTUP_DUR	Real	Minutes
Actual Operating Mode (Stop, Hold, Advancing, Sanitize)	VAL_MODE	Integer	N/A
Actual Header Selection (None, Header A, Header B)	VAL_HEADER	Integer	N/A
Actual Still Temperature Mode (Ambient, Hot)	STS_WFI_TEMP	Boolean	1-Hot
Actual Still Temperature Command (Ambient, Hot)	WFI_TEMP_CMD	Boolean	1-Hot
Actual Current Runtime	TOTAL_RUNTIME	Real	Minutes
Current Hot Runtime	HOT_RUNTIME	Real	Minutes
Current Ambient Runtime	AMBIENT_RUNTIME	Real	Minutes
Current Offline Time	OFFLINE_TIME	Real	Minutes
Time until Sanitization	SAN_TIME_REM	Real	Minutes
Sanitization Temp OK	SAN_OK	Boolean	1-OK
Out of Service Status	STS_OOS	Boolean	1-Out of Service
Offline Status	STS_OFFLINE	Boolean	1-Offline
Ready Status	STS_READY	Boolean	1-Ready
Online Status	STS_ONLINE	Boolean	1-Online
Sanitization Expired Status	STS_SAN_EXPIRED	Boolean	1-Expired

Alarm description	Alarm conditions
"Alert" Alarm	[OFF_SAN_WARN_TIMER] or [SAN_WARN_TIMER] expired
Low Temp while Sanitizing	Sanitization mode active and [STILL_TEMP] < [SAN_TEMP]
Due to Sanitizing Alarm	[OFF_SAN_TIMER] or [SAN_TIMER] expired
Sanitization Complete Alarm	[SAN_DUR_TIMER] expired
Hold Alarm	Hold Mode active
Stop Alarm	Stop Mode active
Maintenance Alarm	Out of Service Active

If any of the following failure conditions are enabled and active, display the description of the failure condition and execute the hold logic.

Condition	Hold description	Hold conditions
1	Controlled Equipment Failure	Failure of any controlled equipment
2	Loss of Instrument Air	[INST_AIR] indicates loss of pressure
3	Low Differential Pressure	Not Sanitize Mode and [STILL_DP_LO]
4	High Conductivity	Not Sanitize Mode and [STILL_COND_HL]
5	Sanitization Timer Expiry	Not Sanitize Mode and [SAN_TIMER] expired
6	Offline Sanitization Timer Expiry	Not Sanitize Mode and [OFF_SAN_TIMER] expired

Parameter	Description
OCMD_MODE	Operator Commanded Operating Mode (Stop, Hold, Advancing, Sanitize)
OCMD_HEADER	Operator Commanded Header Selection (None, Header A, Header B)
OCMD_TEMP_MODE	Operator Commanded Still Temperature Mode (Ambient, Hot)
VAL_MODE	Actual Operating Mode (Stop, Hold, Advancing, Sanitize)
VAL_HEADER	Actual Header Selection (None, Header A, Header B)

Parameter	Description
STS_TEMP_MODE	Actual Still Temperature Mode (Ambient, Hot)
TOTAL_RUNTIME	Current Runtime
HOT_RUNTIME	Current Hot Runtime
AMBIENT_RUNTIME	Current Ambient Runtime
OFFLINE_TIME	Current Offline Time
SAN_TIME_REM	Time until Sanitization
STS_OOS	Out of Service Status
STS_OFFLINE	Offline Status
STS_READY	Ready Status
STS_ONLINE	Online Status
STS_SAN_EXPIRED	Sanitization Expired Status

The following changes will be logged:

Event	Description
Operational Mode Change	Operator Operational Mode Change
Header Selection Change	Operator Header Selection Change
Still Temperature Mode Change	Operator Still Temperature Mode Change
Maintenance Sanitization Duration Change	Maintenance Sanitization Duration Change
Operator ACK All	Acknowledge All Alarms
"Alert" Alarm ACK	"Alert" Alarm ACK
Low Temp while Sanitizing Alarm ACK	Low Temp while Sanitizing Alarm ACK
Due to Sanitizing Alarm ACK	Due for Sanitizing Alarm ACK
Sanitization Complete Alarm ACK	Sanitization Complete Alarm ACK

Example 8.2 ***Functional Requirement Specification***

WFI Silo Equipment Module Class

EM_SILO

This is to define the functional requirements for the Water for Injection (WFI) Silo equipment module class. This type of equipment module is used to manage the Silos for the WFI system. The module may be accessed by the operator directly from the HMI. This document describes the WFI Silo Equipment Module Class for the PAS installed at Client Healthcare Corporation, Marion NC. Batch Control and Data Collection are outside the scope of this document.

Acronym or term	Definition
AUTO	PID Automatic, Automatic
CAS	PID Cascade, Cascade
CM	Control Module
Ethernet	Broadcast networking system that carries digital data packets to local nodes. A 10-Mbps base and local area network (LAN), Ethernet evolved from the IEEE 802.3 standard. Ethernet is the transport vehicle for many upper level protocols, including TCP/IP and Xerox network systems (XNS)
EM	Equipment Module
HMI	Human Machine Interface (Operator Terminal)
I/O	Input/Output signals to/from the Control System and the field devices
MAN	PID Manual, Manual
OEM	Original Equipment Manufacturer
PAS	Process Automation System
PID	PID Control, Proportional-Integral-Derivative control
PIO	Point I/O – Allen Bradley distributed I/O system
P&ID	Piping and Instrumentation Diagram
PLC	Programmable Logic Controller
POU	Point of use
RIO	Remote I/O
SMDS	Software Module Design Specification
SOP	Standard Operating Procedure
URS	User Requirement Specification
WFI	Water for Injection
WFICS	Water for Injection Control System

This type of equipment module is used to manage the Silos for the WFI system. This equipment module class is used to (i) select the operational mode (Stop, Hold, Advance [Equalizing Supply], Drawdown, Sanitize, Fill, Drain), (ii) manage the sanitization timers, including placing the equipment in the required state on expiry, and (iii) monitor and handle equipment faults as required.

Details of the operational modes and fault conditions are described in the WFI Silo Equipment Module Design Specification and instances. This module only operates in AUTO mode, and the operational mode is selected by the operator from the HMI.

Each instance of the module corresponds to an individual Silo. The Module controls the distribution valve path [DIST_VLV] and drain valve [SILO_DRAIN_VLV] (XV_10x09), as well as requests to fill [HDR_A_SUPPLY] and [HDR_B_SUPPLY] based on commands from the status of the supply headers. The module additionally controls a recirculation loop which consists of a recirculation pump [RECIRC_PUMP]. The recirculation loop is active in all modes. The details of these interactions are outlined in the operational modes section of this document.

The SAN_TIMER and SAN_WARN_TIMER monitor the time that the silo temperature as measured by [SILO_TEMP] is below the sanitization temperature [SAN_TEMP]. If the silo temperature rises above [SAN_TEMP], the SAN_TIMER is not reset. When the SAN_WARN_TIMER timer expires, the "Alert" alarm notifies of the pending sanitization expiry. When the SAN_TIMER expires, the Silo will lose its sanitization status, and the module will be placed in hold mode until a sanitization is commanded.

In addition to the sanitization timers, if the silo experiences a rupture disk fault [SILO_BURST], or differential pressure alarm across the Silo heat exchanger [SILO_DP_HIGH], the Silo will lose its sanitization status if the operator does not clear the alarm condition and reset the alarm within the timer reset period. The module will be placed in hold mode until the system receives a sanitization command.

The equipment module also contains a drain timer [DRAIN_TIMER] and a [SILO_DRAIN_PAUSE] request. The [DRAIN_TIMER] starts timing when [SILO_LVL] < [SILO_LVL_LO]. Any time the [SILO_DRAIN_VLV] is open, the operator can execute a [SILO_DRAIN_PAUSE] request. This request overrides any sequence and stops and resets the [DRAIN_TIMER]. When the [SILO_DRAIN_PAUSE] request is cleared, the sequence will resume and the [DRAIN_TIMER] begins timing.

Alias name	Description
EQ_OCMD_SUPPLY	Equalizing Supply Command (Supervisory Control)
EQ_OCMD_Fill	Equalizing Fill Command (Supervisory Control)
HDR_A_STS	Supply Header A Operational Status
HDR_B_STS	Supply Header B Operational Status
INSTR_AIR	Instrument Air Pressure Switch
SILO_FLT_1_TEMP	Silo Filter 1 Temperature (TI_10x01)
SILO_FLT_2_TEMP	Silo Filter 2 Temperature (TI_10x02)
SILO_LVL	Silo Level Transmitter (LI_10x01)

Alias name	Description
SILO_BURST	Silo Rupture Disk Status (PSE_10x01)
SILO_DRAIN_VLV	Silo Drain Valve (XV_10x09)
SILO_OUTLET_VLV	Silo Outlet Valve (XV_10x01)
DIST_VLV	Silo to Distribution Loops Supply Valve Path
DIST_HDR_1_3_SUPPLY_VLV	Distribution Header 1/3 Supply Valve (XV_10015)
DIST_HDR_2_4_SUPPLY_VLV	Distribution Header 2/4 Supply Valve (XV_10016)
DIST_HDR _DRAIN_VLV	Distribution Header Drain Valve (XV_10038)
RECIRC_PMP	Silo Recirculation Pump (SIC_10x01)
HDR_A_SUPPLY	Header A Supply Request
HDR_B_SUPPLY	Header B Supply Request
SILO_HDR_A_SUPPLY_VLV	Silo Header A Supply Valve (XV_10x03)
SILO_HDR_B_SUPPLY_VLV	Silo Header B Supply Valve (XV_10X06)
SILO_SUPPLY_HDR_FLUSH_VLV_A	Silo Supply Header Flush Valve A (XV_10x04)
SILO_SUPPLY_HDR_FLUSH_VLV_B	Silo Supply Header Flush Valve B (XV_10x07)
SILO_STEAM_SUPPLY_VLV	Silo Steam Supply Block Valve (XV_10x11)
RECIRC_PIC	Silo Recirculation Pressure Control (PIC_10x01)
PRIMARY_LOOP_RTRN_VLV	Primary Loop Return Valve (XV_10017, 10018, 10025, or 10026)
SILO_HDR_VLV	Silo Header Valve (XV_10011, 10012, 10013, or 10014),
SILO_DIST_HDR_RTRN_VLV	Silo Distribution Header Return Valve (XV_10035, 10022, 10030, or 10027)
SILO_RTRN_VLV	Silo Return Valve (XV_10x12)
PRIMARY_LOOP_BLK_VLV	Primary Loop Block Valve (XV_10001, 10002, 10003, or 10004)
SILO_RTRN_DRAIN_VLV	Silo Return Drain Valve (XV_10x13)
SILO_DIST_HDR_DRAIN_VLV	Silo Distribution Header Drain Valve (XV_10041, 10042, 10043, or 10042)
SILO_DIST_HDR_RTRN_DRAIN_VLV	Silo Distribution Header Return Drain Valve (XV_10036, 10023, 10031, or 10028)
PRIMARY_LOOP_FLOW	Primary Loop Flow Indicator (FI_10001, 10002, 10003, or 10004)
PRIMARY_LOOP_TEMP	Primary Loop Temperature Indicator (TI_10001, 10002, 10003, or 10004)

Alias name	Description
SILO_PROCESS_AIR_VLV	Silo Process Air Valve (XV_10x14)
SILO_STEAM_BLOCK_VLV	Silo Steam Block Valve (XV_10x11)
SILO_TIC	Silo Temperature Control (TIC_10x01)
DIST_HDR_SILO_HOT_WFI_REC_VLV	Distribution Header Silo Hot WFI Recovery Valve (XV_10034, 10037, 10039, or 10040)
SILO_HOT_WFI_BLOCK_VLV	Silo Hot WFI Block Valve (XV_10711, 10721, 10731, or 10741)
SILO_HOT_WFI_VENT_VLV	Silo Hot WFI Vent Valve (XV_10712, 10722, 10732, or 10742)
SILO_HOT_WFI_COMP_AIR_VLV	Silo Hot WFI Compressed Air Valve (XV_10713, 10723, 10733, or 10743)
RECOVERY_SILO1_BLOCK_VLV	Recovery Silo 1 Block Valve (XV_10705)
RECOVERY_SILO1_SUPPLY_VLV	Recovery Silo 1 Supply Valve (XV_10702)
RECOVERY_SILO2_BLOCK_VLV	Recovery Silo 2 Block Valve (XV_0706)
RECOVERY_SILO2_SUPPLY_VLV	Recovery Silo 2 Supply Valve (XV_10701)

The command source determines which commands and settings will be used for the equipment module. The command sources used for equipment modules have the same look and functionality as the command sources used for the control modules. This section details the command sources that will be configured for the WFI Silo Equipment Module.

The operator controls the device. Operator commands and settings from the HMI will be accepted.

Priority logic controls the device, and operator commands will be superseded.

For the WFI Still Equipment Module, activation of any hold condition will trigger the override command.

Maintenance owns control of the device. Maintenance commands will supersede the override commands and operator commands and settings are accepted from the HMI.

Out of service is initiated by a maintenance personal placing the equipment into maintenance mode and then out of service.

When placed Out of Service, all underlying control modules are released so that they can manipulated as required for the service being performed on individual pieces of equipment (valves/pumps/etc.). The Equipment Module is placed into an idle state.

When a Maintenance Out of Service has been commanded, a sanitization is required before the equipment can be placed in service again. Before the sanitization can be initiated, the following must be filled out by Maintenance Personnel before sanitation is complete:

- Extraordinary Sanitization Number

- Extraordinary Maintenance Number
- Maintenance Sanitization Time (minutes)

The Extraordinary Sanitization Number and Extraordinary Maintenance Number can be changed up until the end of the sanitization and are written to the report at the time of completion.

The controls listed in this section operate continuously regardless of the operational mode. Continuous controls will be disabled when EM is placed Out of Service.

The recirculation loop recirculates the contents of the silo. The silo recirculation pressure control loop (PIC_10x01) [RECIRC_PIC_MODE] = Auto to control given the following conditions:

- [SILO_LVL] (LIT_10 × 01) > [RECIRC_LVL]
- [SILO_OUTLET_VLV] (XV_10 × 01) = Open
- [SILO_DRAIN_VLV] (XV_10 × 09) = Closed

If a drain is required at any point, the operator is required to confirm and electronically sign for the drain to start. Once the operator confirms drain and the silo drain valve has opened, the Drain Timer [DRAIN_TIMER] will be started for [DRAIN_TIME]. If this expires, the Silo Empty Drain Time Done Alarm is triggered. The operator is required to E-Sign to clear the alarm and confirm that the Silo is Confirmed Empty.

Once either silo supply valve has opened, the silo confirmed empty flag [EMPTY] is cleared.

Note that for all equipment modules, underlying control modules are locked in cascade mode before running a mode sequence.

Stop mode is initiated by the operator from the faceplate or when the Silo DP Alert or Silo Rupture Disc Alert alarm is set. While in Stop mode, the silo is placed into an idle condition where the filling of the Silo is disabled, and the water is recirculated until the alarms are cleared and the next operator action. The Stop mode states for equipment controlled by this EM are as follows:

Equipment	State
HDR_A_SUPPLY	Off
HDR_B_SUPPLY	Off
SILO_DRAIN_VLV (XV_10x09)	Closed
Distribution Path	Closed
SILO_RTRN_DRAIN_VLV (XV_10x13)	Closed
RECIRC_PIC_MODE (PIC_10x01)	Auto
SILO_TIC_MODE	Manual
POU_ENABLED	False
SILO_STEAM_SUPPLY_VLV (XV_10x11)	Closed

Hold mode is initiated by a hold condition. Before a hold condition is generated, the system will prompt the operator for a signed Hold confirmation. While in Hold mode, all controlled equipment is placed into a safe state and remains there until all hold conditions have cleared and another mode is initiated by the operator. The safe states for equipment controlled by this EM are as follows:

Equipment	State
HDR_A_SUPPLY	Off
HDR_B_SUPPLY	Off
SILO_DRAIN_VLV (XV_10x09) ([EMPTY] = False and Mode ≠ Drain)	Open
SILO_DRAIN_VLV (XV_10x09) (([EMPTY] = True and Mode ≠ Drain) or ([EMPTY] = False and Mode = Drain))	Closed
Distribution Path	Closed
RECIRC_PIC_MODE (PIC_10x01)	Manual
RECIRC_PMP (SIC_10x01)	Off
SILO_TIC_MODE	Manual
POU_ENABLED	False
SILO_STEAM_SUPPLY_VLV (XV_10x11)	Closed

Advance mode is initiated by the operator. In Advance mode, the silo can both fill and distribute based on commands from the Equalizing Supervisory Control. When a silo is placed into Advance, it is considered “online and ready”. The startup sequence for Advance mode is as follows:

1. Verify [SILO_LVL] > [SILO_LVL_LO]
2. If other two silos are active, open [DIST_HDR_1_3_SUPPLY_VLV] (XV_10015) and
3. Open [SILO_OUTLET_VLV] (XV_10x01).
4. Verify silo recirculation pressure control (PIC_10x01) [RECIRC_PIC_MODE] = Auto. Set [RECIRC_PIC_SP] = [RECIRC_PRESS_SP].
5. Open [SILO_HDR_VLV] (XV_10011, 10012, 10013, or 10014), [PRIMARY_LOOP_RTRN_VLV] (XV_10017, 10018, 10025, or 10026), [SILO_DIST_HDR_RTRN_VLV] (XV_10035, 10022, 10030, or 10027), [SILO_RTRN_VLV] (XV_10x12), and [PRIMARY_LOOP_BLK_VLV] (XV_10001, 10002, 10003, or 10004). Close [SILO_RTRN_DRAIN_VLV] (XV_10x13), [SILO_DIST_HDR_DRAIN_VLV] (XV_10041, 10042, 10043, or 10042), and [SILO_DIST_HDR_RTRN_DRAIN_VLV] (XV_10036, 10023, 10031, or 10028).
6. Set distribution header pressure control valve position [DIST_HDR_PCV] = 0% open.
7. Set [SILO_STS] = “Ready”

8. Wait for [PRIMARY_LOOP_FLOW] (FI_10001, 10002, 10003, or 10004) = [ADVANCE_FLOW_SP]
9. Close [SILO_RTRN_VLV] (XV_10x12), and [PRIMARY_LOOP_RTRN_VLV] (XV_10017, 10018, 10025, or 10026).
10. Open [SILO_DIST_HDR_DRAIN_VLV] (XV_10041, 10042, 10043, or 10042) and energize [SILO_PROCESS_AIR_VLV] (XV_10x14) for [DRY_LINE_TIME].
11. After [DRY_LINE_TIME], close [SILO_DIST_HDR_DRAIN_VLV] (XV_10041, 10042, 10043, or 10042), [SILO_PROCESS_AIR_VLV] (XV_10x14), and [SILO_DIST_HDR_RTRN_VLV] (XV_10035, 10022, 10030, or 10027).
12. Set distribution header pressure control valve position [DIST_HDR_PCV] = [ADVANCE_PCV_SP]
13. Set [POU_ENABLED] = True.

POU valve are still waiting on distribution loop permissive. Once the sequence is complete, ambient recirculation will continue until an alarm is activated or a shutdown or sanitization is initiated.

The supervisory control monitors the status of all silos and issues "Equalizing Fill" [EQ_OCMD_FILL] and "Equalizing Supply" [EQ_OCMD_SUPPLY] commands to the silos based on their status.

The supervisory control determines if all silos are "Equalized" based on whether all silos in Equalizing Supply mode are within ±5% of the highest silo level. If only one silo is in Equalizing Supply Mode, it will be considered "Equalized".

If all silos are "Equalized", the supervisory control will

- Issue an "Equalizing Supply" [EQ_OCMD_SUPPLY] command to all silos in Equalizing Supply mode unless one of the silos is "Hot"
- Issue an "Equalizing Fill" [EQ_OCMD_FILL] command to all silos in Equalizing Supply mode

If any silo that is NOT "Equalized", the supervisory control will

- Issue an "Equalizing Supply" [EQ_OCMD_SUPPLY] command to the silo in Equalizing Supply mode with the HIGHEST level until all silos in Equalizing Supply mode are within ±1.5% of the highest silo level
- Issue an "Equalizing Fill" [EQ_OCMD_FILL] command to the silo in Equalizing Supply mode with the LOWEST level until all silos in Equalizing Supply mode are within ±1.5% of the highest silo level

When an "Equalizing Supply" [EQ_OCMD_SUPPLY] command is issued by the supervisory control, the distribution valve path [DIST_VLV] will open if:

- No other silo is in active "Drawdown" mode
- [SILO_LVL] is greater than [LOW_TRIP_LVL]

When a "Equalizing Fill" [EQ_OCMD_FILL] command is issued by the supervisory control, a fill request ([HDR_A_SUPPLY] or [HDR_B_SUPPLY]) is issued if:

- The header temperature status matches the commanded temperature mode [OCMD_TEMP_MODE]
- The High-Level trip [SILO_LVL]. Hi Alarm is not active

The resulting request ([HDR_A_SUPPLY] or [HDR_B_SUPPLY]) is used by the Supply Header Equipment Modules to request the corresponding Header EM to open the supply valve to the Silo.

The sanitization sequence is initiated by the operator from the HMI. The sanitization sequence is as follows:

1. If [SAN_EXPIRED] is True
   a. Open the silo drain valve [SILO_DRAIN_VLV] (XV_10x09).
      i. Opening the silo drain valve will stop the [RECIRC_PMP].
   b. Start [DRAIN_TIMER] when [SILO_LVL] < [SILO_LVL_LO].
   c. Close [SILO_DRAIN_VLV] (XV_10x09) when [DRAIN_TIMER] expires and an empty confirmation is received from the operator.
   d. Set [HDR_A_SUPPLY] or [HDR_B_SUPPLY]
   e. When [SILO_LVL] ≥ [SAN_LVL] set [HDR_A_SUPPLY] or [HDR_B_SUPPLY] = False.
2. If [SAN_EXPIRED] is False and [SILO_LVL] > [SAN_LVL]
   a. Open the silo drain valve [SILO_DRAIN_VLV] (XV_10x09) until [SILO_LVL] = [SAN_LVL]
   b. Close the silo drain valve [SILO_DRAIN_VLV] (XV_10x09).
3. If [SAN_EXPIRED] is False and [SILO_LVL] < [SAN_LVL]
   a. Set [HDR_A_SUPPLY] or [HDR_B_SUPPLY]
   b. When [SILO_LVL] = [SAN_LVL] set [HDR_A_SUPPLY] or [HDR_B_SUPPLY] = False.
4. Disable primary loop point of use valves [P1_POU_VLV] = Disabled.
5. Enable [GROUP_SAN].
6. Close [DIST_HDR_1_3_SUPPLY_VLV] (XV_10015), [DIST_HDR_2_4_SUPPLY_VLV] (XV_10016), and [DIST_HDR_DRAIN_VLV] (XV_10038).
7. Close the distribution valve path.
8. Open the silo outlet valve [SILO_OUTLET_VLV] (XV_10x01).
9. Open [SILO_STEAM_BLOCK_VLV] (XV_10x11). Set silo temperature control (TIC_10x01) [SILO_TIC_MODE] = AUTO and [SILO_TIC_SP] = [SILO_SAN_TEMP_SP].
10. Open [SILO_HDR_VLV] (XV_10011, 10012, 10013, or 10014).
11. Set [SILO_SAN_RDY] = TRUE.
12. Wait for [DIST_SAN_RDY] = True
    a. Start Sanitization timer [SAN_DUR_TIMER] for [SAN_DUR].

13. When [SAN_DUR_TIMER] is complete. Close [SILO_STEAM_SUPPLY_VLV] (XV_10x11). Set silo temperature control (TIC_10x01) [SILO_TIC_MODE] = Manual. Set [SILO_TIC_OP] = 0% open.
14. Begin the Drain Hot Water Recovery Sequence.
    a. Open [DIST_HDR_SILO_HOT_WFI_REC_VLV] (XV_10034, 10037, 10039, or 10040), [SILO_HOT_WFI_BLOCK_VLV] (XV_10711, 10721, 10731, or 10741), and close [SILO_DIST_HDR_RTRN_VLV] (XV_10035, 10022, 10030, or 10027).
    b. Open [RECOVERY_SILO1_BLOCK_VLV] and [RECOVERY_SILO1_SUPPLY_VLV] or [RECOVERY_SILO2_BLOCK_VLV] and [RECOVERY_SILO2_SUPPLY_VLV].
15. When [SILO_LVL] < [SILO_COOLDOWN_LVL] set [HDR_A_SUPPLY] or [HDR_B_SUPPLY] = True.
16. When [PRIMARY_LOOP_TEMP] < [LOOP_COOLDOWN_TEMP_SP]
    a. Open [SILO_DIST_HDR_RTRN_VLV] (XV_10035, 10022, 10030, or 10027) and close [DIST_HDR_SILO_HOT_WFI_REC_VLV] (XV_10034, 10037, 10039, or 10040).
    b. Enable [P1_POU_VLV] valves.
17. Begin Hot Water Recovery Air Dry Sequence.
    a. Open [SILO_HOT_WFI_VENT_VLV] (XV_10712, 10722, 10732, or 10742) and Start [HOT_WATER_REC_VENT_TIMER].
    b. When [HOT_WATER_VENT_TIMER] elapses, close [SILO_HOT_WFI_VENT_VLV] (XV_10712, 10722, 10732, or 10742) and activate [SILO_HOT_WFI_COMP_AIR_VLV] (XV_10713, 10723, 10733, or 10743) to dry the hot water recovery line.
    c. Start [HOT_WATER_REC_DRY_TIMER].
    d. When [HOT_WATER_REC_DRY_TIMER] elapses close [RECOVERY_SILO1_BLOCK_VLV] (XV_10705) and [RECOVERY_SILO1_SUPPLY_VLV] (XV_10702) or [RECOVERY_SILO2_BLOCK_VLV] (XV_0706) and [RECOVERY_SILO2_SUPPLY_VLV] (XV_10701), [SILO_HOT_WFI_VENT_VLV] (XV_10712, 10722, 10732, or 10742), and [SILO_HOT_WFI_BLOCK_VLV] (XV_10711, 10721, 10731, or 10741).
18. Set Sanitization Complete Alarm.
    a. Reset [SAN_TIMER]

Fill mode can be initiated either by the operator or by the completion of the sanitization sequence. In Fill mode, the WFI temperature [OCMD_TEMP_MODE] is set by the operator from the faceplate, and if either header has a matching temperature mode, the resulting request ([HDR_A_SUPPLY] or [HDR_B_SUPPLY]) is used by the Supply Header Equipment Modules to request the corresponding Header EM to open the supply valve to the Silo.

1. Set [HDR_A_SUPPLY] or [HDR_B_SUPPLY] = True.
2. Open [SILO_SUPPLY_HDR_FLUSH_VLV_A] (XV-10X04) or [SILO_SUPPLY_HDR_FLUSH_VLV_B] (XV-10X07).
3. Wait for flush to occur for [FILL_FLUSH_HDR_TIME].
4. After [FILL_FLUSH_HDR_TIME], open [SILO_HDR_A_SUPPLY_VLV] (XV_10x03) or [SILO_HDR_B_SUPPLY_VLV] (XV_10X06) and close [SILO_SUPPLY_HDR_FLUSH_VLV_A] or [SILO_SUPPLY_HDR_FLUSH_VLV_B] (XV-10X04 or XV-10X07).
5. Once [SILO_LEVEL] ≥ [SILO_LO_LVL_SP], open [SILO_OUTLET_VLV] (XV_10x01) and [SILO_DRAIN_VLV] (XV_10x09).
6. Wait for flush to occur for [FILL_FLUSH_SILO_TIME].
7. After [FILL_FLUSH_TIME], close [SILO_DRAIN_VLV] (XV-10X09).
8. The system will continue to fill and recirculate until [SILO_LVL] ≥ [SILO_HIGH_LVL_SP]. Once a high level is sensed in the silo, set [HDR_A_SUPPLY] or [HDR_B_SUPPLY] = False.
9. Close [SILO_HDR_A_SUPPLY_VLV] (XV_10x03) or [SILO_HDR_B_SUPPLY_VLV] (XV_10X06).

If at any time during Fill the High-Level trip [SILO_LVL]. Hi Alarm, is activated, the request ([HDR_A_SUPPLY] and [HDR_B_SUPPLY]) will be overridden to Off. The Silo remains in Fill mode until another mode is selected.

In drain, the filling of the Silo is disabled [HDR_A_SUPPLY] off and [HDR_B_SUPPLY] off. The drain valve [DRAIN_VLV] is opened, and the distribution valve path [DIST_VLV] is closed. A [SILO_DRAIN_PAUSE] will allow the operator to stop the drain step. The drain sequence is as follows:

1. Disable filling of the Silo ([HDR_A_SUPPLY] and [HDR_B_SUPPLY] off).
2. Close the distribution valve path.
3. Open the silo outlet valve [SILO_OUTLET_VLV] (XV_10x01).
4. Open the drain valve [SILO_DRAIN_VLV] (XV_10x09).
   a. Opening the drain valve will stop the [RECIRC_PMP].
   b. If the silo is confirmed empty [EMPTY], the drain valve will close.

In Drawdown mode, the silo selected is the only silo to supply the distribution loops. In Drawdown mode, the distribution valve path [DIST_VLV] will open if:

- No other silo is in active "Drawdown" mode
- [SILO_LVL] is greater than [LOW_TRIP_LVL]

The sequence for Drawdown mode is as follows:

1. Disable filling of the Silo ([HDR_A_SUPPLY] and [HDR_B_SUPPLY] off).
2. Verify [SILO_LVL] > [LOW_TRIP_LVL].
3. If other two silos are active, open [DIST_HDR_1_3_SUPPLY_VLV] (XV-10015) and [DIST_HDR_2_4_SUPPLY_VLV] (XV-10016) and close [DIST_HDR _DRAIN_VLV] (XV_10038).

4. Open [SILO_OUTLET_VLV] (XV_10x01).
5. Verify silo recirculation pressure control (PIC_10x01) [RECIRC_PIC_MODE] = Auto. Set [RECIRC_PIC_SP] = [RECIRC_PRESS_SP].
6. Open [PRIMARY_LOOP_RTRN_VLV] (XV_10017, 10018, 10025, or 10026), [SILO_DIST_HDR_RTRN_VLV] (XV_10035, 10022, 10030, or 10027), [SILO_RTRN_VLV] (XV_10x12), and [PRIMARY_LOOP_BLK_VLV] (XV_10001, 10002, 10003, or 10004). Close [SILO_RTRN_DRAIN_VLV] (XV_10x13), [SILO_DIST_HDR_DRAIN_VLV] (XV_10041, 10042, 10043, or 10042), and [SILO_DIST_HDR_RTRN_DRAIN_VLV] (XV_10036, 10023, 10031, or 10028).
7. Set distribution header pressure control valve position [DIST_HDR_PCV] = 0% Open.
8. Set [SILO_STS] = "Ready".
9. Wait for [PRIMARY_LOOP_FLOW] (FI_10001, 10002, 10003, or 10004) = [ADVANCE_FLOW_SP].
10. Close [SILO_RTRN_VLV] (XV_10x12), and [PRIMARY_LOOP_RTRN_VLV] (XV_10017, 10018, 10025, or 10026).
11. Open [SILO_DIST_HDR_DRAIN_VLV] (XV_10041, 10042, 10043, or 10042) and energize [SILO_PROCESS_AIR_VLV] (XV_10x14) for [DRY_LINE_TIME].
12. After [DRY_LINE_TIME], close [SILO_DIST_HDR_DRAIN_VLV] (XV_10041, 10042, 10043, or 10042), [SILO_PROCESS_AIR_VLV] (XV_10x14), and [SILO_DIST_HDR_RTRN_VLV] (XV_10035, 10022, 10030, or 10027).
13. Set [DISABLE_SILO_DIST] to true for remaining Silos in "Advance" mode.

If [SILO_LVL] is less than [LOW_TRIP_LVL] the mode is set to "Drain" and set [DISABLE_SILO_DIST] to false for remaining Silos in "Advance" mode.

Equipment	State
RECIRC_PUMP	Running
DIST_HDR_PCV (PCV_10001, 10002, 10003, or 10004)	Open to [ADVANCE_PCV_SP]
SILO_HDR_VLV (XV_10011, 10012, 10013, or 10014),	Open
PRIMARY_LOOP_BLK_VLV (XV_10001, 10002, 10003, or 10004)	Open
DIST_HDR_SILO_HOT_WFI_REC_VLV (XV_10034, 10037, 10039, or 10040)	Closed
SILO_DIST_HDR_RTRN_VLV (XV_10035, 10022, 10030, or 10027)	Open
PRIMARY_LOOP_RTRN_VLV (XV_10017, 10018, 10025, or 10026)	Open

Parameter description	Parameter tag name	Parameter type	Engineering units
Command Source (Operator, Override, Maintenance, Out of Service)	CMD_SRC	PlantPAx Command Source Module	N/A
Operating Mode (Stop, Advance, Sanitize, Fill, Drain, Drawdown)	OCMD_MODE	Integer	N/A
Silo Temperature Mode (Ambient, Hot)	OCMD_TEMP_MODE	Boolean	N/A
Silo Recirculation Pressure Control Mode (Manual/Auto)	RECIRC_PIC_MODE	Integer	N/A
Silo Recirculation Pressure Control Setpoint	RECIRC_PIC_SP	Real	psig
Silo Temperature Control Mode (Manual/Auto)	SILO_TIC_MODE	Integer	N/A
Silo Temperature Control Setpoint	SILO_TIC_SP	Real	°F

Engineering Parameters

Parameter description	Parameter tag name	Parameter type	Engineering units
Silo Empty Status	EMPTY	Boolean	N/A
Disable Silo Distribution Request	DISABLE_SILO_DIST	Boolean	N/A
POU Valves Enable	POU_ENABLED	Boolean	N/A
Group Sanitization Status	GROUP_SAN	Boolean	N/A
Primary Loop Distribution Sanitization Ready Status	DIST_SAN_RDY	Boolean	N/A
Silo Sanitization Ready Status	SILO_SAN_RDY	Boolean	N/A
Primary Loop Valves Enabled	P1_POU_VLV	Boolean	N/A
Sanitization Time	SAN_TIME	Real	Hours
Sanitization Warning	SAN_TIME_WARN	Real	Hours
No Recirculation Time	NO_RECIRC_TIME	Real	Hours
Sanitization Temperature	SAN_TEMP	Real	°F

Engineering Parameters

Parameter description	Parameter tag name	Parameter type	Engineering units
Sanitization Duration	SAN_DUR	Real	Minutes
Post-Maintenance Sanitization Duration	MAINT_SAN_DUR	Real	Minutes
Post-Maintenance Sanitization Temperature	MAINT_SAN_TEMP	Real	°F
Empty Drain Time	DRAIN_TIME	Real	Minutes
Minimum Recirculation Level	RECIRC_LVL	Real	%
Low-Level Trip Point	LOW_TRIP_LVL	Real	%
Silo Capacity	SILO_CAPACITY	Real	gal
Silo Low-Level Alarm Limit	SILO_LVL_LO	Real	%
Silo High-Level Setpoint	SILO_HIGH_LVL_SP	Real	%
Silo Low-Level Setpoint	SILO_LO_LVL_SP	Real	%
Recirculation Pressure Setpoint	RECIRC_PRESS_SP	Real	psig
Advance Flow Setpoint	ADVANCE_FLOW_SP	Real	gpm
Dry Line Time	DRY_LINE_TIME	Real	Minutes
Advance Pressure Control Valve Position Setpoint	ADVANCE_PCV_SP	Real	%
Minimum Level for Sanitization	SAN_LVL	Real	%
Silo Sanitization Temperature Setpoint	SILO_SAN_TEMP_SP	Real	°F
Loop Cooldown Temperature Setpoint	LOOP_COOLDOWN_ TEMP_SP	Real	°F
Hot Water Recovery Vent Timer	HOT_WATER_REC_ VENT_TIMER	Real	Minutes
Silo Cooldown Level	SILO_COOLDOWN_ LVL	Real	%
Hot Water Recovery Dry Timer	HOT_WATER_REC_ DRY_TIMER	Real	Minutes
Fill Flush Header Time	FILL_FLUSH_ HDR_TIME	Real	Minutes

Engineering Parameters

Parameter description	Parameter tag name	Parameter type	Engineering units
Fill Flush Silo Time	FILL_FLUSH_ SILO_TIME	Real	Minutes
Actual Operating Mode (Stop, Hold, Advance, Drawdown, Sanitize, Fill, Drain, Drawdown)	VAL_MODE	Integer	N/A
Actual Supply Header (None, Header A, Header B, Both Headers)	VAL_HEADER	Integer	N/A
Actual Silo Temperature Command (Ambient, Hot)	TEMP_MODE_CMD	Boolean	1-Hot
Actual Silo Temperature Mode	STS_TEMP_MODE	Boolean	1-Hot
Recirculation Drain Override	STS_RECIRC_OVR	Boolean	N/A
Current Runtime	TOTAL_RUNTIME	Real	Minutes
Current Hot Runtime	HOT_RUNTIME	Real	Minutes
Current Ambient Runtime	AMBIENT_RUNTIME	Real	Minutes
Current Offline Time	OFFLINE_TIME	Real	Minutes
Recirc Time	RECIRC_TIME	Real	Minutes
Time until Sanitization	SAN_TIME_REM	Real	Minutes
Sanitization Temp OK	SAN_OK	Boolean	1-OK
Out of Service Status	STS_OOS	Boolean	1-Out of Service
Offline Status	STS_OFFLINE	Boolean	1-Offline
Ready Status	STS_READY	Boolean	1-Ready
Online Status	STS_ONLINE	Boolean	1-Online
Out of Service Status	STS_OOS	Boolean	1-Out of Service
Offline Status	STS_OFFLINE	Boolean	1-Offline
Silo Empty	STS_EMPTY	Boolean	1-Empty

Individual instances of this equipment module class will alarm whenever faulted.

Alarm description	Alarm conditions
Temperature Alarm active for >24 h alarm	[SILO_FLT_1_TEMP].Alarm or [SILO_FLT_2_TEMP].Alarm or [SILO_FLT_3_TEMP].Alarm for >24 h
Low Temp while Sanitizing alarm (<176 °F)	Sanitizing Mode and [SILO_TEMP]. Lo Alarm
Silo Empty Drain Time Done Alarm	Silo Empty Flag [EMPTY]
“Alert” Alarm	SAN_WARN_TIMER expired
Due to Sanitizing Alarm	NO_RECIRC or SAN_TIMER expired
Sanitization Complete Alarm	SAN_DUR_TIMER expired
Silo Heat Exchanger DP High	[SILO_HX_DP_HIGH]
Hold Alarm	Hold Mode Active

If any of the following failure conditions are enabled and active, display the description of the failure condition and execute the stopping logic.

Condition	Stop description	Stop conditions
1	Silo Heat Exchanger DP High	[SILO_HX_DP_HI_ALERT]
2	Silo Heat Exchanger DP Low	[SILO_HX_DP_LO_ALERT]
3	Silo Rupture Disk	[SILO_BURST_ALERT]
4	High High Silo Level Alarm	[SILO_LVL].HiAlarm
5	Silo High High-Level Switch Active	[SILO_LVL_HIHI]
6	Silo High High-Recirculation Temperature	[SILO_RECIRC_TEMP].HiHiAlarm
7	High Conductivity	[SILO_COND_HI]
8	Low Conductivity	[SILO_COND_LO]

If any of the following failure conditions are enabled and active, display the description of the failure condition and execute the hold logic.

Condition	Hold description	Hold conditions
1	Controlled Equipment Failure	Failure of any controlled equipment
2	Rupture Disk Burst	[SILO_BURST_ALARM]
3	High differential pressure	[SILO_HX_DP_HI_ALARM]
4	Low differential pressure	[SILO_HX_DP_LO_ALARM]
5	Sanitization Timer Expiry	Not Sanitize Mode and [SAN_TIMER] expired
6	Silo Low Level	[SILO_LVL_LO]

Alarm horns and amber beacons reside in the Boiler Room to provide notice based on the monitoring of the silo rupture disks, silo differential pressures, and instrument air for the WFICS silos. If any of the critical silo conditions are activated, the silo horns sound. If the silo rupture disk condition is no longer active or an operator acknowledges the alarm through the HMI, the horns will silence. If the silo differential pressure or instrument air conditions are no longer active, the horns will continue to sound until an operator acknowledges the alarm through the HMI. Any new alarm will sound the horns.

Parameter	Description
OCMD_MODE	Current Operating Mode
OCMD_SUPPLY	Current Supply Mode Command
VAL_MODE	Actual Operating Mode (Stop, Hold, Advance, Sanitize)
VAL_HEADER	Actual Header Filling (None, Header A, Header B)
STS_TEMP_MODE	Actual Still Temperature Mode (Ambient, Hot)
STS_OOS	Out of Service Status
STS_OFFLINE	Offline Status
STS_READY	Ready Status
STS_ONLINE	Online Status
STS_SAN_EXPIRED	Sanitization Expired Status
STS_EMPTY	Current Silo Empty Status

Event	Description
Operational Mode Change	Operator Operational Mode Change
Header Selection Change	Operator Header Selection Change
Maintenance Sanitization Temperature Change	Maintenance Sanitization Temperature Change
Maintenance Sanitization Duration Change	Maintenance Sanitization Duration Change
Operator ACK All	Acknowledge All Alarms
"Alert" Alarm ACK	"Alert" Alarm ACK
Low Temp while Sanitizing Alarm ACK	Low Temp while Sanitizing Alarm ACK
Due to Sanitizing Alarm ACK	Due to Sanitizing Alarm ACK
Sanitization Complete Alarm ACK	Sanitization Complete Alarm ACK
Silo Empty Alarm ACK	Silo Empty Alarm ACK

Example 8.3 ***Functional Requirement Specification***

WFI Primary Loop Equipment Module Class

EM_PRIM_LOOP

This is to define the functional requirements for the WFI Primary Loop equipment module class. This type of equipment module is used to manage the WFI Primary Loops for the WFI system. The module may be accessed by the operator directly from the HMI.

This document describes the WFI Primary Loop Equipment Module Class for the PAS installed at Client Healthcare Corporation, Marion NC. Batch Control and Data Collection are outside the scope of this document.

Acronym or term	Definition
AUTO	PID Automatic, Automatic
CAS	PID Cascade, Cascade
CM	Control Module
Ethernet	Broadcast networking system that carries digital data packets to local nodes. A 10-Mbps base and local area network (LAN), Ethernet evolved from the IEEE 802.3 standard. Ethernet is the transport vehicle for many upper level protocols, including TCP/IP and Xerox network systems (XNS)
EM	Equipment Module
HMI	Human Machine Interface (Operator Terminal)
I/O	Input/Output signals to/from the Control System and the field devices
MAN	PID Manual, Manual
OEM	Original Equipment Manufacturer
PAS	Process Automation System
PID	PID Control, Proportional-Integral-Derivative control
PIO	Point I/O – Allen Bradley distributed I/O system
P&ID	Piping and Instrumentation Diagram
PLC	Programmable Logic Controller
POU	Point of use
RIO	Remote I/O
SMDS	Software Module Design Specification
SOP	Standard Operating Procedure
URS	User Requirement Specification
WFI	Water for Injection
WFICS	Water for Injection Control System

This type of equipment module is used to manage the WFI Primary Loops for the WFI system. This equipment module class is used to (i) select the operational mode (Stop, Maintenance, Hold, Advance [Hot/Ambient], Sanitize, Flush), (ii) manage the sanitization timer, including placing the equipment in the required state on expiration, and (iii) monitor and handle equipment faults as required.

Details of the operational modes, specific alias tables, and fault conditions are described in the WFI Primary Loop Equipment Module Design Specification and instances.

This module only operates in AUTO mode, and the operational mode is selected by the operator from the HMI. Each instance of the module corresponds to an individual Primary Loop. The details of these interactions are outlined in the operational modes section of this document.

There are main sanitization timers that are monitored by this equipment module. The SAN_TIMER monitors the time that the skid is not in HOT mode. When the SAN_TIMER expires, an alarm alerts the operator about the sanitization expiry.

The equipment module also contains a flush timer [FLUSH_TIMER] and a [FLUSH_REQUIRED] request. Any time the [FLUSH_PATH] is open, the operator can execute a [FLUSH_PAUSE] request. This request overrides any sequence and stops and resets the [FLUSH_TIMER]. When the [FLUSH_REQUIRED] request is cleared, the sequence will resume and the [FLUSH_TIMER] begins timing.

The equipment module also handles WFI requests from Mix Tank POU valves based on the distribution loop flow capacity and the draw rates of POU valves requesting WFI. The WFI request control logic monitors the current total draw rate of WFI supplied to the mix tanks and prevents WFI requests from being granted if granting will exceed loop capacity. The HMI displays a queue of the equipment requesting WFI from the Primary Distribution Loop. The requests will be granted first in/first out method however, the operator will be able to switch any two queue positions on the HMI.

The WFI Request Control Continuous Logic will constantly check for the following two conditions:

1. If a WFI Request to the Primary Distribution Loop is Initiated
   a. Get the equipment ID for the request
      i. Check if equipment ID is valid. If valid:
         1. Store request to the next available spot in queue
         2. If the request is next in queue for approval
            a. Check if request draw rate will cause the TOT_CALC_DRAW_RATE to exceed [LOOP_CAPACITY]
               i. If the TOT_CALC_DRAW_RATE > [LOOP_CAPACITY], do not grant request

ii. If the TOT_CALC_DRAW_RATE ≤ [LOOP_CAPACITY], grant the WFI request

ii. If equipment ID is not valid, do not store request in queue.

2. If a WFI Request is cleared
   a. Remove the request from the queue and move subsequent requests up
   b. Calculate TOT_CALC_DRAW_RATE using the current granted requests and the request next in line.
      i. If the TOT_CALC_DRAW_RATE > [LOOP_CAPACITY], do not grant request
      ii. If the TOT_CALC_DRAW_RATE ≤ [LOOP_CAPACITY], grant the WFI request

The Primary Loop can have multiple pumps consecutively supplying the Secondary loops.

Alias name	**Description**
SILO_VLV_STS	Silo Valve Status (can monitor multiple Silos)
SILO_HOT_STS	Silo Hot Status (can monitor multiple Silos)
SILO_MINLEVEL_STS	Silo Level Above MIN_SAN_LVL Status (can monitor multiple Silos)
DIST_PUMP_A	Pump Control Module
DIST_PUMP_B	Pump Control Module
FLUSH_VLV	Flush/Recirculate Valve Status
RECIRC_TEMP	Recirculation Temperature
PUMP_HEX	Heat Exchanger for Loop
DIST_FIC	Distribution Flow Control
DIST_PIC	Distribution Pressure Control (PIC_20x01)
DIST_STEAM_SUPPLY_VLV	Distribution Steam Supply Valve (XV_20x07)
DIST_TIC	Distribution Temperature Control (TIC_20x01)
DIST_LOOP_TEMP	Distribution Loop Temperature (TI_20x01)
DIST_HDR_TEMP	Distribution Header Temperature (TI_10001, 10002, 10003, or 10004)
SILO_HOT_WFI_VENT_VLV	Silo Hot WFI Vent Valve (XV_10712, 10722, 10732, or 10742)
SILO_HOT_WFI_BLOCK_VLV	Silo Hot WFI Block Valve (XV_10711, 10721, 10731, or 10741).
SILO_HOT_WFI_COMP_AIR_VLV	Silo Hot WFI Compressed Air Valve (XV_10713, 10723, 10733, or 10743)
RECOVERY_SILO1_BLOCK_VLV	Recovery Silo 1 Block Valve (XV_10705)

Alias name	Description
RECOVERY_SILO1_SUPPLY_VLV	Recovery Silo 1 Supply Valve (XV_10702)
RECOVERY_SILO2_BLOCK_VLV	Recovery Silo 2 Bock Valve (XV_0706)
RECOVERY_SILO2_SUPPLY_VLV	Recovery Silo 2 Supply Valve (XV_10703)
PRIMARY_LOOP_RTRN_VLV	Primary Loop Return Valve (XV_10017, 10018, 10025, or 10026)
SILO_DIST_HDR_RTRN_VLV	Silo Distribution Header Return Valve (XV_10035, 10022, 10030, or 10027)
SILO_RTRN_VLV	Silo Return Valve (XV_10x12)
PRIMARY_LOOP_BLK_VLV	Primary Loop Block Valve (XV_10001, 10002, 10003, or 10004)
SILO_RTRN_DRAIN_VLV	Silo Return Drain Valve (XV_10x13)
SILO_DIST_HDR_DRAIN_VLV	Silo Distribution Header Drain Valve (XV_10041, 10042, 10043, or 10042)
SILO_DIST_HDR_RTRN_ DRAIN_VLV	Silo Distribution Header Return Drain Valve (XV_10036, 10023, 10031, or 10028)
PRIMARY_LOOP_SUPPLY_ SAMPLE_VLV	Primary Loop Supply Sample Valve (XV_20x05)
DIST_HDR_1_3_SUPPLY_VLV	Distribution Header 1/3 Supply Valve (XV_10015)
DIST_HDR_2_4_SUPPLY_VLV	Distribution Header 2/4 Supply Valve (XV_10016)
DIST_HDR_DRAIN_VLV	Distribution Header Drain Valve (XV_10038)
DIST_HDR_PCV	Distribution Header Pressure Control Valve (PCV_10001, 10002, 10003, or 10004)
PRIMARY_LOOP_SAMPLE_VLV	Primary Loop Sample Valve (XV_10007, 10008, 10009, or 10010)
DIST_HDR_SILO_HOT_WFI_ REC_VLV	Distribution Header Silo Hot WFI Recovery Valve (XV_10034, 10037, 10039, or 10040)

The command source determines which commands and settings will be used for the equipment module. The command sources used for equipment modules have the same look and functionality as the command sources used for the control modules. This section details the command sources that will be configured for the Primary Loop Equipment Module.

The operator controls the device. Operator commands and settings from the HMI will be accepted.

Priority logic controls the device, and operator commands will be superseded.

For the WFI Primary Loop Equipment Module, activation of any hold condition will trigger the override command and transition the equipment module to the Hold Operational Mode.

Maintenance owns control of the device. Maintenance commands will supersede the override commands and operator commands and settings are accepted from the HMI.

Out of service is initiated by a maintenance personal placing the equipment into maintenance mode and then out of service.

When placed Out of Service, all underlying control modules are released so that they can be manipulated as required for the service being performed on individual pieces of equipment (valves/pumps/etc.). The Equipment Module is placed into an idle state.

When a Maintenance Out of Service has been commanded, a sanitization is required before the equipment can be placed in service again. Before the sanitization can be initiated, the following must be filled out by Maintenance Personnel:

- Extraordinary Sanitization Number
- CWR/EMP Number
- Maintenance Sanitization Temperature (°F)
- Maintenance Sanitization Time (minutes)

Note that for all equipment modules, underlying control modules are locked in cascade mode before running a mode sequence.

Stop mode is initiated by the operator from the faceplate. The safe states for equipment controlled by this EM are as follows:

Equipment	**State**
RECIRC_PATH	Enable
STS_LOOP	Offline
POU_ENABLED	Disabled

Hold mode is initiated by a hold condition. Before a hold condition is generated, the system will prompt the operator for a signed Hold confirmation. While waiting on the confirmation, the system will be recirculating without being supplied from the Silo and POU valves are disabled. While in Hold mode, all controlled equipment is placed into a safe state and remains there until all hold conditions have cleared and another mode is initiated by the operator. The safe states for equipment controlled by this EM are as follows:

Equipment	**State**
FLUSH_REQUIRED	True
STS_LOOP	Offline
POU_ENABLED	Disabled

A flush is required upon restarting from the Hold mode. Flush will continue for the duration of [FLUSH_DURATION_SP].

Advance mode is when any silo is in advance or by the completion of Sanitization Mode. When in advance mode, the [DIST_FLUSH_PATH] is disabled. Once a silo is online ([STS_SILO] is “Ready”), the distribution pumps [DIST_PUMP_A] and [DIST_PUMP_B] are permitted to start, and (Loop Status [STS_LOOP] set to “Ready”). Once any distribution pumps are running, [STS_LOOP] is updated to “Online”. If no silo is attached to the loop, then pumps are inhibited. When a silo becomes advanced the loop will automatically be set to advance if it is not in the middle of a flush or sanitization.

1. Wait for [STS_SILO] = “Ready”.
2. Energize both recirculation pumps [DIST_PUMP_A] and [DIST_PUMP_B].
3. Set distribution flow control (FIC_20x01) [DIST_FIC_MODE] = Auto.
   a. Set [DIST_FIC_SP] = [DIST_RECIRC_FLOW_SP].
4. Set distribution pressure control (PIC_20x01) [DIST_PIC_MODE] = Auto.
   a. Set [DIST_PIC_SP] = [DIST_RECIRC_PRESS_SP].
5. Set [STS_LOOP] = “Online”
6. Set [POU_ENABLED] = True.[1]
   a. POU valve are still waiting on Silo permissive

When a hot water request is received, set [POU_ENABLED] = False.

1. Open [DIST_STM_SUPPLY_VLV] (XV_20x07). Set distribution temperature control (TIC_20x01) [DIST_TIC_MODE] = Auto with [DIST_TIC_SP] = [HOT_WTR_SUPP_TEMP_SP]
2. Wait for distribution loop temperature (TI_20x01) [DIST_LOOP_TEMP] = [HOT_WTR_SUPP_TEMP_SP] open requested POU valve.
3. Once the required volume of hot water is delivered to the selected mix tank, set distribution temperature control (TIC_20x01) [DIST_TIC_MODE] = Manual.
4. Wait until distribution loop temperature (TI_20x01) [DIST_LOOP_TEMP] = [WTR_SUPP_TEMP_SP].
5. Set [POU_ENABLED][1] = True.
6. Wait until [HOT_WATER_REQUEST] = False.
7. Begin Hot Water Recovery Air Dry Sequence.
   a. Open [SILO_HOT_WFI_VENT_VLV] (XV_10712, 10722, 10732, or 10742) and Start [HOT_WATER_REC_VENT_TIMER].
   b. When [HOT_WATER_VENT_TIMER] elapses, close [SILO_HOT_WFI_VENT_VLV] (XV_10712, 10722, 10732, or 10742) and activate [SILO_HOT_WFI_COMP_AIR_VLV] (XV_10713, 10723, 10733, or 10743) to dry the hot water recovery line.

---

1 Use of POU valves will be enabled if POU_ENABLED = True for both the Silo EM and Primary Distribution Loop EM.

c. Start [HOT_WATER_REC_DRY_TIMER].
d. When [HOT_WATER_REC_DRY_TIMER] elapses close [RECOVERY_SILO1_BLOCK_VLV] (XV_10705) and [RECOVERY_SILO1_SUPPLY_VLV] (XV_10702) or [RECOVERY_SILO2_BLOCK_VLV] (XV_0706) and [RECOVERY_SILO2_SUPPLY_VLV] (XV_10701), [SILO_HOT_WFI_VENT_VLV] (XV_10712, 10722, 10732, or 10742), and [SILO_HOT_WFI_BLOCK_VLV] (XV_10711, 10721, 10731, or 10741).

The sanitization sequence is initiated by the operator from the HMI. Primary Loop PN1 will be sanitized at the same time as Primary Loop South PS1. Primary Loop PN2 will be sanitized at the same time as Primary Loop South PS2. The sanitization sequence is as follows:

1. Disable [P1_POU_VLV] valves.
2. Enable [GROUP_SAN].
3. Wait for [SILO_SAN_RDY] = True.
4. Open [PRIMARY_LOOP_RTRN_VLV] (XV_10017, 10018, 10025, or 10026), [SILO_DIST_HDR_RTRN_VLV] (XV_10035, 10022, 10030, or 10027), [SILO_RTRN_VLV] (XV_10x12), and [PRIMARY_LOOP_BLK_VLV] (XV_10001, 10002, 10003, or 10004). Close [SILO_RTRN_DRAIN_VLV] (XV_10x13), [SILO_DIST_HDR_DRAIN_VLV] (XV_10041, 10042, 10043, or 10042), and [SILO_DIST_HDR_RTRN_DRAIN_VLV] (XV_10036, 10023, 10031, or 10028).
5. Enable the distribution pumps to start and run (set loop status [STS_LOOP] to "Ready"). Set distribution flow control (FIC_20x01) [DIST_FIC_MODE] = Auto. Set [DIST_FIC_SP] = [SAN_FLOW]
   a. These pumps will maintain flow at the required flow rate. Both pumps will run at the same speed to meet the flow setpoint.
6. Open [DIST_STEAM_SUPPLY_VLV] (XV_20107). Set distribution temperature control (TIC_20x01) [DIST_TIC_MODE] = Auto. Set [DIST_TIC_SP] = [SAN_TEMP].
7. When [DIST_HDR_TEMP] = [SAN_TEMP] start Sanitization timer [SAN_DUR_TIMER] for [SAN_DUR].
   a. Cycle [PRIMARY_LOOP_SUPPLY_SAMPLE_VLV] (XV_20x05) for [VLV_CYCLE_TIME]
   b. Cycle [DIST_HDR_1_3_SUPPLY_VLV] (XV_10015) or [DIST_HDR_2_4_SUPPLY_VLV] (XV_10016) for [VLV_CYCLE_TIME]
   c. Cycle [DIST_HDR_DRAIN_VLV] (XV_10038) for [VLV_CYCLE_TIME]
   d. Cycle [DIST_HDR_PCV] (PCV_10001, 10002, 10003, or 10004) for [VLV_CYCLE_TIME]
   e. Cycle [PRIMARY_LOOP_SAMPLE_VLV] (XV_10007, 10008, 10009, or 10010) for [VLV_CYCLE_TIME]

   f. If the temperature drops below [SAN_TEMP], [SAN_DUR_TIMER] is reset and starts over once the temperature has returned to [SAN_TEMP]
   g. Set [DIST_SAN_RDY] = True.
8. After [SAN_DUR], close [DIST_STEAM_SUPPLY_VLV]] (XV_20x07). Set distribution temperature control (TIC_20x01) [DIST_TIC_MODE] = Manual and [DIST_TIC_OP] = 0% open.
9. Upon [SAN_DUR] timer completion, set Sanitization Complete Alarm,
   a. Reset [SAN_TIMER].

Flush mode is initiated when transitioning from a Hold state or an operator initiating a flush via the HMI pushbutton. The flush sequence is as follows:

1. Set [POU_ENABLED] = False.
2. Wait for [STS_SILO] = "Ready"
3. Enable [DIST_FLUSH_PATH].
4. Start Flush timer [FLUSH_DUR_TIMER] for [FLUSH_DUR].
   a. The timer will reset if there are no silos online ([STS_SILO] indicates all closed) or no distribution pumps are running.
5. Upon [FLUSH_DUR_TIMER] completion,
   a. Disable [DIST_FLUSH_PATH].
   b. Enable [DIST_RECIRC_PATH].
   c. Clear the [STS_FLUSH] flag.
   d. Reset FLUSH_TIMER.

At the end of the flush sequence, [FLUSH_TIMER] is reset, and the distribution loop is set to Advance.

Equipment	**State**
DIST_PUMP_A	Running
DIST_PUMP_B	Running
DIST_HDR_PCV (PCV_10001, 10002, 10003, or 10004)	Closed (0%)
PRIMARY_LOOP_RTRN_VLV (XV_10017, 10018, 10025, or 10026)	Open
PRIMARY_LOOP_SAMPLE_VLV (XV_10007, 10008, 10009, or 10010)	Open
DIST_HDR_SILO_HOT_WFI_REC_VLV (XV_10034, 10037, 10039, or 10040)	Closed
SILO_DIST_HDR_RTRN_VLV (XV_10035, 10022, 10030, or 10027)	Open
SILO_DIST_HDR_RTRN_DRAIN_VLV (XV_10036, 10023, 10031, or 10028)	Open
SILO_RTRN_VLV (XV_10x12)	Closed

Equipment	State
SILO_RTRN_DRAIN_VLV (XV_10x13)	Open
DIST_PUMP_A	Running
DIST_PUMP_B	Running
DIST_HDR_PCV (PCV_10001, 10002, 10003, or 10004)	Open to [ADVANCE_ PCV_SP][2]
PRIMARY_LOOP_RTRN_VLV (XV_10017, 10018, 10025, or 10026	Open
PRIMARY_LOOP_SAMPLE_VLV (XV_10007, 10008, 10009, or 10010)	Closed
DIST_HDR_SILO_HOT_WFI_REC_VLV (XV_10034, 10037, 10039, or 10040)	Closed
SILO_DIST_HDR_RTRN_VLV (XV_10035, 10022, 10030, or 10027)	Open
SILO_DIST_HDR_RTRN_DRAIN_VLV (XV_10036, 10023, 10031, or 10028)	Closed
SILO_RTRN_VLV (XV_10x12)	Open
SILO_RTRN_DRAIN_VLV (XV_10x13)	Closed

Parameter description	Parameter tag name	Parameter type	Engineering units
Command Source (Operator, Override, Maintenance, Out of Service)	CMD_SRC	PlantPAx Command Source Module	N/A
Operating Mode (Stop, Advance, Sanitize, Flush)	OCMD_MODE	Integer	N/A
Distribution Flow Control Setpoint	DIST_FIC_SP	Real	gpm
Distribution Flow Control Mode	DIST_FIC_MODE	Integer	N/A
Distribution Pressure Control Setpoint	DIST_PIC_SP	Real	psig
Distribution Pressure Control Mode	DIST_PIC_MODE	Integer	N/A
Distribution Temperature Control Setpoint	DIST_TIC_SP	Real	°F

2 [ADVANCE_PCV_SP] anywhere in the sequences with the comment [ADVANCE_PCV_SP] varies depending on the current demand of WFI from the Primary Distribution Loop.

Parameter description	Parameter tag name	Parameter type	Engineering units
Distribution Temperature Control Mode	DIST_TIC_MODE	Integer	N/A
Sanitization Time	SAN_DUR	Real	Hours
Sanitization Temperature	SAN_TEMP	Real	°F
Minimum Silo Level for Sanitization	MIN_SAN_LVL	Real	%
Flush Time	FLUSH_DUR	Real	Minutes
Sanitization Time	SAN_DUR	Real	Minutes
Distribution Flush Path Enabled	DIST_FLUSH_PATH	Boolean	N/A
Distribution Recirculation Path Enabled	DIST_RECIRC_PATH	Boolean	N/A
Flush Pause Request	FLUSH_PAUSE	Boolean	N/A
Advance Flow Setpoint	DIST_RECIRC_FLOW_SP	Real	gpm
Advance Pressure Setpoint	DIST_RECIRC_PRESS_SP	Real	gpm
Hot WFI Supply Temperature Setpoint	HOT_WTR_SUPP_TEMP_SP	Real	°F
Hot Water Request	HOT_WATER_REQUEST	Boolean	N/A
Hot Water Recovery Vent Timer	HOT_WATER_REC_VENT_TIMER	Real	Minutes
Ambient Water Supply Temperature Setpoint	WTR_SUPP_TEMP_SP	Real	°F
Group Sanitization Request	GROUP_SAN	Boolean	N/A
Silo Sanitization Ready	SILO_SAN_RDY	Boolean	N/A
Sanitization Flow Target	SAN_FLOW	Real	gpm
Valve Cycle Time during Sanitization	VLV_CYCLE_TIME	Real	Minutes
Distribution Loop Sanitization Ready	DIST_SAN_RDY	Boolean	N/A
POU Enabled Status (True/False)	POU_ENABLED	Boolean	N/A

Parameter description	**Parameter tag name**	**Parameter type**	**Engineering units**
Primary Loop POU Enabled Status	P1_POU_VLV	Boolean	N/A
Flush Required Status	FLUSH_REQUIRED	Boolean	N/A
Loop Status	STS_LOOP	Boolean	N/A
Silo Status	STS_SILO	Boolean	N/A
Primary Loop Flow Capacity	LOOP_CAPACITY	Real	gpm
Advance Pressure Control Valve Position Setpoint	ADVANCE_PCV_SP	Real	%
Actual Operating Mode (Stop, Hold, Advance, Sanitize)	VAL_MODE	Integer	N/A
Current Runtime	TOTAL_RUNTIME	Real	Minutes
Current Hot Runtime	HOT_RUNTIME	Real	Minutes
Current Ambient Runtime	AMBIENT_RUNTIME	Real	Minutes
Time until Sanitization	SAN_TIME_REM	Real	Hours
Time until Flush	FLUSH_TIME_REM	Real	Hours
Sanitization Temp OK	SAN_OK	Boolean	1-OK
Actual Loop Temperature Mode	STS_TEMP_MODE	Boolean	1-Hot
Flush Expired Status	STS_FLUSH_EXPIRED	Boolean	1-Expired
Flush Active	STS_FLUSH	Boolean	1-Active
Loop Status (Offline, Ready, Online)	STS_LOOP	Integer	N/A
Out of Service Status	STS_OOS	Boolean	1-Out of Service
Offline Status	STS_OFFLINE	Boolean	1-Offline
Online Status	STS_ONLINE	Boolean	1-Online
Sanitization Expired Status	STS_SAN_EXPIRED	Boolean	1-Expired
Post-Maintenance Sanitization Duration	MAINT_SAN_DUR	Real	Minutes
Post-Maintenance Sanitization Temperature	MAINT_SAN_TEMP	Real	°F

Individual instances of this equipment module class will alarm whenever faulted.

Alarm description	Alarm conditions
Due for Flush Alarm	[FLUSH_SAN_TIMER].DN
Due for Sanitization Alarm	[SAN_TIMER].DN
Pump Offline Alarm	Flush or Sanitization Mode and no distribution pump [PUMP_n] running
Low Sanitize Temperature Alarm	Sanitize Active And [STS_FLUSH]and loop not in hot mode
Sanitize without Hot Silo Alarm	Sanitization Mode Started and there are no attached Silos that are "Hot" and Level > 15%
Sanitization Complete Alarm	[SAN_DUR_TIMER] expired
Maintenance Mode Alarm	Out of Service Active
Hold Mode Alarm	Hold Mode Active
Stop Mode Alarm	Stop Mode Active

If any of the following failure conditions are enabled and active, display the description of the failure condition and execute the hold logic.

Condition	Hold description	Hold conditions
1	Controlled Equipment Failure	Failure of any controlled equipment
2	Loss of Instrument Air Pressure	[INST_AIR] indicates loss of pressure.
3	Sanitization Timer Expiry	[SAN_TIMER] expired
4	High Return Conductivity	[RETURN_COND_HI]
5	High Supply Conductivity	[SUPPLY_COND_HI]
6	Low Supply Conductivity	[SUPPLY_COND_LO]
7	High Return Temperature	[RETURN_TEMP_HI]
8	High Return Pressure	[RETURN_PRESS_HI]
9	Low Return Pressure	[RETURN_PRESS_LO]
10	High Return Flow	[RETURN_FLOW_HI]
11	Low Return Flow	[RETURN_FLOW_LO]
12	High Return TOC	[RETURN_TOC_HI]
13	Low Return TOC	[RETURN_TOC_HI]
14	High Supply TOC	[SUPPLY_TOC_HI]
15	Low Supply TOC	[SUPPLY_TOC_LO]

Parameter	Description
OCMD_MODE	Operator Commanded Operating Mode (Stop, Advance, Sanitize, Flush)
STS_TEMP_MODE	Current Temperature Mode
STS_FLUSH	Flush Active
TOTAL_RUNTIME	Current Runtime
HOT_RUNTIME	Current Hot Runtime
AMBIENT_RUNTIME	Current Ambient Runtime
SAN_TIME_REM	Time until Sanitization
FLUSH_TIME_REM	Time until Flush
STS_TEMP_MODE	Actual Loop Temperature Mode
STS_FLUSH_EXPIRED	Flush Expired Status
STS_FLUSH	Flush Active
STS_LOOP	Loop Status (Offline, Ready, Online)
STS_OOS	Out of Service Status
STS_OFFLINE	Offline Status
STS_ONLINE	Online Status
STS_SAN_EXPIRED	Sanitization Expired Status
Event	**Description**
Operational Mode Change	Operational Mode Change
Maint. Sanitization Temperature Change	Maint. Sanitization Temperature Change
Maint. Sanitization Duration Change	Maint. Sanitization Duration Change
Operator ACK All	Acknowledge All Alarms
Low Temp while Sanitizing Alarm ACK	Low Temp while Sanitizing Alarm ACK
Due for Sanitizing Alarm ACK	Due for Sanitizing Alarm ACK
Sanitize without Hot Silo Alarm ACK	Sanitize without Hot Silo Alarm ACK
Sanitization Complete Alarm ACK	Sanitization Complete Alarm ACK

Example 8.4 ***Functional Requirement Specification***

WFI Supply Header Equipment Module Class

EM_HEADER

This is to define the functional requirements for the Water for Injection (WFI) Supply Header equipment module class. This type of equipment module is used to manage the WFI Supply Headers for the WFI system. This module controls a header supply permissive [STS_HDR_OK] as well as the header supply and header drain valves for each of the attached silos. The module may be accessed by the operator directly from the HMI.

Acronym or term	Definition
AUTO	PID Automatic, Automatic
CAS	PID Cascade, Cascade
CM	Control Module
Ethernet	Broadcast networking system that carries digital data packets to local nodes. A 10-Mbps base and local area network (LAN), Ethernet evolved from the IEEE 802.3 standard. Ethernet is the transport vehicle for many upper level protocols, including TCP/IP and Xerox network systems (XNS)
EM	Equipment Module
HMI	Human Machine Interface (Operator Terminal)
I/O	Input/Output signals to/from the Control System and the field devices
MAN	PID Manual, Manual
OEM	Original Equipment Manufacturer
PAS	Process Automation System
PID	PID Control, Proportional-Integral-Derivative control
PIO	Point I/O – Allen Bradley distributed I/O system
P&ID	Piping and Instrumentation Diagram
PLC	Programmable Logic Controller
POU	Point of use
RIO	Remote I/O
SMDS	Software Module Design Specification
SOP	Standard Operating Procedure
URS	User Requirement Specification
WFI	Water for Injection
WFICS	Water for Injection Control System

This equipment module is used to manage the WFI Headers for the WFI system. This equipment module class is used to (i) select the operational mode (Stop, Maintenance, Hold, Advance [Hot/Ambient], Sanitize), (ii) manage the sanitization timer, including placing the equipment in the required state on expiry, and (iii) monitor and handle equipment faults as required.

Details of the operational modes and fault conditions are described in the WFI Header Equipment Module Design Specification and instances. This module only operates in AUTO mode, and the operational mode is selected by the operator from the HMI.

Each instance of the module corresponds to an individual WFI Header. The Module controls a header supply permissive [STS_HDR_OK], silo supply valves [SILO_n_ SUPPLY_VLV], and header drain valves [SILO_n_DRAIN_VLV] based on inputs from the operator, commands from the silos, and status of the supply WFI Stills.

There are sanitization timers which are monitored by this equipment module. The OFF_SAN_TIMER and OFF_SAN_WARN_TIMER monitor the time that the header is offline ([STILL_VLV_STS] indicating no stills attached). When the OFF_SAN_WARN_TIMER timer expires, a message is displayed to the operator to notify of the pending sanitization expiry. When the OFF_SAN_TIMER expires, the module will be placed in hold mode, and the silo supply valves will be closed, and all header drains valves will be opened until a sanitization is commanded.

The SAN_TIMER and SAN_WARN_TIMER monitor the time since the last sanitization. When the SAN_WARN_TIMER timer expires, a message is displayed to the operator to notify of the pending sanitization expiry. When the SAN_TIMER expires, the module will be placed in hold mode, and the silo supply valves will be closed, and all header drains valves will be opened until a sanitization is commanded. The Header can supply multiple silos consecutively from multiple stills.

Alias name	Description
INSTR_AIR	Instrument Air Pressure Switch
STILL_VLV_STS	Still Header Valve Status (rollup of all stills)
SILO_n_TEMP	Header Temp at Silo "n" Drain
SILO_n_HDR_CMD	Silo "n" Header Supply Request
SILO_n_SUPPLY_VLV	Silo "n" Header Supply Valve
SILO_n_DRAIN_VLV	Silo "n" Header Drain Valve

The command source determines which commands and settings will be used for the equipment module. The command sources used for equipment modules have the same look and functionality as the command sources used for the control modules. This section details the command sources that will be configured for the WFI Still Equipment Module.

Maintenance owns control of the device. Maintenance commands will supersede the override commands and operator commands and settings are accepted from the HMI. Out of service is initiated by a maintenance personal placing the equipment into maintenance mode and then out of service.

When the header is placed Out of Service, all underlying control modules are released so that they can be manipulated as required for the service being performed on individual pieces of equipment (valves/pumps/etc.). The Equipment Module is placed into an idle state.

When a Maintenance Out of Service has been commanded, a sanitization is required before the equipment can be placed in service again. Before the sanitization can be initiated, the following must be filled out by Maintenance Personnel:

- Extraordinary Sanitization Number
- Extraordinary Maintenance Number
- Maintenance Sanitization Time (minutes)

The Extraordinary Sanitization Number and Extraordinary Maintenance Number can be changed up until the end of the sanitization and are written to the report at the time of completion.

Note that for all equipment modules, underlying control modules are locked in cascade mode before running a mode sequence. Stop mode is initiated by the operator. While in Stop mode, the permissive for Stills to supply (STS_READY) is removed, all supply valves to the silos (SILO_x_SUPPLY_VLV) are closed, and all drain valves (SILO_x_DRAIN_VLV) are opened. Hold mode is initiated by a hold condition. While in Hold mode, all controlled equipment is placed into a safe state and remains there until all hold conditions have cleared and another mode is initiated by the operator. The safe states for equipment controlled by this EM are as follows:

Equipment	State
SILO_x_SUPPLY_VLV	Closed
SILO_x_DRAIN_VLV	Open
STS_READY	Not Ready
STS_HDR_OK	Not OK

Advance mode is initiated either by the operator or by the completion of the Flush sequence. In Advance mode the stills feeding the header are selected based on the operator selection of the following:

- OCMD_TEMP_MODE (Ambient/Hot)

The actual temperature of the water entering must match the header commanded temp unless Still sanitization is active, and no header is hot. While in Advance mode and any Still (as indicated by STILL_VLV_STS) is actively feeding the header, the header will set the [STS_HDR_OK] status and begin to monitor supply requests from the silos. When the header supply permissive [STS_HDR_OK] has been set, if an operator has selected the header for supply [SILO_n_HDR_CMD], the header will open the silo's supply valve [SILO_n_SUPPLY_VLV] and close all the drain valves [SILO_x_DRAIN_VLV].

If the header supply permissive [STS_HDR_OK] has not been set, or no silo is issuing a command for water from the header (SILO_n_HDR_CMD), then all of the drain valves [SILO_x_DRAIN_VLV] will be opened.

The sanitization sequence is initiated by the operator from the HMI. When a header is placed into Sanitization (or advanced hot) mode, any Still feeding it is also placed into sanitization mode. The sanitization sequence is as follows:

1. Check the number of available stills and verify the minimum sanitization flow rate will be achieved.
2. Clear header supply permissive [STS_HDR_OK], close all supply valves [SILO_x_SUPPLY_VLV] and open all the drain valves [SILO_x_DRAIN_VLV].
3. Set [WFI_TEMP_CMD] to Hot.
   a. Set all Still's that are currently attached to the header [WFI_TEMP_CMD] to Hot.
4. Wait for at least one Still (as indicated by STILL_VLV_STS) to actively be feeding the header.
5. Wait for all Silo drain temperatures [SILO_x_TEMP] to reach [SAN_TEMP]
   a. Set [SAN_OK] sanitization permissive as long as this condition is true.
6. Start Sanitization Duration timer [SAN_DUR_TIMER] for [SAN_DUR] or [MAINT_SAN_DUR].
   a. [MAINT_SAN_DUR] is used if the previous mode was Out of Service (Maintenance).
   b. If any Silo drain temperature ([SILO_n_TEMP]) is less than [SAN_TEMP], the timer will restart
7. Upon timer completion, set Sanitization Complete Alarm and reset OFF_SAN_TIMER, OFF_SAN_WARN_TIMER, SAN_TIMER, SAN_WARN_TIMER.
   a. If temperature drops below the San temp, timers do not start again until temp is hot.
8. Set Sanitization Complete Alarm.
9. If the [SAN_OK] permissive is set and Sanitization is not required (STS_SAN_EXPIRED is false), the header will honor any silo requests for supply [SILO_n_HDR_CMD] and will open that silo's supply valve [SILO_n_SUPPLY_VLV] and close all the drain valves [SILO_x_DRAIN_VLV]. At the end of the sanitization sequence, the mode is set to Advance.

Flush mode is initiated when no stills are attached for [FLUSH_TIME]. The Flush sequence is as follows:

1. Clear header supply permissive [STS_HDR_OK] and open all the drain valves [SILO_x_DRAIN_VLV]
2. Start Flush Duration timer [FLUSH_TIMER] for [FLUSH_DUR]. At the end of the flush sequence, the mode is set to Advance.

Parameter description	**Parameter tag name**	**Parameter type**	**Engineering units**
Command Source (Operator, Override, Maintenance, Out of Service)	CMD_SRC	PlantPAx Command Source Module	N/A
Operating Mode (Stop, Advance, Sanitize, Flush)	OCMD_MODE	Integer	N/A
Header Temperature Mode (Ambient, Hot)	OCMD_TEMP_MODE	Boolean	N/A
Offline Sanitization Time	OFF_SAN_TIME	Real	Hours
Offline Sanitization Warning	OFF_SAN_TIME_WARN	Real	Hours
Sanitization Time	SAN_TIME	Real	Hours
Sanitization Warning	SAN_TIME_WARN	Real	Hours
Sanitization Temperature	SAN_TEMP	Real	°F
Sanitization Duration	SAN_DUR	Real	Minutes
Post-Maintenance Sanitization Duration	MAINT_SAN_DUR	Real	Minutes
Offline Flush Time	FLUSH_TIME	Real	Hours
Flush Duration	FLUSH_DUR	Real	Minutes
Actual Operating Mode (Stop, Hold, Advance, Sanitize)	VAL_MODE	Integer	N/A
Actual Header Temperature Command (Ambient, Hot)	WFI_TEMP_CMD	Boolean	1-Hot
Actual Still Temperature Mode (Ambient, Hot)	STS_WFI_TEMP	Boolean	1-Hot
Current Runtime	TOTAL_RUNTIME	Real	Minutes
Current Hot Runtime	HOT_RUNTIME	Real	Minutes
Current Ambient Runtime	AMBIENT_RUNTIME	Real	Minutes
Current Offline Time	OFFLINE_TIME	Real	Minutes
Time until Sanitization	SAN_TIME_REM	Real	Minutes

Parameter description	Parameter tag name	Parameter type	Engineering units
Sanitization Temp OK	SAN_OK	Boolean	1-OK
Header Ready for Supply	STS_READY	Boolean	1-Ready
Header OK Status	STS_HDR_OK	Boolean	1-Header OK
Out of Service Status	STS_OOS	Boolean	1-Out of Service
Offline Status	STS_OFFLINE	Boolean	1-Offline
Online Status	STS_ONLINE	Boolean	1-Online
Sanitization Expired Status	STS_SAN_EXPIRED	Boolean	1-Expired

Individual instances of this equipment module class will alarm whenever faulted.

Alarm description	Alarm conditions
"Alert" Alarm	[OFF_SAN_WARN_TIMER] or [SAN_WARN_TIMER] expired
Low Temp while Sanitizing Silo alarm	Sanitization mode active and [SILO_n_TEMP] < [SAN_TEMP]
Due for Sanitizing Alarm	[OFF_SAN_TIMER] or [SAN_TIMER] expired
Sanitization Complete Alarm	[SAN_DUR_TIMER] expired
Hold Alarm	Hold Mode Active
Stop Alarm	Stop Mode active
Maintenance Mode Alarm	Out of Service Active

If any of the following failure conditions are enabled and active, display the description of the failure condition and execute the hold logic.

Condition	Hold description	Hold conditions
1	Controlled Equipment Failure	Failure of any controlled equipment
2	Loss of Instrument Air Pressure	[INST_AIR] indicates loss of pressure
3	Sanitization Timer Expiry	[SAN_TIMER] expired
4	Offline Sanitization Timer Expiry	[OFF_SAN_TIMER] expired

Data	Description
OCMD_MODE	Operator Commanded Operating Mode (Stop, Advance, Sanitize, Flush)
OCMD_TEMP_MODE	Operator Commanded Temperature Mode (Ambient, Hot)
VAL_MODE	Actual Operating Mode (Stop, Hold, Advance, Sanitize, Flush)
STS_WFI_TEMP	Actual Temperature Mode (Ambient, Hot)
TOTAL_RUNTIME	Current Runtime
HOT_RUNTIME	Current Hot Runtime
AMBIENT_RUNTIME	Current Ambient Runtime
SAN_TIME_REM	Time until Sanitization
STS_READY	Header Ready for Supply
STS_HDR_OK	Header OK Status
STS_OOS	Out of Service Status
STS_OFFLINE	Offline Status
STS_ONLINE	Online Status
STS_SAN_EXPIRED	Sanitization Expired Status
Event	**Description**
Operational Mode Change	Operator Operational Mode Change
Temperature Mode Change	Operator Temperature Mode Change
Maintenance Sanitization Duration Change	Maintenance Sanitization Duration Change
Operator ACK All	Acknowledge All Alarms
"Alert" Alarm ACK	"Alert" Alarm ACK
Low Temp while Sanitizing Alarm ACK	Low Temp while Sanitizing Alarm ACK
Due for Sanitizing Alarm ACK	Due for Sanitizing Alarm ACK
Sanitization Complete Alarm ACK	Sanitization Complete Alarm ACK

Electrical – Controls Equipment Specifications

Component	Brand name	Notes
Buttons, Push	Allen-Bradley	800T series, non-textured in cleanroom applications, LEDs for lighted buttons, sealed contacts in wet areas
Circuit Breakers	Allen-Bradley, Phoenix	Use circuit breakers in place of fuses when possible

Component	Brand name	Notes
Conditioner, Line Power	Islatrol, Allen-Bradley	Drive power source should be supplied through a Power Line Conditioner
Conduit, Flexible	Sealtite or Liquatite	Refer to section titled
Contactors	Allen-Bradley	IEC series with finger-safe cover
Cords, Quick Disconnects	Hubbell, Meltric	Twist lock plug or de-contactor types
Data Collection	Allen-Bradley	FactoryTalk Historian Card
Disconnects	Allen-Bradley	Refer to section titled "*Enclosures*"
Drives	Allen-Bradley	Refer to section titled "*Drives/Motors*"
Enclosures	Hoffman, Rittal	Refer to section titled "*Enclosures*"
Fuses	Bussman, Shawmut	Fuses are to be used only if Circuit Breakers do not fit the application, review with client engineering before implementation
HMI Industrial PC	B&R, Hope Industrial, Advantech, Allen-Bradley	Windows 10 or higher
HMI Software	Allen-Bradley	FactoryTalk View SE. Specific applications may be required to be CFR 21-Part 11 compliant. Reference URS for this requirement
HMI thin clients	Allen-Bradley, Advantech, Hope Industrial, B&R	
Instrumentation	Rotronic – Dew point Ashcroft – DP Rosemount- – Conductivity Endress Houser – Level Vega – Level Vaisala – GMP Temp, RH, DP Honeywell – Temp/pressure IFM – Pressure Mettler Toledo – pH Meters Pyromation – RTDs	Application specific
Interlocks, Guard Door	Allen-Bradley, Schmersal	Each door switch is to provide input to PLC so that HMI can indicate which door is open. Must be locking type

Component	Brand name	Notes
Lights, Pilot	Allen-Bradley	800T series, LED
Light Curtains/Scanners	Sick or Keyence	
Motors, AC	Baldor, Emerson, Nord	Refer to section titled "*Drives/Motors*"
Motor, Starters	Allen-Bradley	IEC Rated
Power Supply – 24 V	Sola, Allen-Bradley, Phoenix	Use DIN-Rail mounting
Programmable Logic Controller	Allen-Bradley	Refer to section titled "*Programmable Logic Controllers*"
Robots	Yaskawa (Motoman), Fanuc, or Staubli	Specific application
Relays, Safety Relays	Allen-Bradley, Phoenix, or Sick	Use DIN-Rail mounted, use solid-state where possible
Safety Controllers	Allen-Bradley or Sick	May be integrated into the PLC where applicable
Switches, Limit	Allen-Bradley, Sick	
Switches, Selector	Allen-Bradley	800T Series
Switches/ Routers, Ethernet	Allen-Bradley, Phoenix, Hirschman, Antaira	
Sensors	Banner, Keyence, Allen-Bradley, Turck, Sick, IFM	Must use 24-V DC source when possible with quick disconnects. Use I/O remote blocks where possible. I/O Link devices may be acceptable
Servo Motors and Drives	Allen-Bradley	See Drives/Motors section
Transformer: Control	Allen-Bradley, Sola	Equivalent transformer may be used
Transformer: Power	Acme	Equivalent transformer may be used
Voltage	—	480-V AC 3 Phase nominal, 60 Hz 120/240-V AC 1 Phase nominal, 60 Hz
Wiring, Color	Follow NFPA-79	Refer to "*Color Codes*"
Wiring, Components		All components are to be marked with permanent labels securely fastened to the mounting surface, not the component itself

Component	Brand name	Notes
Wiring, Service		Wire type: THHN or European equivalent
Wiring, Markings	Sleeves or heat shrink are preferred. Stickers are acceptable for components	All components are to be marked with permanent labels securely fastened to the mounting surface, not the component itself
Wiring, Panel		Wire type: MTW
Wiring, Splices		Wire Nuts and Butt splices are not permitted outside of motor connections. Use terminals in junction boxes where splices are needed
Wiring, Terminations	Allen-Bradley, Phoenix, or WAGO	IEC screw type or ferrules are preferred

Documentation

Electrical documentation shall include:

(1) One hardcopy set of electrical prints in the control cabinet.
(2) Electronic copy of electrical prints on flash drive in AutoCAD 14 or higher dwg or dxf format.
(3) Electronic copy of ladder logic on flash drive.
(4) Electronic copy of any support equipment programs (such as an HMI) on flash drive.

Drives/Motors

AC Motors:

Standard frame, SS enclosure in cleanroom applications. SS gearboxes preferred. No through shaft gearboxes.

Variable Frequency Drives:

Allen-Bradley PowerFlex 527 series, 525 series for stand-alone applications

Servo Motors:

Allen-Bradley MP-Series, TL-Series, or Kinetix VP

Servo Drives:

Allen-Bradley Kinetix series

Notes:

- Each VFD Controller shall have an operator interface display installed. If there are more than one VFD Controllers present in a panel or project, each VFD will have an operator interface display installed.

- If parameters are controlled from a ControlLogix processor, adjustments should be made through HMI over Ethernet.

DC Motors:

None unless required for the application and pre-approved

Wire/Cable Management:

Cleanroom Specific

1. Wash downs areas require enclosed SS and drainable troughs.
2. Rigid conduit must be stainless steel in ISO Class C and D rooms or higher.
3. NMT 6″ of flex for fixed connections, e.g. motors, valves, etc.
4. Any application that requires greater than 6″ of flex must be pre-approved.

General Requirements

1. Outside of clean room, ladder-style wire ways are acceptable.
2. Use cable pass-through devices for servo and other large cables/hoses (e.g. Murrplastik).
3. Metallic conduit is preferred for applications outside of cleanrooms.
4. Wireways and conduit must be NMT of 20% filled.
5. When conduit has 2+ conductors, must be NMT 40% filled. Reference NEC 310.15(B)(3)a for ampacity adjustments for multiple conductors.

Enclosures

1. NEMA 4× rating for wash down applications and NEMA 12 for all others. Components must be (IP20) finger safe.
2. Control enclosures shall have a minimum of 20% spare space for future expansion.
3. Ethernet switches should have a minimum of 20% room for expansion.
4. 120-V AC electrical outlets must be installed in enclosures with programmable devices.
5. Isolate high voltage (>50 V) from low voltage.
6. Include shelf for laptop.
7. External programming (Grace ports) port and 110 V outlet (water tight where necessary).
8. Disconnects – Allen-Bradley. Disconnect must be remote from main electrical panel and meet NEC and NFPA standards
9. Cooling units are required for applications outside of controlled temperature spaces.

Circuit Breakers and Fuses (only if breakers are not an option)

1. Fuses are to be used only if Circuit Breakers do not fit the application
2. Circuit breakers must be equipped with a contact that will allow a fault to be displayed on the HMI if it is tripped.

Programmable Logic Controllers

1. Allen-Bradley ControlLogix and CompactLogix.
2. Refrain from using specialty I/O cards unless they are required for their unique functionality.
3. When analog cards are used, 4–20 mA signals are preferred. Use signal splitters if two output sources are needed, e.g. for multiple displays.
4. When using intelligent devices such as Variable Frequency Drive Controllers or Weight Scales use an Ethernet IP communication protocol.
5. Each input and output of the PLC shall have only one device connected. Switches or sensors connected in series to a single input point are not acceptable.
6. Each type of digital I/O used in an application (DC input, DC output, AC input… etc.) shall have spare points left on the cards. There shall be at least 10% points of each type available for future use.
7. All spare I/O points must be labelled and landed on terminals blocks.
8. The PLC rack shall have spare slots for future expansion. Leave at least two slots free.
9. Point I/O od Flex I/O shall have panel space for 10% additional modules.
10. Ethernet IP valve manifolds shall have 10% additional valves, and 10% additional I/O.
11. Ethernet IP valve manifolds shall have panel space for 10% additional valves, and 10% additional I/O.
12. The PLC shall have at least 20% spare memory.
13. There shall be no passwords used within the programs.
14. There shall be no password or protection of the source code.
15. SQIs/SQOs are not acceptable.
16. PLC tag names should be descriptive on the device or parameter. As an example Target Amount versus Parameter [19].

Note: Stand Alone Controllers (HVAC/Doors/etc.) rule of thumb is if the controller requires a program to be loaded to operate then it must meet the requirements above.

Human Machine Interface (HMI)

1. Screen navigation shall be clear and consistent from screen to screen.
2. Screen navigation buttons should appear in a consistent location in the screens.
3. Start or On buttons shall be green.
4. Stop or Off buttons shall be red.
5. Reset buttons shall be amber or yellow.
6. Emergency Stop buttons shall be red with a yellow round label.
7. Emergency Stop buttons must be rated for a minimum of Cat 3 Sil 2

8. Other button functions colors shall be pre-approved.
9. Parameter entry from the HMI shall be limited to parameters that change frequently.
10. Graphic representations of the machine shall be displayed on HMI screens.
11. Graphic representations shall include locations of input and output devices on the HMI screen.
12. A list of parameters entered from the HMI shall be approved before HMI development begins.
13. HMI screens shots and Fault List shall be provided four weeks prior to FAT.
14. Machines that have direct impact on product, or fillers shall be 21 CFR Part 11 compliant.
15. HMIs that store 21 CFR Part 11 compliant data must store to a local SQL or SQLExpress database unless the HMI is run from a server.
16. HMIs that store 21 CFR Part 11 compliant data shall require electronic signature.
17. Preliminary HMI screen design shall be pre-approved prior to full development of the HMI application.

Software

Software versions must be up to the current level/version in use

Wiring – Color Codes

208-V AC

- Single Phase Line (labeled as L): black or red (in case of a second active)
- Phase 1 (labeled as "L1"): black
- Phase 2 (labeled as "L2"): red
- Phase 3 (labeled as "L3"): blue
- Neutral Conductor (labeled as "N"): gray
- Protective Ground (labeled as "PG"): green, green-yellow, or bare copper.

480-V AC

- Single Phase Line (labeled as L): black or red (in case of a second active)

480-V AC Equipment Wiring

- Phase 1 (labeled as "L1"): black
- Phase 2 (labeled as "L2"): red
- Phase 3 (labeled as "L3"): blue

480-V AC Feed

- Phase 1 (labeled as "L1"): brown
- Phase 2 (labeled as "L2"): orange
- Phase 3 (labeled as "L3"): yellow
- Neutral Conductor (labeled as "N"): gray
- Protective Ground (labeled as "PG"): green

Use VFDs cables specifically designed for VFDs.

Notes:

(1) Use IEC terminals only for special applications. Use Standard as a rule.

(2) All electrically common points are to be marked with the same number on the wire and termination points.

(3) All wires are to be marked with the line number on the print per drawing standards.

CSV - References

1. Gough and Nettleton [20, 21]
2. Liscouski [22]

9

Personal Protective Equipment (PPE) and Process Flow

Sterile manufacturing standard operating procedures (SOPs) should emphasize personnel gowning requirements, qualifications, and behaviors inside clean rooms and suites (ISO 5, 6, 7, and 8). Clean room design, barrier technologies, process design and qualifications, process control, environmental monitoring, data integrity, and production batch record accuracies should all be well understood and adhered to by following procedures to facilitate team integration and seamless operations in this regulated environment. ISO classifications are easily understood as it relates to particle counts in cascading clean rooms to reach the aseptic core. Training of personnel to understand these classifications and their importance in terms of various gowning levels will help personnel awareness in compliance to required procedures. Supervisory roles will focus on assuring conformance to written SOPs. In addition, personnel movements and monitoring will be observed in aseptic processing areas.

As companies evolve in moving from manual operations to automated ones, interventions during manufacturing operations will be reduced and this will lead to a reduction in deviations to justify these interventions. In this regard, unidirectional airflow patterns' analysis needs to be confirmed due to operator interventions. Besides air handler unit (AHU) overall flow metered measurements, airflow velocity measurements should be taken in the proximity of high-efficiency particulate air (HEPA) filters and at the filling surface levels. Continuous particle monitoring systems (CPMS) should be utilized with the aid of computers for nonviable particle counts. Air sampling should be through fixed cones close to the filling and capping station heads in sensitive locations under unidirectional airflow. Settling microbiological plates should be used to collect samples in sensitive locations to measure viable counts for the environmental monitoring program. Continuous monitoring of pressure differentials between

Sterile Processing of Pharmaceutical Products: Engineering Practice, Validation, and Compliance in Regulated Environments, First Edition. Sam A. Hout.

the aseptic core (ISO 5) and the surrounding (ISO 6) cascading to ISO 7 should be balanced to ensure no pressure reversals as doors are opened to exit aseptic areas. Interlocking doors are important when doors are opened, the outward airflow in the airlocks should minimize potential ingress of contaminants. Therefore, the timing of open doors is critical and should be minimized.

10

Sterile Aseptic Processing

There are three stages to consider in sterile processing. First, process development involves the design and execution of engineering runs to confirm process parameter targets and limits, and formulation and mixing of chemicals to evaluate dispersion and mixing speed and times followed by pH titration curves. Filling requires setup fill parameters, proposed tubing set design and aseptic connections, and length/ID of tubing. A dosing study must be completed to determine ability to dose required amounts within time constraints to meet machine speed requirements.

Second, batch processing tanks require IOQ validation, then steam in place (SIP) timing and readout of temperatures and kill factors (F_0) are determined as part of SIP validation. Bulk materials' sterilization validation must be confirmed using aseptic simulations and bulk holding studies. SIP validation confirms bioburden reduction and claims sterility after SIP. Bulk sterilization validation claims sterility after bulk sterilization. Pressurization post sterilization is done via an autoclaved valve with a 0.22-μm filter that was isolated post autoclave. Pressurization after sterility claims – maintain ~10 psi 0.22 μm filtered N_2 when possible, positive pressure to prevent microbial ingress [23].

SIP process development ensures lethality can be met throughout the vessel. In addition, tank must be isolated prior to disconnection from SIP utilities to maintain sterility for aseptic formulation. Bulk sterilization development verifies proposed approach will work and predetermined lethality can be met in worse-case locations. In addition, tank must be isolated prior to disconnection from bulk sterilization utilities to maintain sterility. Bulk holding studies confirm hold of sterile medium following process steps after sterility is claimed. The uses of all sterilized formulation equipment require a male Kleenpak (Pall) connector or equivalent and sterile tubing pathway to filter sterile medium after hold time that is based on filling requirements, usually 2-day fill ≥ 3-day hold.

An aseptic simulation is a bulk holding study to demonstrate that sterility can be maintained during the formulation process where multiple aseptic manipulations occur. This simulation is representative of the full aseptic process after sterility

Sterile Processing of Pharmaceutical Products: Engineering Practice, Validation, and Compliance in Regulated Environments, First Edition. Sam A. Hout.

is claimed. Aseptic manipulations include sterile transfer from one vessel into another presterilized vessel. Sterile recirculation of formulation for filling or process operations requires a male Kleenpak connector and a sterile tubing pathway to filter sterile medium after hold time.

Container Closure Integrity (CCI) test demonstrates that the proposed container closure system can form a sterile boundary to prevent microbial ingress after the product leaves the filling lines [24]. Filling and closing should be adjacent under ISO 5 conditions. Equipment filling parameters microbial challenge is required prior to media fills [25]. Dye ingress is foregone depending on risk, e.g. not usually done for vials with stoppers. CCI testing is required to mitigate any risk to media fill from the container.

Third, filling sterility involves filler IOQ, CCI using dye ingress, and microbial challenge [26]. Following confirmation of process integrity through media fills, current Good Manufacturing Practice (cGMP) batches, stability batches, and submission batches can be produced. Media fills demonstrate that sterile media can be aseptically filled and closed in the filling room. Non-aseptic-formulated media is sterile filtered into the filling room. Formulation uses generic media fill dedicated equipment, and sterility is claimed after sterile filtration of media into the fill room. Any media fill failures indict all products aseptically filled in that fill room at a minimum. Aseptic process is identical to media fills.

There are limitations and gaps in aseptic validation [27]. One risk factor is the assumption that product is sterile prior to entry into the fill room. Therefore, any process steps that generate a sterile product need to be validated to demonstrate process sterility. Another risk is the SIP validation of aseptic formulation tank, and for example bulk sterilization validation of Cipro solution. Other risk factors can be summoned in the areas of aseptic transfer of bulk sterilized Cipro solution, sterile filtration of poloxamer solution into formulation tank, aseptic formulation of bulk product, aseptic transfer into filling surge vessel, and aseptic recirculation of bulk product in fill tubing set during aseptic simulation.

Cleaning verification only is performed for dedicated equipment. Cleaning verification verifies that bioburden and endotoxin levels meet internal requirements for United States Pharmacopeia (USP) Highly Purified Water. Cleaning validation requirement is based on the evaluation of various products to use the equipment.

11

Integrated Facility Design

In general connotations, the aseptic core where filling and capping occur should be controlled and classified to ISO 5 standards [28]. Around the filling and capping and outside the restricted-access barrier systems (RABS), ISO 6 standards might be applied. Outside the filling and capping suites, an aseptic corridor with ISO 6 classification is appropriate. All material and personnel flow adjacent rooms must be designed with dedicated paths for in and out. Materials can be transferred through path-through design. These adjacent rooms can usually cascade from ISO 6 to ISO 7 toward gowning rooms. Exit to ISO 8 aseptic corridor and out to controlled not classified (CNC) space (sanitary). All component preparation suites must be in ISO 7 classified rooms. All compounding and mixing of product formulations must be in ISO 7–classified suites [29].

The manufacturing process includes the cascading space described earlier and other work-in-progress (WIP) areas for handling and storage or staging for product fill/finish. To list a few process steps, these will include weighing of excipients, active pharmaceutical ingredient (API) weighing and dosing, milling, mixing formulation to add solids to liquids, which might include gas purge to blanket product. Following compounding, a sterile filtering might occur before the transfer of sterilized materials and components such as stoppers or caps into the filling suite.

Lyophilization incorporates the filling and partial stoppering of vials before loading into lyophilizers, whereby water is removed from the product under specific vacuum conditions. This is followed by unloading sealing and capping, washing in the case of high potency drugs, drying, labeling, and tray off, then serialization and packaging.

Sterile Processing of Pharmaceutical Products: Engineering Practice, Validation, and Compliance in Regulated Environments, First Edition. Sam A. Hout.

A typical and efficient setup of a sterile drug manufacturing building in a two-level construction will typically house all the manufacturing suites on the first level, including all necessary utilities, bathrooms, break rooms, personal protective equipment (PPE) rooms, and visual corridors for line observations. The second floor might be split into an equipment mezzanine to house air handling units (AHU) for heating, ventilation, and air conditioning (HVAC) that supports the clean rooms, other critical utilities installations, control cabinets, and maintenance support space. Chemistry and microbiology laboratories might be on the second floor in support of manufacturing operations' sampling and testing requirements, including environmental monitoring of clean space and critical utilities measurements and testing to ensure efficient operations. A wraparound sanitary corridor might provide for ease of access to all these labs and support equipment needed.

Manufacturing classified rooms are controlled through high-efficiency particulate air (HEPA) filtration and room pressurization, whereby the aseptic core has the highest pressure and cascades down to the wraparound corridor. The intent is to mitigate risk by minimizing product contamination. The use of PPE protective clothing and separate gowning and de-gowning rooms plus the sterilization of all components needed in the fill room utilizing various pretreatment methods is the recommended practice to support sterile transfers into the aseptic core.

CNC designation implies that the space is cleanable, but with access control and air is filtered for ventilation. Personnel transfers and required garment is discretionary. No environmental monitoring is required, but optional, to control ingress of mold.

Capping areas must be close and adjacent to filling, but with proper barriers to prevent particulate matter transfer into open containers. All design criteria and selection of capping equipment must emphasize capping and crimping systems that are designed to mitigate shedding of particles. Additional measure such as vacuum suction systems might be employed very close to the capping head to ensure that any generated particles would be collected through the vacuum system.

Major advances in aseptic/sterile manufacturing focus on processes that allow for minimal personnel interventions through further automation and robotic manipulations. Furthermore, advances in gating and automated access tools in specific critical locations might allow for personnel interventions but with minimal intervention impact.

Drug substance/ API processing	Drug product/fill-finish
Dispensing	Dispensing
Compounding	Compounding
Sterile filtering	Sterile filtering
Crystal seed	Container/cap preparations (depyrogenation/sterilization)
Crystallization	Aseptic transfer of components and materials
Drying	Aseptic filling/capping
Milling	Lyophilization – partial stoppered vials transfer
Sizing	Complete capping and crimping
Container packing	Terminal sterilization if not aseptic
Closure	Manual or automated visual inspection
Packaging	Serialization and labeling
	Packaging

Aseptically processed products that cannot be sterile filtered, e.g. sterile powders must be dispensed in an ISO Class 5 environment. Liquids can be metered or weighed. Vessels might be placed on load cells or floor balance to achieve desired weights. All necessary calibrations must be in place and paying attention to interference from auxiliary cables or hoses that might be attached to the equipment such that there is no impact on the accuracy of weighing.

Every precaution must be taken to prevent cross-contamination during dispensing. This could be due to the carryover of residual materials from a previous batch, or parallel dispensing of different products. Proper good manufacturing practice (GMP) implementations would minimize risk of contamination, for example, established validated cleaning procedure between batches. Clearance procedure must be followed by trained operators to achieve correct, clean, and efficient dispensing [30].

Double-door batch ovens, path-through steam sterilizers, or chain and conveyor depyrogenation tunnels can directly transfer sterile components into the aseptic filling environment. If this type of sterile transfer is not available, components might be double wrapped in sterilizable sheets that permit air/steam ingress while maintaining sterility of contents. Components that are γ irradiated or ethylene oxide gas sterilized can be aseptically transferred to the ISO 5 clean room.

Precautions must be taken through validated steps to remove all disinfected outer wraps without introducing risk of biocontamination into the ISO 5 environment.

Water for injection (WFI) is using as the basic solvent for aqueous formulations, including emulsions. All processing equipment must be sterilized in validated cycles. Liquid mixing tanks are SIP sterilized. Aseptic powder transfers are usually challenging and must be validated. In addition, appropriate expertise should be sought to prevent dust explosions, as the understanding of angle of repose, static electricity, grounding and bonding of equipment, powder density, void factors, and moisture content effects on ignition/explosion in the presence of oxygen.

Critical process utilities systems such as WFI, clean steam (CSM), clean compressed air (CCA), clean N_2, and clean air in classified areas are considered as direct impact to product quality and safety. These critical systems are designed to prevent product contamination in pharmaceutical sterile manufacturing. Materials of construction of all storage and distribution systems are very important to specify such that they will contribute to transport of fluid safely without any ingress of contaminants. Besides the chemical nature of materials, continuous cleaning and sanitization as validated systems require are crucial aspect of maintainability against microorganism growth. Other process utilities support systems are important for operations reliability. Critical utilities must not be reactive or absorptive and must withstand chemical and sanitization repeated cleaning.

The impact of power failure in sterile manufacturing aseptic facilities must be of major design consideration. All critical controls must be backed up with uninterrupted power supply (UPS) units. The main classified areas and all critical stability chambers, and critical chemistry and microbiology storage systems must be backed up with a power generator (either natural gas or diesel) and these systems must be very reliable and automated such that they are always on standby and ready to start in the case of power failures. Putting the aseptic building on different power grid supplies from the electric company must prove to be useful with allowances for automatic switching from one grid to another. In all cases, the impact of power interruptions must be properly evaluated and documented as it might affect the sterility of the products.

The advancement in lighting moving toward LED lighting provides for more luminacins and should be measured at least 500 lx 3 ft from the floor. Amber lights might be used in ISO 5 suites to protect the product from UV light interference with product quality. Lighting heat generation must be incorporated in the design of clean rooms to control product temperature. All light fixtures must be sealed and cleanable to be free from particulate accumulation.

Materials of construction of light fixture must be compatible with classified room requirements. Special provision for inspection lighting and backgrounds are usually specified for such applications. Automated visual inspection must have neutral lighting that will not interfere with camera function during inspection. All wiring, outlets, and associated equipment must meet the US electrical code as well as the permits for the city in which equipment and facilities are being prepared for manufacturing operations.

Case Study

Issue

AHU-1 was initially balanced in support of commissioning and qualification of the air handler and the supplied clean rooms C-1 and C-2. During this balancing activity, all acceptance criteria for room air change per hour rate and differential pressurization were met.

Upon balance data review, client engineering and contractor installation construction personnel determined that the level of outside air needed to balance the system could create a situation, during low outside temperature conditions, that would lead to colder than acceptable temperatures, per the freeze stat alarm that alerts at 38 °F, inside the air handler. This situation could lead to actuation of the freeze stat protection interlock shutting down the air handler, or in extreme cases, damage to the chilled water coil.

Root Cause

AHU-1, in its final design state, will service multiple clean rooms with air supplied and returned from conditioned space. During the initial phase of installation, only two rooms are supplied with air from this air handler, C-1 and C-2. And since all air from C-1 is exhausted out of the building by design, only room C-2 returns air to the air handler through return ductwork.

The initial phase airflow specification that fed the balance assumed a room pressure of the adjoining material air lock (MAL) room C-3 of 0.09″ WC (differential to building base). The current condition of this room was upward of 0.2″ WC to base at the time of balancing. The reason for this high existing condition is the state of the pressure cascade from adjoining spaces and cannot be modified at this time without effecting other critical differential pressures. Because of this deviation between design conditions and actual existing conditions, room C-2 had to be pressurized at a much higher differential pressure to building base than was intended in design to maintain the relative differential pressure to the adjoining MAL room.

This pressure was met, but only by reducing the return air by closing the mixed air damper and deactivating the return fans. This caused a deficit in makeup air to supply the air handler. This deficit was overcome by opening the minimum outside air damper allowing more outside air to feed the supply fans.

Corrective Action

The following steps were taken to maintain the acceptable balance conditions while reducing outside air needs:

1. Recruit AHU makeup air from a return duct intended for future room return. This duct is already fabricated, mechanically complete to the return header for AHU-1, and will draw air from inside the existing plant at typical room temperature conditions.
   a. A volume damper will be installed allowing for more precise balancing using this return air source.
   b. A filter media will be placed at the inlet of this return duct to augment the existing filters that are in the air supply between the AHU and the clean rooms it serves. This additional filter will serve to keep general plant particulates from entering the ductwork.
2. The AHU/rooms will be rebalanced reducing the outside airflow that is supported by design and psychrometric calculations intended to protect the coil and prevent a freeze stat alarm condition.
3. The balance data from this rebalance, along with all AHU device settings resulting, will be documented and turned over in support of commissioning, qualification, and acceptance of the air handler and rooms.

Preventive Actions

Balance conditions for future phased turnover will consider the existing plant pressure conditions.

12

Barriers and Isolators

Restricted-access barrier systems (RABS) and isolators are now important considerations in the design and construction of filling and capping lines. These systems can be fully or partially automated to include interlocks, alarms, and complete processing, including clean-in-place (CIP) and steam-in-place (SIP) (vaporized hydrogen peroxide [VHP]). Mainly, these setups are intended to reduce and mitigate the risk of operator interventions, which might impact product integrity during aseptic filling. These systems can be small scale for filling clinical trial formulations or large scale to entertain commercial high-speed automated lines [31].

Isolators might be placed in an ISO 8 zone [32]. RABS are built around filling lines in ISO 6 zone with target ISO 5 aseptic core inside the barrier. Design considerations for optimized operations are critical to the use of these systems. Accessibility, intervention, cleaning, sanitization, and sterilization are all process requirements that are used to ensure functionality of these barriers. Unidirectional flow and pressure inside these systems are very controlled to ensure that particulate matter is as specified.

Active RABS use an integrated high-efficiency particulate air (HEPA) filtered air supply over the filling and capping stations. Cleaning and disinfection are introduced through the use of active agents such as VHP. Gloves and transfer path-throughs are used for materials movements with the aid of specialty tools.

Passive RABS airflow is provided by room ceiling HEPA filters. Unidirectional airflow is directed over the filling and capping stations inside the RABS and balances to flow inside and outside barrier walls to control particulate from concentrating in the surrounding environment. Gloves and gauntlets are used and controlled by standard operating procedure (SOP) as they are attached and removed from glove ports in the critical locations on the RABS or Isolator systems. Gloves are sterilized and disinfected or changed as established by validation protocols.

Sterile Processing of Pharmaceutical Products: Engineering Practice, Validation, and Compliance in Regulated Environments, First Edition. Sam A. Hout.

RABS and Isolator systems are superior to just cleanrooms. They require a higher ceiling and accessible footprint around these self-contained units. Isolators can be built with complex installations on the inside and thereby require extensive validation protocol to ensure validity of repeatable operational procedures. There are trade-offs for operating costs and manufacturing hurdles using these systems. Environmental reproducible decontamination with automated cycles should be validated, but as operators are removed from the direct filling/processing environment, products are more protected, and operator safety is enhanced with the use of potent compounds. Once line operation is established, cleaning and handling complexities become routine. Nonetheless, as line manufacturing steps are assembled inside an isolator, systems integration and control can be complex as transfer occurs from one station to another inside of the isolator. Line changes involving size, change parts, and component change will require product change cleaning and line clearance, which might not be as straightforward, but becomes easier as procedural methods are applied.

When products are manufactured inside an isolator, the gowning level of operating personnel would be eased to a lower classification level, which increases operator comfort under strict temperature conditions. Access to aseptic processing areas where isolators are put in operation does not require sterile gowning, but more like gloves, goggles, hair net, beard covers, scrubs, and shoe covers. RABS do not offer relaxation of sterile gowning procedures.

Dust and dust explosions are of concern in the use of powder transfer systems. Isolators and other containment methods will mitigate these effects as the isolator will shield operators from the processing environment. KST values must be calculated as part of the design while handling powders that might be susceptible to explode depending on oily contents, fine milling particle size, surface tension, angle of repose, and other ignition factors that must be vetted to prevent catastrophic failures. However, isolators can be useful for processing hazardous powders or biological potent materials when operated inside the closed system. Toxic compounds can also be handled inside an isolator but are limited while processing in RABS.

Isolator Design Considerations

As with confined space, heat can be generated due to space limitation. Heat loads inside an isolator should be carefully calculated to manage temperature control. Besides ergonomics concerns, the positioning of glove ports is critical to specific process transfer of containers inside the isolator if operator interface interactions are required, including any sampling. The use of sterilant and generated vapors and effects on both materials of construction and operators should be well defined.

Sterilants such as chlorine dioxide or hydrogen peroxide and peracetic acid are all aggressive oxidizers and their use must be documented stepwise in SOP. Rapid transfer ports should be built with flexible connectors to ensure that the aseptic core remains intact. Laminar flow inside an isolator needs to be controlled for recirculation through built-in HEPA filtration system. If powders are handled inside the isolator, special design consideration must plot the flow regime and the accumulation of powder on the various surfaces. All considerations for dust control and decontamination for batch start-up requirements are crucial to clean operations. Positive pressure inside the isolator prevents ingress of any contamination from the surrounding environment. Nonetheless, balanced design pressure should consider specific application when dealing with high potency drugs that must not exit the isolator into the surrounding environment.

13

Guidelines for Statistical Procedure

This is to establish guidelines for the use of statistical techniques in protocols and other official studies [33].

1. Alpha (α): The probability of rejecting a true hypothesis (typically referred to as producer's risk)
2. Beta (β): The probability of accepting a false hypothesis (typically referred to as customer's risk)
3. DPM: Defects per million [34]
4. CPM: Complaints per million
5. USL: Upper specification limit
6. LSL: Lower specification limit
7. MSA: Measurement system analysis

Basic guidelines [35] will be set forth for each procedure; however, the instruction for a procedure may not contain sufficient information to actually execute. If this is the case, a statistics text such as the Quality Control Handbook should be consulted. Minitab statistical software may be used for analysis. If Minitab is used and data is not stored on the server, the data will be included with the study results.

Hypothesis testing involves procedures that compare two groups to identify if there is a statistical difference between them. This procedure will cover three different hypothesis tests:

1. Z-test: Comparison of means, standard deviation known
2. T-test: Comparison of means, standard deviation unknown
3. F-test: Compares the variance of two samples

The basic assumption or null hypothesis used in hypothesis testing is that the samples are from the same population. The purpose of the test is to determine if

Sterile Processing of Pharmaceutical Products: Engineering Practice, Validation, and Compliance in Regulated Environments, First Edition. Sam A. Hout.

there is enough evidence to state that a statistical difference exists. The procedure cannot prove two samples are the same only that there is not a statistical difference. Hypothesis testing parameters are selected based on two types of risks:

1. Alpha (α): The probability of rejecting a true hypothesis (typically referred to as producer's risk)
2. Beta (β): The probability of accepting a false hypothesis (typically referred to as customer's risk)

The α and β parameters should be chosen based on the criticality of the analysis and the effect of each instance occurring. Reducing the parameter enhances the sensitivity of the test to that particular occurrence. As a general guideline, α is normally set at 0.05 and β at 0.10. These parameters are used to determine the acceptance region for the test. Tables for this value are found in most statistics texts or Quality Control Handbook [36].

A one- or two-tailed evaluation must be specified to choose the acceptance region. A one-tailed test is used when the average must fall on one side of the mean. If the test results must be either higher or lower than the current process, a one-tailed test is used. The two-tailed test is used if the test mean may fall on either side of the control mean. The number of tails is important when choosing the acceptance region for the test. A two-tail test will require the α level to be divided by two to capture equal areas on either side of the mean.

DPM or CPM; Complaint Incidents Per Million (CIPM) [37]

DPM is calculated by the following formula:

$$\text{Detects/Units Produced} \times 1\,000\,000$$

CIPM is calculated by the following formula:

$$\text{Complaint Incidents/Units Produced} \times 1\,000\,000$$

It should be noted that DPMs cannot be added. To recalculate, simply add all defects or complaint incidents together and divide by the sum into the total units produced.

Process Capability Analysis

A process capability analysis is used to predict process performance. This study is powerful as it allows the use of variable data to better predict process performance. The analysis allows predictions to be made on the process or product. The example table next is a guide as a normal distribution model to understand process performance.

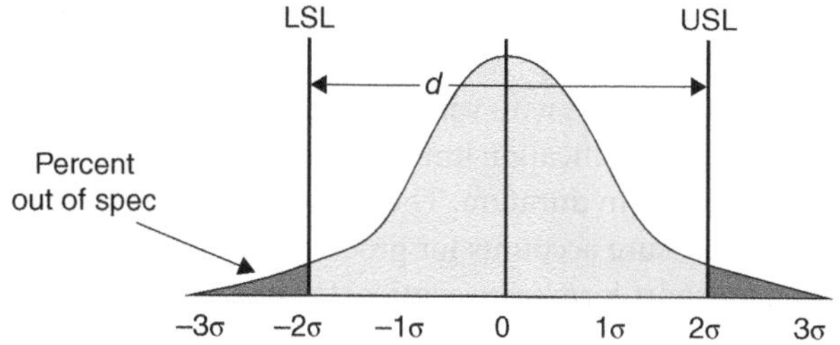

Distance from average	Percentage within	Percentage outside
±0.5 σ	38.3%	61.7%
±1 σ	68.3%	31.7%
±1.5 σ	86.6%	13.4%
±2 σ	95.4%	4.6%
±2.5 σ	98.8%	1.2%
±3 σ	99.7%	0.27% or 2700/million
±3.5 σ	99.95%	465/million
±4 σ	99.994%	63/million
±4.5 σ	99.9993%	6.8/million
±5 σ	99.99994%	0.6/million

By using the abovementioned reference, we can determine that the bell shape distribution would cover 99.73% of the data. The location and the spread can easily determine the capability of the process. The examples next give the histogram when used with upper and lower specification limits.

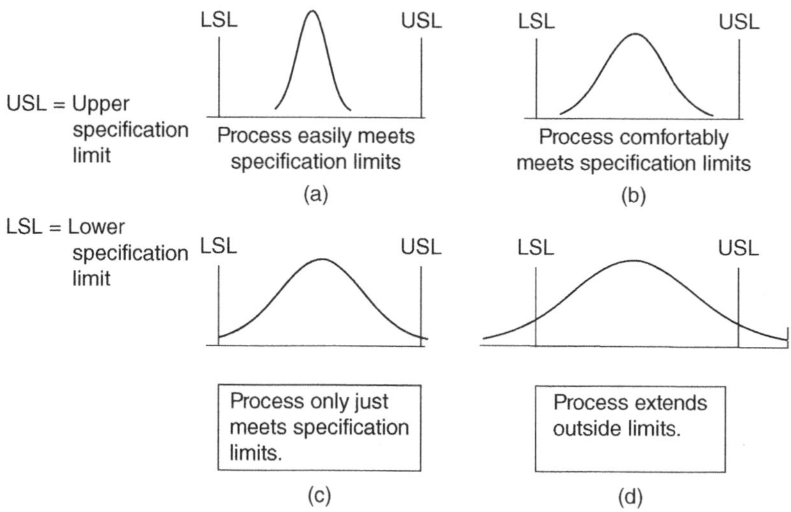

Process capability refers to the natural variation of a process that occurs due to common causes. This is calculated by comparing a measure of the variation to the specification limits for the characteristic of interest. In this way, the ability of a process to meet the product requirements can be quantified. There are two types of studies: process potential (Pp) and process performance (Ppk). Pp and Ppk may be used to qualify a process with variable data. The Pp measure is a ratio of the process variation to the specification limits. Pp is normally performed under optimal conditions and is short in duration. The data is typically not maintained in a time order. The Ppk measure accounts for process centering and stability. Pp studies are appropriate for short-term (consecutive) testing. *Note*: Ppk can be used to review the processes centering ability.

1. Study may be performed during actual production or under conditions that simulate actual production.
2. For a snapshot of the process capability, a minimum of 30 samples can be produced continuously. Any sample size greater than 30 is acceptable as it improves the data reliability. No process adjustments will be made, and the same operator will be used.
3. Measure the characteristic of interest using a calibrated instrument(s) and record this number of the datasheet.
4. Calculate Pp and Ppk. Desired Pp and Ppk measurement is 1.33 with no out-of-spec values. *Note*: other variable plans may be used, such as Mil Std 414 or other variable sampling methodologies.

Note: Central limit theorem – "For most populations, the sampling distribution of the mean can be approximated closely by a normal distribution, provided the sample size is sufficiently large." "Approximate normality is clearly evident at $N = 30$."

Minitab will present the overall capability as Pp and Ppk. The overall capacity (Pp and Ppk) is to be used for all studies with a sample size of one in which the data is not tested in sequence. This overall evaluation calculates the standard deviation based on all samples.

Long-term Studies

Cpk studies are long-term studies, which samples are taken at different time periods. These studies should be performed as follows:

1. Study may be performed during actual production or under conditions that simulate actual production.
2. Samples should be pulled at regular intervals over a specified period of time to yield a true process picture. Example: Pull samples every hour for three shifts of production.
3. Measure the characteristic of interest using a calibrated instrument(s) and record this number of the datasheet.

4. For long-term studies place the 50 samples into 10 subgroups of 5 units each. A standard subgroup size is five samples.
5. Calculate the Cp and Cpk of the process.
6. Review the control chart for points out of control or out-of-specification.
7. Desired Cp and Cpk measurement is 1.33 with no out-of-spec values. *Note*: Other variable plans may be used such as Mil Std 414 or other variable sampling methodologies.

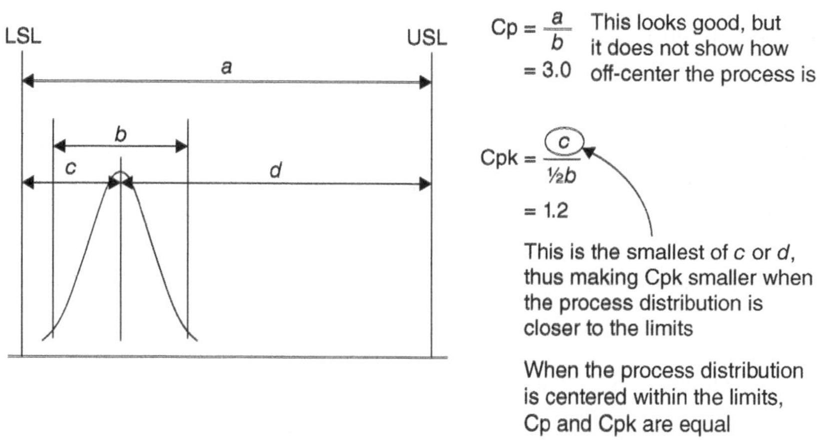

Minitab may be used and is the preferred method for performing capability and gauge repeatability and reproducibility studies. Minitab will also present the within capability study results as Cp and Cpk. These calculations are based on longer term studies in which the data is collected in time sequence. These studies typically use average subgroup's range or subgroup's standard deviation to calculate the overall standard deviation. These within calculation from Minitab (Cp or Cpk) will only be used with time sequence–based data.

Acceptance Sampling

Attribute and Variable Sampling Plans

Sample plans are to be selected based on acceptable quality limit (AQL) and lot tolerance percent defective (LTPD). In order to replace a current plan, follow the steps outlined next. The AQL/LTPD should be used for finished product. Component level or in-process sampling plans should be tighter than those used for finished product.

1. Calculate the AQL and LTPD of the current plan.
2. Find other methodologies that have similar AQLs and LTPDs.
3. Choose the lower cost or easiest to maintain.
   LTPD – Lot tolerance percent defective
   AQL – Acceptable quality limit

Sampling plans may be calculated using Wayne Taylor's sample size software or other standards.

Variable Sampling Plans – ANSI Z1.9

Variable sampling plans may be substituted for attribute plan provided they represent the AQL's and LTPD's needed. Variable plans use the sample size and a *K* constant to establish confidence levels. The average and standard deviation from the analyzed data is compared to the *K* values to determine acceptance. The *K* value represents the number of standard deviations from the mean is from the limit.

The formula for calculation is listed below.

$$(\text{USL} - \text{Average})\ /\text{Standard deviation} = X$$

$$(\text{Average} - \text{LSL})\ /\text{Standard deviation} = Y$$

The minimal value of X and Y must be greater than K (standard) for the data to be acceptable.

Normality

The normality assumption must be assessed when using variable sampling plans with standard deviation unknown for nonstandard processes. The method to test for normality can be selected from the list next and should be conducted using sample sizes of minimum 30 units, and a confidence level of 90%. These normality tests are available in Minitab.

1. Probability plotting of individual values and achieving an adequate linear fit of the data.
2. Performing a statistical test of normality; such as:
   a. A Shapiro–Wilk W-Test
   b. A Chi-Square test
   c. The Anderson–Darling (AD) test for normality
   d. Any test having an equivalent statistical power to the previous tests

The test rejects the hypothesis of normality when the *P*-value is less than or equal to 0.1.

Failing the normality test allows you to state with 90% confidence the data does not fit the normal distribution. Passing the normality test only allows you to state no significant departure from normality was found.

The following graphs display normal distributed data and an example of the application of the AD test for normality. The test confirms that the data meets the normality assumption at a confidence level of 90%.

No data perfectly matches a normal distribution. The normal distribution is typically more conservative than other models and therefore is sufficient for most analysis (see reference next).

> Three sigma limits do, indeed filter out virtually all of the routine variation and they do this regardless of the shape of the model used. This means that we do not have to have normally distributed data in order for the limits to work. Three sigma limits are completely general and will work with all types of routine variation. By filtering out virtually all of the routine variation they will yield very few false alarms.
>
> *Reference "Normality and the Process Behavior Chart" Donald J. Wheeler, SPC Press.*

Attention will be paid to the tails of the data and potential outliers (>4 standard deviations). There is a reasonable chance that points may be beyond the three standard deviation limit of a normal distribution, especially as the sample size increases. For example, 50 samples from a normal distribution have a 13.5% chance ([100% – 99.73%] × 50) that a point will lie outside three standard deviations.

Outliers are not to be removed unless the cause is identified, and the testing is repeated.

Under unique circumstances, data may yield a distribution that will not fit the normal distribution assumption. Extreme deviations from the normal distribution may require normalization or other statically sound methods.

The following graph is an example of data that could require transformation or other statistical analysis to be able to predict confidence levels.

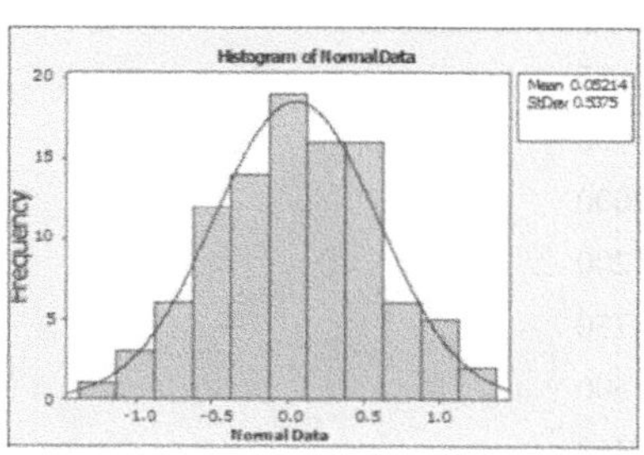

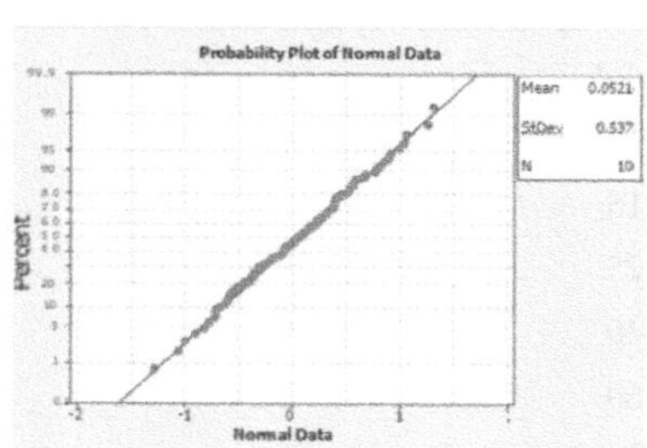

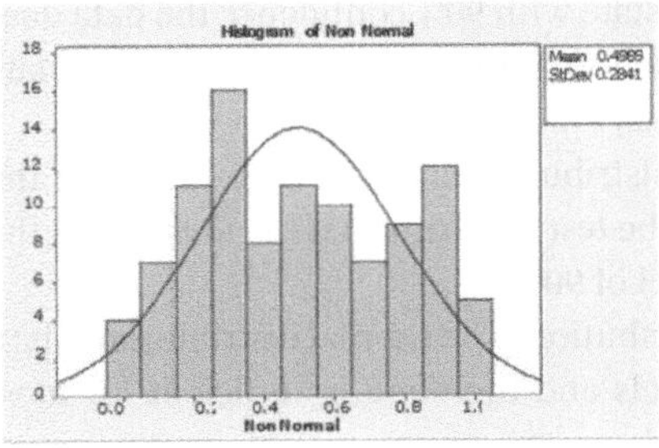

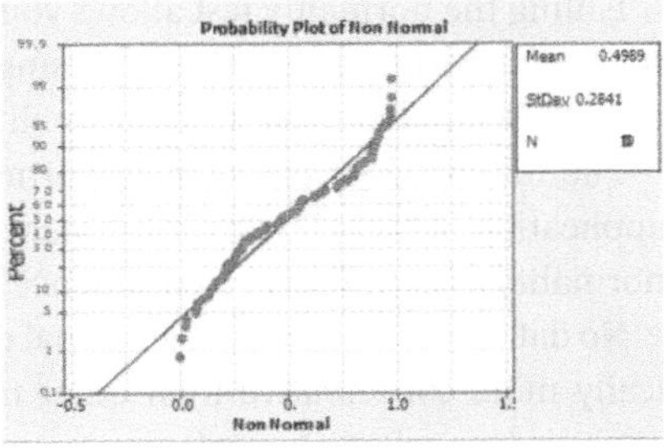

Transformation of Non-normal Data (Normalization)

If the data to be analyzed are not normally distributed, transform the data by taking the natural logarithm of each value and then perform a second normality test on the transformed values to verify that the data behave normal. If the transformed data are not yet normally distributed, consult a person trained on statistical data analysis (i.e. a Black Belt, or a statistician) to review the data and provide rationale for the analysis performed.

Protocol Sampling

Acceptance sampling will be applied in two different ways during a protocol. During the installation qualification (IQ) or operational qualification (OQ) phases, a zero accept plan may be used to qualify a process or a production lot while a 90% LTPD plan may be used in performance qualification (PQ) to evaluate finished goods batches.

A zero accept plan is chosen based on the AQL for the quality characteristic(s) affected. These plans ensure a 95% confidence that the process of defect level is at or below the specified AQL. Samples may be produced continuously or pulled randomly.

Zero Accept Plan (Typical OQ Plan)

AQL (%)	Accept	Minimum sample size
0.065	0	4615
0.10	0	3000
0.15	0	2000
0.25	0	1200
0.40	0	750
0.60	0	500
0.65	0	462

AQL (%)	Accept	Minimum sample size
1.00	0	300
1.50	0	200
2.50	0	120
4.0	0	75

This table was developed using the formula 3 (constant)/AQL expressed as fraction defective. This formula may yield slightly higher sample sizes. A more precise method is to use Sampling Software version 1.0 or higher from Taylor Enterprises.

LTPD plans to be used in the PQ phase are modified versions of ANSI Z1.4 double sampling plans. The accept–reject levels have been modified to achieve an LTPD equal to the desired AQL. These plans will be selected independent of lot size based on the AQL of the defect(s). The stated AQL and LTPD of the plan are based on a process and not individual lots.

90% LTPD Plan (Typical PQ Plan)

Defect(s) AQL	Sample 1	Accept/ Reject	Sample 2	Accept/ Reject	AQL p(a) = 0.95	LTPD p(a) = 0.10
Less than 0.065	3850	0/2	3850	1/2	0.054	0.0647
0.65	385	0/2	385	1/2	0.054	0.653
1.0	385	1/3	385	2/3	0.14	1.02
2.5	150	1/3	150	2/3	0.35	2.59

Failure Mode and Effect Analysis (FMEA)

For the assessment and management of risk, the use of analysis techniques is recommended [38]. They will serve as prevention tools to manage and reduce risks associated with the failure of machines, processes, and systems. Techniques like Fault Tree Analysis (FTA), Hazard and Operability Analysis (HAZOP), Failure Mode and Effects Analysis (FMEA), or other sound methods may be used.

The goal of an FMEA is to define and eliminate problems prior to or after implementation. The tool may be applied to machines, product, process, computer system, etc. This risk analysis will be applied to change control processes. The Quality Engineer will determine if an update of the FMEA is required based on input from the change owner.

The standardized risk priority number (RPN) approach will allow different processes to be compared in the plant. The FMEA may also define key risk areas that require corrective action or increased validation. The tool is a common-sense approach to risk identification based on subject-matter expert's (SME) empirical knowledge and experience. The occurrence of the defect is related to the criticality and ability to remove the defect. As each step in the process is reviewed, a risk value is assigned based on the occurrence times criticality times inspection level.

An attachment to the FMEA will be created and will include the following items:

1. The process steps will be recorded if applicable.
2. The potential failure mode will be listed for each process step. Multiple failure modes for each process step may exist.
3. The potential cause for the failure mode will be identified. Multiple causes for the failure mode may exist.
4. The potential effect of the failure will be listed. If available, this should align with the AQL defect classifications.
5. Occurrence: Chance of the potential cause occurring and resulting in the failure mode. A rating of 1–5 will be assigned. Use scrap rates and other applicable information if known.
6. Severity: Severity of failure. A rating of 1–5 will be assigned. The product specifications should be used if applicable. Special attention is to be given to scores or 4 or 5 (quality items).
7. Detection: Detection rate of failure. A rating of 1–5 will be assigned. A review of product checks, machine controls, and downstream inspection should be included in this review.
8. Current controls: A listing of current control may be listed or referenced from the process flow chart if provided.
9. RPN: Calculate the RPN. Multiply occurrence × severity × detection to determine RPN. High RPN ratings need to be addressed to minimize risk and that documentation reflects the corrective action(s). Reference the table next for comments, rationale, and approval requirements related to RPN values.

RPN ($O \times S \times D$)	Comments and rationale required?	Approvals required
≥45	Yes	Routine approvals, Quality Manager, Quality Director
28–44	Yes	Routine approvals, Quality Manager
13–27	Yes	Routine approvals
1–12	No[a]	Routine approvals

a) For failure modes with a severity rating of 4 or 5, comments or rationale are required. When developing FMEAs for new products, processes, or equipment, all failure modes with a severity of 4 or 5 must be reviewed to ensure that some control or detection is in place.

Comments: List comments, rationale, and/or corrective actions related to the step, if needed. The Reference table mentioned earlier is to determine when required.

An example of an FMEA format is as follows:

(1) Process step	(2) Potential failure mode	(3) Potential cause	(4) Potential effect of failures	(5)	(6)	(7)	(8) Current controls	(9) RPN $O \times S \times D$	(10) Comments

Applicable approvals must be obtained from appropriate areas as related to the FMEA. Approvals indicate agreement with the analysis. For FMEAs with RPN values 28–44, approval from the area affected Quality Manager is required. For FMEAs with RPN values ≥45, Quality Manager and Quality Director approval are required. After the document has been assigned an FM number, the FMEA has been created and approved, QE will update the FMEA master file.

Environmental FMEAs are used for risk assessment of controlled environmental rooms and processes. Environmental FMEAs are performed and documented and are controlled in addition to the SOP change control process.

Calculating or Recalculating Control Limits

Control limits are special upper and lower limits used to monitor the performance of a process. These limits are calculated from actual process data and are generally positioned as three standard deviations from the process mean. Theoretically, this range of ±3 standard deviations will contain 99.7% of the process data; therefore, any data point outside these limits is indicative of a significant process shift. Limits must be reviewed periodically to detect changes in process performance which may require narrowing or widening the limits.

Procedure

1. Limits are to be calculated based upon the controlling plant procedure.
2. Limits should be established based on historical data with the limits calculated for ±3 standard deviations. Limits may be derived with a spread other than ±3 standard deviations but these exceptions must be noted, and the reason documented in the applicable SOP.
3. Limits may not be established outside of spec limits. If the control limits are greater than or equal to the spec limits, control limits cannot be used for that process.
4. Limits should be recalculated using a minimum of 30 data points pulled randomly from historical data. If available data for the period reviewed does not contain more than 30 data points, then all data should be used.

14

Calibration

Routine calibration of equipment and documentation of the calibration results is outlined by reflecting on methods of calibrating and documentation practices of miscellaneous devices, which measure products or processes [39]. All inspection, measuring, and test equipment that can affect product quality or processes are to be calibrated against Standard Measurement and Test Equipment (M&TEs) that are traceable to the National Institute of Standards and Technology (NIST). If NIST or national traceable standards are not available for the parameter being measured, an independent reproducible standard shall be used. The following is to establish approach to the Control of Measuring and Test Equipment Calibration Program.

Calibration: Set of operations that establish, under specified conditions, the relationship between values of quantities indicated by a measuring asset or measuring system, or values represented by a material measure or a reference material, and the corresponding values realized by Standard M&TEs.

CMMS: Computerized Maintenance Management System is a centralized computer system controlled by client's Global IT department, with all system servers residing in client's outsourced data center. CMMS comprises a Web-based computer enterprise planning system application developed by IBM that can be accessed by any computer in the client network using Internet. All asset records are maintained electronically using computer enterprise planning system. Computer enterprise planning system is also used for all maintenance records.

Measurement Standard/M&TE: Material measuring asset, reference material, or measuring system intended to define, realize, conserve, or reproduce a unit or one or more values of a quantity to serve as a reference.

Sterile Processing of Pharmaceutical Products: Engineering Practice, Validation, and Compliance in Regulated Environments, First Edition. Sam A. Hout.

Test accuracy ratio (TAR): Ratio of the accuracy of the unit under test (UUT) and the reference standard used to calibrate the UUT. A test accuracy ratio of 4 : 1 is required for measurement standards where TAR can be applied.

Test uncertainty ratio (TUR): Ratio of the accuracy tolerance of the UUT to the uncertainty of the measurement standard (M&TE) used.

Tolerance: Extreme values of an error permitted for a given asset.

As new assets are received, they will be reviewed by Plant Engineering and Quality Management departments to determine any calibration requirements. For replacement of existing equipment, the area QM supervisor or appropriate manufacturing supervisor should always notify the calibration supervisor or calibration technician as these assets are obtained. Calibration requirements are reviewed as follows:

1. Analyze the use of all assets to determine whether or not the asset should be calibrated.
2. It is the owner's responsibility to assure that the calibrated equipment is suitable for the application in which it will be used. Suitability includes assuring that the range; tolerance, accuracy (uncertainty), and capacity of the asset/equipment are sufficient for the application.
3. Evaluate manufacturer's recommended calibration process and provide calibration procedures in conjunction with the calibration department.
4. Whenever an asset Out-of-Tolerance (OOT) Report is assigned, the owner will notify all users of the asset since its last calibration and will complete the OOT Report.
   a. It will be the responsibility of the owner and/or area supervisor to notify the calibration lab of new assets and changes in usage to any assets. When an operation limit or calibration frequency is changed, documentation to support that change will be recorded on the calibration work order (i.e. protocol, critical work numbers or rationale for the change) and approved by quality management.
   b. Obvious changes to the operation limit such as temperature or mass unit designations (i.e. kg to lb) will not need supervisor approval.
   c. All assets being made inactive or decommissioned must be turned in to the calibration lab for a final verification and status update in the calibration system.
   d. It will be the responsibility of the owner and equipment users to assure that the equipment is used only within the calibrated range, that assets are not used beyond the calibration expiration date and that all equipment is properly identified.
   e. Provide calibration procedures.

Calibration Lab Responsibilities:

1. Perform calibrations.
2. Maintain NIST Traceable Standard M&TEs. The calibration lab must remain locked at all times to limit access to M&TEs for security and to prevent damage of standards.
3. Review and file calibration paperwork generated by the calibration lab.
4. Issue OOT Reports, as required.
5. Maintain the records in computer enterprise planning system.
6. Generate a daily overdue calibration report.
7. Generate a daily active asset/active PM report.
8. Generate weekly computer enterprise planning system reports for supervisor review.
9. Generate monthly calibration reports.
10. Issue asset status notification forms as required.

Assets calibrated will show an accuracy ratio of 4 : 1 or better. The test ratio is determined using the manufacturer's stated accuracy of the equipment being used against the process tolerance of the UUT. A list of temperature calibrations performed by the calibrations department and all of the equipment available to perform the calibration is maintained and updated as needed. It shows that all equipment maintained in the calibration departments can be combined in any combination without dropping below the required 4 : 1 ratio. Procedures will be used for determining manufacturer accuracies and the uncertainty or combined uncertainty of any component(s) used to perform a calibration. Procedures are required to be updated whenever changes occur to Standard M&TEs or when adding or deleting a Standard M&TE from service.

Hard copies of instrumentation manuals, showing where all accuracies for all Standard M&TE equipment were derived, are filed in the Calibration Lab in the Manuals Section of the Standards File Section.

Note: The Test of Uncertainty Ratio and Test Accuracy Ratio for Standard M&TE equipment used in sterilization calibrations is governed by individual specifications and plant standard operating procedures (SOPs) and meets the required TAR/TUR of 4 : 1 or better.

The calibration departments will manage their calibrations generated by computer enterprise planning system. The chemistry, micro, and environmental labs are responsible for their department calibrations. Their calibrations are completed on paper, and those departments enter a Pass/Fail status in computer enterprise planning system. Other departments that have not been trained on computer enterprise planning system calibrations will return their completed paper calibrations to the calibration department, who will enter the Pass/Fail results into computer enterprise planning system.

Note: Assets will also be inspected for physical damage during calibration and production operation. If physical damage is noted during calibration or manufacturing operation, manufacturing supervisor, quality, and calibration will be notified to determine the appropriate corrective action. If visual potential microbial growth is noted, Micro lab will also be notified, and the appropriate documentation and an investigation will be initiated, including all testing needed.

Whenever the asset status is being changed to decommissioned or inactive, verification must be performed prior to removal from service. If an asset is made inactive or decommissioned following an acceptable final calibration, computer enterprise planning system calibration records will be updated to reflect the status change. When the calibration department identifies an asset as nonoperational, missing or is notified of a change of asset status, a nonconformance form will be initiated to complete the status change and the updating of the computer enterprise planning system calibration records. An assessment will be performed to determine if an OOT Report will be issued if the asset is damaged to the extent of being nonoperational.

When it is necessary to change the status of a calibrated asset to inactive or decommissioned, all associated computer enterprise planning system records i.e. PM record, and Job Plan record must reflect the change and become inactive. The asset assigned to the data sheet record must be deleted if the asset is decommissioned. Conversely, when returning an asset to service all associated Records, i.e. PM Record and Job Plan Record will reflect the status change and become Active. The Asset will be added to the Data Sheet Record under the Asset tab. When a calibration Standard (M&TE) is being removed from service for recertification, the asset status should be changed to reflect: Pending Vendor Service and the PM Record inactive. When returning the M&TE to Active status, the PM Record will be returned to Active.

Standard M&TEs to be decommissioned must be verified prior to decommissioning if they have been used to perform any calibration since the Standard M&TE's last calibration. The status of a Standard M&TE or Asset should be changed to inactive if it is being sent out to an approved vendor for re-calibration and is not expected to be returned to service prior to the calibration due date. The computer enterprise planning system calibration work order status will be changed to in progress until the Standard M&TE or asset is received, inspected, and the Certification has been verified. Then, the calibration work order may be completed and closed.

Equipment owners are responsible for verifying current calibration labels are present. When an Asset is removed from service, it will be identified so as not to be used for calibration or equipment verification. Gauges and other monitoring devices that do not require calibration should be labeled as calibration not required. An OOT Report will be initiated for inoperative Standard M&TEs that are placed out of service. A record of all assets, gauges, and devices being routinely calibrated will be maintained in computer enterprise planning system for notification when those assets are due for calibration.

The number of calibration points selected for testing is determined by the usage of the instrument under test, the characteristics of the test unit, and the characteristics of available calibration standards. A minimum of three calibration points is recommended unless the UUT cannot be tested at three points due to limitations in accessing the internal measuring device(s) for test point manipulation or the units have fixed calibration points. For example, light intensity, conveyor speeds, and ambient room temperature/humidity are considered real-time fixed-point calibrations that are nonadjustable and the UUT remains in the area/system being monitored; therefore, multiple test points are not permissible and one-point verification checks are used. Calibrated equipment and/or instruments cannot be operated past the calibration due date and must be removed from service/operation.As-found calibration data results will be documented to the same number of significant digits as the tolerance in the appropriate calibration procedure. When comparing as-found data to a procedure tolerance, rounding to the specified number of decimal places prior to judging pass/fail is required. When the digit to the right is significant, digit is equal to or greater than 5, round to the next highest number. If the digit is less than 5, the preceding digit is unchanged.

EXAMPLE

25.462 = 25 if tolerance is in whole numbers
25.462 = 25.5 if tolerance is in tenths
25.462 = 25.46 if tolerance is in hundredths

Computer enterprise planning system rounds based on the highest resolution value entered. For example, a calibration data sheet can be defined with a tolerance of two decimal places, but if readings from a Standard M&TE are obtained with a tolerance of three decimal places, computer enterprise planning system will round the data values with a tolerance of three decimals when determining if the values are OOT. Computer enterprise planning system calculates the calibration set point acceptance tolerance from the initial as-found Input value entered. For example, if an instrument's set point check is at 15.00 ± 1.00 and the as-found Input value observed is 15.24, then the acceptance tolerance is calculated to 14.240–16.240. Computer enterprise planning system adds a third significant figure (a zero) to be able to perform the required calculation rounding back to the tolerance of two significant figures. The same applies for instruments with one, three, or more significant figures. This system design ensures the correct rounding to the number of significant figures listed in the calibration tolerance is achieved so that an accurate determination for in/out of tolerance is made.

Individual specifications, standard operating procedures will outline the specific procedure for each asset to be calibrated. Calibration frequencies are established in regard to one or more of the following considerations:

- Frequencies stated in specifications
- Manufacturer's recommended recertification frequency
- Stability
- Degree of usage

Calibration Standards or M&TE may be recertified by an outside standards laboratory. Calibration M&TEs are normally cleaned and serviced as part of the recertification process. All M&TEs should be stored in their respective container or storage fixture when not in use. All M&TEs should be stored in a laboratory or office environment. Calibration M&TEs should not be stored in manufacturing areas or warehouse. Manufacturer's instructions should be followed for care and handling of equipment. Standard M&TEs are to be used only for calibration and verification. If Standard M&TEs remain in the possession of an outside vendor to perform contract calibrations, a copy of their certifications will be reviewed and maintained by the calibration department.

Standard M&TEs will be reviewed to verify the presence of NIST traceable certifications. If the certification of the Standard M&TE has as-found data within NIST tolerance but has a less than 4 : 1 TUR, justification will be initiated to provide acceptance of the M&TE by recalculating the TUR using process tolerances. If the as-found data is "out of tolerance," an OOT Report will be issued unless one of the following applies:

a. The Standard M&TE has not been used since its last calibration.
b. The Standard M&TE has not been used at the set point(s) that were found to be OOT.
c. The Standard M&TE has not been used at the set point(s) that were found to have a less than 4 : 1 TUR.
d. The 4 : 1 TUR is maintained. This is determined by dividing the process tolerance through recalculating the uncertainty of the calibration system by using each system component.
e. The error of the Standard M&TE is not detectable by the under test asset's resolution.
f. The Standard M&TE was verified to be within tolerance just prior to shipment, then the OOT situation will be attributed to damage occurring during transport.
g. Correction factors in the amount of the Standard M&TE's error were applied to the calibrated assets and all test points were found within tolerance.
h. The instrument has been calibrated with a different Standard M&TE since the last calibration and found within tolerance.

If for reasons a Standard M&TE does not have a TUR of 4 : 1 or better, other methods maybe used to ensure adequacy of measurement. Such methods may include:

a. Uncertainty analyses
b. Guard Banding of the UUT
   Example: Subtract the Standard M&TE's tolerance from the UUT process tolerance. This accounts for the Standard M&TE error.
   A temperature controller has an assigned tolerance of ±5 °C, the best available Standard has a tolerance of ±2.5 °C. This would only provide a TUR of 2 : 1. By assigning a new tolerance to the controller of ±2.5 °C, the error of the standard is accounted for.
c. Widening of UUT tolerance limits
   In some instances (e.g. UV radiometer), the calibration vendor is not capable of providing uncertainties for the Standard M&TE due to the nature of the test equipment and this is adequate provided uncertainty readings are not obtainable.

Calibration Standards (M&TEs)	Frequency
AMP. Meter ac probe	1 yr
Barometers	Initially
Bubble-O-Meter	Initially
Conductivity cells/conductivity meter	1 yr
dc power supply	Initially
Deadweight testers (pneumatic and hydraulic)	1 yr
Decade boxes	1 yr
Digital current/voltage meter	1 yr
Digital force-torque indicator	1 yr
Digital pressure indicator/gauge	1 yr
Digital thermometers	1 yr
Gauge blocks/gauge blocks visual inspection	1 yr
Gloss meter	1 yr
Hygrometer indicator	1 yr
Light meter	1 yr
Multimeters/process meters	1 yr
Particle counters	6 mo
Pin/plug gauge	2 yr
Polarimeter quartz control plate	Initially

Calibration Standards (M&TEs)	Frequency
Portable airflow hood	1 yr
Portable electronic manometer	1 yr
Precision voltage standards (standard voltage)	1 yr
Profilometer roughness standard	2 yr
PRT monitor	1 yr
PRT probes	1 yr
Resistors	1 yr
Salt solutions for dew point monitor calibrations	Replace after 1 yr
Secondary spectrometric STD	1 yr
Smoke photometer	6 mo
STAGE micrometer (microscope)	Initially
Standard steel rulers	Initially
Stopwatches	6 mo
Tachometers	1 yr
Thermocouple probe (omega)	1 yr
UV radiometer	6 mo
Weights – up to 500 lb	1 yr
Weights – 500 lb and over	1 yr
Weights used for analytical balances	1 yr

Calibration frequencies for standards use specifications where defined manufacturer and external calibration service recommendations, instrument stability, and degree of usage are determined. Calibration standards are sent to external calibration service vendors on or before the defined frequencies listed in the above-mentioned table and any instrument deviation is investigated and dispositioned through the Out-of-Limit Investigation and/or Calibration Addendum process. Deadweight testers can be verified in house to avoid damage incurred during shipment.

Note: If the gauge blocks are defaced, bent, or altered in such a way to affect the integrity of the certification, they must be returned for NIST recertification.

Note: If quartz control plate has been subjected to extremes in temperature or has been mishandled: dropped, broken, cracked, etc., plate must be returned to the vendor immediately for calibration check.

Each asset will be assigned and identified with a calibration number at the time it is placed in the calibration system. If possible, the asset ID number will be engraved into the asset or identified by other suitable means such as label tape if engraving is impractical. Additionally, the asset will be tagged with a calibration label to indicate the asset number, the calibration date, asset description and location, the calibration expiration date, and the calibrated range. The calibration technician will verify the calibration label and asset match prior to every calibration. When available, the manufacturer, model, and serial number will be entered into the asset record.

Care and handling of measurement and test equipment: All measurement and test equipment shall be handled per manufacturer's recommendations or individual SOP. When equipment is not in use, it should be stored in its respective container or designated storage area or cabinet. Manufacturer's instructions should be followed for care and handling of test equipment. A thermograph will monitor the temperature and humidity of the laboratory environment. If calibrations are performed onsite, any environmental conditions that affect or may affect the measurement results shall be recorded.

A calibration overdue report will be generated monthly and sent to management for review. The calibration status of any asset exceeding the due date and its preventative maintenance record will be changed to inactive in the computer enterprise planning calibration system and removed from service. If an asset was in production beyond the assigned calibration due date and not made inactive, a Calibration Addendum will be issued, and the asset immediately calibrated. If the calibration results are acceptable, then justification of no impact will be documented on the Calibration Addendum. If an OOT result is obtained, additionally an OOT Report will be issued, and a product impact disposition performed per procedure and any required escalations to a nonconformance report (NCR) will be addressed in the OOT investigation. If a Calibration Addendum is not required, a comment shall be made on the calibration record and approved by manufacturing and quality management.

A calibrations trend report will be generated monthly. This provides an overview of the calibrations performed and any resulting OOT conditions found by type of instrument. This report is compiled with a minimum of the following computer enterprise planning system-generated information:

- Number of routine calibrations performed
- Unscheduled calibrations performed
- Calibrations performed within due date
- Number of missing and decommissioned assets
- OOT calibrations
- Number of active and operating assets with an inactive PM Record

The calibration lab trend report will be filed in the Document Center after review by management. Computer enterprise planning system calibration records are continually being updated and information generated for the compilation of the monthly trend reports will be signed and dated and filed in the calibration lab for a rolling year.

Immediately upon awareness of an OOT result, the person performing the calibration will initiate a tracking number and notify the area manufacturing director and quality section manager. If the OOT result may impact product or process, an NCR will be initiated per procedure. In these cases, the NCR must be initiated immediately in case a field alert report (FAR) needs to be submitted to Food and Drug Administration (FDA) within three days.

If an item of M&TE is found to be OOT, malfunctions or does not conform to specified or assigned requirement, corrective action shall be taken. The OOT investigative report shall include at minimum:

- M&TE ID number
- M&TE description/type
- M&TE owner/department
- M&TE location
- Date of calibrations when OOT was detected
- Measurement data, including tolerance limits and maximum deviation or magnitude of detected error
- Actions taken
- Quality impact assessment to any product or process that used the M&TE

The assessment must determine the impact on any product or process that used the M&TE between the date the OOT occurred and the previous calibration.

For assets found to be outside the specified tolerance limit, the rationale listed next will be used to determine if an OOT Report is required. No OOT report is required if one of the following applies:

a. The asset has not been used since its last calibration.
b. The asset has not been used at the set point(s) that were found to be OOT. Additional set points above or below the operational range must be verified to be within tolerance.
c. The error between the M&TE and the asset under test is not detectable by the asset's resolution.
d. If the asset was verified to be within tolerance just prior to shipment, then the OOT situation will be attributed to damage occurring during transport.

The OOT investigation will include a review of the two previous calibrations. If a second out of calibration is found in this review, action will be taken which may include:

- Decrease interval between calibrations
- Broaden the asset tolerance (if the process allows)
- Replace asset with a more suitable one

An assessment will be made by the quality supervisor/delegate for the impact on product or components due to the out-of-limit condition.

Adjustments may be documented on the original calibration record where applicable or a new calibration record referencing an adjustment reading. Instruments capable of being adjusted should be adjusted to as close to nominal as possible. Adjustable instruments are considered instruments in which there is an electrical or mechanical means of adjusting the instrument for the purpose of reducing measurement error.

When calibrations are performed electronically using computer enterprise planning system, the electronic calibration record stored by computer enterprise planning system will serve as the master record of the original data. A completed work order details report can be generated to show the results of the calibration. When calibrations are performed on paper and are not a Pass/Fail or an initial calibration, the paper record is the Master Record of the original data.

The calibration supervisor or quality department manager's approval is required when:

- The use of a correction factor was applied: Calibration Addendum or Out-of-Calibration Report will be initiated using a control number issued to the calibration record.
- The asset's due date was exceeded: When an asset has been in service for production use and exceeded the calibration due date, a nonconformance, or comment shall be issued and referenced on the calibration record.
- A frequency change: Calibration Addendum or Out-of-Calibration Report will be initiated using a control number issued to the calibration record.

When an asset is moved to a new location, the calibration label for that asset will have to be reprinted with the new location value. Prior to printing the new label, the asset must be moved within the asset record of computer enterprise planning system to its new location so the correct location value will be displayed on the label.

Contingency Plan/Disaster Recovery

The CMMS system (computer enterprise planning system) is a Web-based application that is hosted in client's outsourced data center. In the event that the system is unavailable, a contingency plan is provided to prevent impacting the ability to maintain calibration records. Problems with or changes to the computer enterprise planning system application (configuration changes, reports, workflow, etc.) will be managed per change control procedures. Computer enterprise planning system will be audited annually per procedure. All operations at client's outsourced data center, including backup and recovery of CMMS data, are controlled by IT SOPs.

15

Cleaning Validation

Based on Food and Drug Administration (FDA) audits regarding establishing the basis for scientific approach, besides coverage and recovery studies, a matrix study based on design of experiments (DOE) (Hi–Lo; crevices; difficult to reach, internal structures; tiers; nozzle connections; etc.) to establish justifiable critical swabbing points (sampling locations) is recommended to ensure equipment repeatable cleanliness (batch-to-batch; between-batches). This is usually covered under HACCP (hazard analysis critical control points) risk assessment program, which covers cleaning validation.

The validation of cleaning processes for process equipment in a dedicated and/or multiuse environment, both manual and automated, for indirect and direct product contact surfaces applies to entities, functions, and personnel that perform or support cleaning validation activities [19]. This does not apply to product decontamination processes for electromechanical devices and/or nonproduct contact surfaces.

Term	Definitions
Acceptable daily intake (ADI)	A measure of the amount of a specific substance that can be consumed on a daily basis over a lifetime without an appreciable health risk
Chemical constituent	A substance or ingredient found in the product
Clean hold time (CHT)	The maximum time that clean equipment can be held without requiring additional recleaning prior to use
Continued process verification (CPV)	Assuring that during routine production the process remains in a state of control
Coupon	A tab representative of the surface material being cleaned used for surface testing

Sterile Processing of Pharmaceutical Products: Engineering Practice, Validation, and Compliance in Regulated Environments, First Edition. Sam A. Hout.

Term	Definitions
Critical quality attributes (CQA)	Are chemical, physical, biological, and microbiological attributes and/or variables that can be defined, measured, and continually monitored to ensure final product outputs remain within acceptable quality limits
Critical process parameters (CPP)	Key variables affecting the production process and/or attributes that are monitored to detect deviations in standardized production operations and product output quality or changes in CQA
Dedicated equipment	Equipment that is used to process a single product
Dirty hold time (DHT)	The maximum time that dirty equipment can be held following production and prior to cleaning
Hard to clean areas	Areas identified as more difficult to be cleaned
Limulus amebocyte lysate (LAL)	Is an aqueous extract of blood cells (amoebocytes) from the horseshoe crab, *Limulus polyphemus*. LAL reacts with bacterial endotoxin or lipopolysaccharide (LPS), which is a membrane component of Gram-negative bacteria. This is used in test to quantify endotoxins
Non-dedicated equipment	Equipment that is used to process multiple products
Process qualification plan (PQP)	Plan to define and manage the process qualification
Process performance qualification protocol (PPQ)	A written protocol that specifies the manufacturing condition, controls, testing, and expected outcomes
Rinsate	(also referred to as "rinse water/rinse solvent") A representative aliquot of rinse water/solvent that is used to determine the adequacy of cleaning and rinsing procedures
Swab	An absorbent pad or piece of material used for cleaning validation sampling operations
Validation	The collection and evaluation of data, which establishes scientific evidence that a process is capable of consistently delivering quality products
Verification	Confirmation, through the provision of objective evidence, that specified requirements have been fulfilled
Worst case	A condition or set of conditions encompassing upper and lower processing limits and circumstances, within standard operating procedures, which pose the greatest chance of product or process failure when compared to ideal conditions. Such conditions do not necessarily induce product or process failure

Each facility shall establish procedures that describe the design and operation of its cleaning validation program as outlined in this procedure. Local procedures shall include the types of equipment (e.g. non-dedicated or dedicated), cleaning agents, and cleaning processes (e.g. manual or automated) validated and maintained at the facility. The number of runs performed within the validation shall be based on knowledge (process understanding) as well as the overall risk to product, including process monitoring. The rationale to justify the number of runs must be documented. The rationale can be based but not limited, on the following: development studies, the equipment under validation, risk controls in place, and relative product risk. Minimally, cleaning procedures shall be validated for all product contact equipment that is used to produce one or more commercial or validation lots. This applies to dedicated and non-dedicated equipment. The "test until clean" method shall not be used. Where appropriate, cleaning verification without validation is allowable but shall be justified and documented. This may be appropriate in situations where validation is in process, during refinement of a cleaning procedure, or for clinical material other than validation lots. If a product is not produced frequently, it is expected that validation may take a longer timeframe to complete [40].

Until such time as there are sufficient runs completed, cleaning verification shall be performed following every production lot [41]. Analytical methods used to verify equipment cleanliness and release of equipment shall be validated. In some cases, it may be that the method is developed and tested but not validated. Use of not validated method shall be justified. Where accessible, equipment shall be visually examined for cleanliness following each cleaning validation run. Cleaning validations must be performed on each product/code with either a matrix approach or risk assessment. For equipment with a defined lifespan (e.g. chromatography columns used in manufacturing operations, filter cassettes), cleaning validations shall be performed throughout the equipment use lifespan.

Non-dedicated equipment shall be cleaned between production of different products to prevent cross-contamination. For non-dedicated equipment being cleaned to allow production of different products, acceptable limits are based on the most conservative of the dose and/or an appropriate calculation for the circumstance (e.g. weight percent calculation). Health-based exposure limits (e.g. ADI) should be used when available.

Where dedicated equipment or campaign production of successive batches of the product is used, equipment shall be cleaned at appropriate intervals to prevent build up and carry-over of contaminants; intervals depend on microbial growth and/or product degradation. The maximum time and/or number of cycles between cleanings shall be established and validated [42].

Note: It may not be necessary to remove residue to the same level as in a product changeover, since carry-over may not represent adulteration of the subsequent lot. When a product campaign includes identical formulations of different potencies, testing shall be performed when sequencing from a higher to a lower concentration to ensure that the quantity of residue of the higher potency formulation will not significantly impact the potency of the subsequent formulation unless the cleaning sequence is validated.

New Products and Product Changes

Any new products introduced to a manufacturing facility shall be evaluated with respect to cleaning validation. Any new chemical ingredient or a change in the concentration of a previously validated component or ingredient shall be evaluated with respect to cleaning validation.

Refer to the following table (Cleaning validation decision table for products) for guidance on when cleaning validation and/or cleaning verification runs shall be performed.

Cleaning Validation Decision Table for Products

If it is determined that...	Then...
A new product contains a chemical or biological component that represents a new worst-case condition to an existing product group…	…Cleaning validation is required per the requirements in this procedure
A new product falls within the validated existing worst-case condition(s)…	…A formal cleaning validation is not required. However: • The rationale and justification shall be documented • Cleaning verification confirmatory run or runs may be performed
An existing product or production process has a change per criteria identified in the Risk Assessment/Matrix Approach and Worst-Case Identification sections	...Evaluate if the change constitutes a new worst-case condition • If so, validate in the new conditions • If not, document the justification. Cleaning verification confirmatory run or runs may be performed

Cleaning Processes and Changes

All newly developed cleaning processes require validation. Any changes to existing cleaning products, process, or equipment require a thorough documented evaluation with respect to cleaning validation. Any changes determined to potentially alter the cleanability per worst-case conditions require a cleaning validation be performed. Any changes that are determined to fall within the existing worst-case conditions require a documented justification but may not require formal cleaning validation.

Risk Assessment/Matrix Approach

For both new and revised cleaning methods, a documented risk assessment shall be performed. Risk assessments are an important element of risk management and need to be included early in the process of cleaning development and validation. This process is best driven by a team of subject matter experts that have multiple experiences from operations, technical services, engineering, quality, and regulatory affairs. There are multiple tools that can be used to complete the risk assessment, and it is up to the team to decide on which are most appropriate for cleaning design. At minimum, the following shall be considered within the risk assessment [43]:

- Equipment design.
- Complexity of the cleaning method.
- Frequency of equipment use (production throughput).
- Equipment location in the production process.
- Single product as compared to multiproduct equipment (e.g. non-dedicated).

Regardless of the tools used, the overall approach is to aid in understanding risks, their root cause(s), and potential mitigations. This activity can also help drive the determination of CPP and CQA.

Matrix Development

A matrix approach should include but not limit to an evaluation of the following:

- Product families/categories
- Equipment families/categories
- Cleaning procedure similarities and differences

The rationale for including a particular product in any grouping shall be properly documented. Each facility may divide its products into product categories/families. Each category/family represents a major division that is based on chemical constituents. Validation is conducted on a representative product which then covers the remaining products in that group.

Each facility may divide its equipment into categories/families. Categories/families may be based on similar characteristics including but not limited to:

- Tank configuration;
- Solution transmission system;
- Instrumentation;
- Jacketing;
- The presence or absence or agitators.

Groupings may be created for related equipment deemed equivalent for the purposes of cleaning. Validation is conducted on representative equipment which then covers the remaining equipment in that group (a process known as bracketing).

Cleaning Processes (Manual and Automated)

Cleaning validations shall consider the different cleaning procedures being used at a facility. All cleaning procedures shall be evaluated for similarities/differences such as, but not limited to, the following:

- Surface coverage demonstrating complete wetting of all surfaces.
- Detergents/solvents used, including concentration and quantity.
- Parameters (e.g. temperatures and flow rates monitored at predefined process locations, time durations).
- Manual processes [44].
- Automated process.
- Pattern of load configurations.

CPP/CQA

Cleaning validation requirements are determined from process parameters based on process knowledge, soiling agents, and equipment cleaning aids (mechanical and chemical). Product and process knowledge and quality attributes should be used to define technologies that best support cleaning activities. Since cleaning validation is another form of process validation with a specialized scope, CPP and CQA should be defined for development of cleaning cycles, verified during validation and monitored to provide assurance cleaning maintains a validated state. Examples of CPP and CQA are as follows:

Critical process parameters	Critical quality attributes
• Process temperature • Process pressure/flow • Process time • Cleaning agent concentration • Clean hold time • Dirty hold time	• Visual inspection/limit • Chemical residue limits • Microbiological residue limits • Drain ability/drying

Cleaning Validation Life Cycle – Cleaning Method Development

Cleaning methods should be developed prior to validation with an equal focus on a robust cleaning method design and the cleaning validation approach. Knowing that the purpose of pre- and post-cleaning is to remove residual product, soils, cleaning agents, and microbial contaminants to an acceptable level, teams designing and developing manufacturing processes and associated equipment need to consider cleaning method development as important as the manufacturing process itself, with heavy emphasis on process knowledge.

Parameters that characterize the cleaning process such as cleaning agents, temperature, physical–chemical properties, contact time, water rinses, and specifics about the equipment such as material, sequence of cleaning steps, pathways, and flow rates should be established prior to validation.

Tools such as quality by design (QbD) should be considered to assist in identifying requirements, defining cleaning processes and optimizing development during validation design. During the design of a cleaning process, it is essential that thorough documentation is created, not only for capturing data and resulting analysis, but equally important is what rationale is used to render decisions. There should be efforts to optimize and continuously improve the cleaning process and be able to adapt to changes in the process.

All cleaning methods developed shall include equipment draining and/or drying as final step.

Strategy for Process Controls

Regardless if the process employs manual cleaning or solely comprises automated cleaning steps, there needs to be a strategy for each step in the cleaning process. This should focus on reducing variation and steps to take when manufacturing variation occurs.

In automated systems, critical process parameters should be monitored and set with action/alert alarms when limits are approached. In some cases, advanced strategies such as process analytical technology (PAT) could be used to control process conditions and show cleaning effectiveness has been reached (e.g. in-line pH and conductivity, or on-line total organic carbon [TOC]).

Worst-Case Identification – Product/Component

Validation activities shall be carried out using the worst-case product (component) as this is used as an indicator of cleaning method robustness. Considerations in the selection of the worst-case residual indicators should include (as applicable):

- Highest concentration seen in a particular code;
- Hardest to clean (physical attributes that make it difficult to remove);
- Pharmacological potency of a particular indicator/component;
- The number of ingredients;
- Solubility;
- Batch size;
- Formulation temperature;
- Viscosity;
- Toxicity;
- Routine microbial content and growth potential.

Justification shall be documented in local cleaning validation procedures and/or identified within the specific equipment cleaning validation under performance.

Equipment

Validation activities shall be carried out using the worst-case equipment. Considerations in the selection of the representative equipment should include (as applicable):

- Areas more difficult to clean;
- Equipment material types (stainless steel, plastic, etc.);
- Size or surface area;
- Dedicated versus non-dedicated equipment;
- Equipment configuration/geometry;
- Product contact time;
- Cleaning method.

Justification for the worst-case equipment selected shall be documented in local cleaning validation procedures and/or identified within the specific equipment cleaning validation under performance.

Validation Tests/Inspections – Visual Inspection

Where accessible, equipment shall be visually examined for cleanliness following the cleaning validation cycle. It will be considered visually acceptable based on the evident absence of foreign material (product, degradants, soiling agents, etc.), for example no visual discoloration, clumps, particulates, coating, or film. Visual inspection should include verification of draining and/or drying of equipment.

Chemical Testing

Rinse and direct surface sampling shall be performed to assess chemicals and analytes. Acceptable chemical assays include, but are not limited to, the following:

- Component-specific assays
- TOC
- Total solids
- pH
- HPLC
- Conductivity
- Resistivity
- Osmolarity
- Ultraviolet scans/reads
- Infrared scans

The selection of a method must be explained in the protocol.

Microbiological Testing

Rinse and/or direct surface sampling shall be performed to assess microbial content. Acceptable microbiological assays include, but are not limited to, the following [43, 45]:

- Total microbial count
- Total yeast
- Mold count
- Bioluminescence
- Coliform count

Endotoxin Testing

For equipment endotoxin rinse, samples should be collected at the completion of the cleaning process. An acceptable endotoxin assay would include, but is not limited to, LAL.

Sampling Methods

There are two general types of sampling that are considered to be acceptable: direct surface sampling (swab or coupon method) and indirect sampling (use of rinse solutions). A justification shall be provided if only one sampling method is used. Local procedures shall define the specific sampling methods to be used.

Sampling methods shall be appropriate for the analyte being measured. Analytical method validation and recovery studies shall be established for all worst-case product residues and chemical cleaning agents.

Direct Swab Sampling

The direct swab test can be applied to both water-soluble materials and water insoluble raw materials (or poorly soluble materials).

For swabbing techniques, a representative area of the equipment is swabbed using a swab or wipe and then tested for recovery of the analyte.

Rinse Sampling

Rinse sampling incorporates the use of a solvent to contact all product contact surfaces of sampled items to quantitatively remove the target residue (if present). Rinse sampling may be performed on the final process rinse or on a separate sampling rinse performed directly after the process rinse is complete.

Coupon Testing

Coupon testing for residues is limited to water-soluble materials or materials that are made water soluble by converting the material into an appropriate salt form.

Coupons representative of the all surface materials being cleaned shall be used.

Coupons are inoculated with the target/equivalent material, placed in predetermined hard to clean locations, recovered at the end of the cleaning process, and quantitatively analyzed.

Sampling Sites

Sampling sites shall include worst-case sites and be identified in a local procedure and documented in the established protocol for the equipment being validated.

Sampling sites shall represent all material types in the equipment (e.g. rubber, stainless steel, glass, etc.).

Acceptance Criteria

Acceptance criteria shall be predetermined based on a documented scientific rationale.

Residual Levels

Where inspection is possible, no quantity of residue shall be visible on the equipment or utensil after cleaning procedures are performed. When agents regulated by local agencies have defined residual limits (e.g. Environmental Protection Agency, Occupational Safety and Health Administration), these shall be followed. However, if product and agents do not have pre-defined limits, residual limits should be determined. The following factors may be used for this determination:

- Acceptable levels shall be predetermined based on scientific rationale not to exceed 0.001 (0.1%) of a dose of any product.
- For any pharmaceutical inactive component (e.g. dextrose, inorganic salts) that has no dosage data, the ADI should be used to calculate the acceptable carry over. If the value obtained is greater than 10 ppm, use 10 ppm as the limit.

- For appropriate applications, at least a three log reduction of carry over in protein concentration between batches/campaigns/or groupings should be demonstrated.
- For appropriate applications, toxicity data may be used to calculate limits (e.g. calculation based on PDE as per EMA).

Residual level calculations may be used as a starting point during the development of new cleaning methods or revised cleaning methods. If during cleaning method development, the preliminary results indicate that the method is more capable than the acceptance criteria calculated, the final acceptance criteria should reflect the capability of the cleaning method developed.

Endotoxin Levels

Where applicable, the endotoxin limit shall be no more than the endotoxin limit of the product and/or equipment in contact with the product.

Microbiological Levels

The bioburden limit shall not be greater than the established lowest bioburden level for the product, product family, and/or equipment in contact with the product.

Cleaning Agents/Sanitizer Validation Studies

All sanitizers used shall be validated.

Cleaning agents shall be chosen based on appropriate criteria such as the following:

- Interaction with other cleaning agents in use.
- Facility surface materials to be cleaned.
- Equipment to be cleaned including its materials of construction.
- Any facility-specific processes or interactions that may affect the effectiveness of the sanitizer.

The rationale shall be documented in the sanitizer validation study.

Acceptance criteria for the specific sanitizer being validated shall be justified based on use including adhering to local, national, and regulatory requirements.

At a minimum, sanitizer validation studies shall include the following:

- Validation of the neutralization method for the selected sanitizer shall be performed.
- Testing in vitro including the use of positive and negative controls.
- Testing in situ including the use of negative controls.

Testing shall be performed on representative surfaces within environmental conditions for which the sanitizer will be used.

Effectiveness studies for which bioburden reduction is the primary acceptance criterion shall be performed.

Effectiveness studies for cleaning agents for which the physical removal of gross contamination is the primary acceptance criterion (e.g. detergents used to remove gross debris immediately following use) are not required.

Challenge organisms used in the validation of sanitizers shall be selected based on local flora organism trending for the area of use. The rationale for the organisms selected shall be documented within the validation.

Dependent on the sanitizing agent under validation (e.g. agent with sporicidal properties, cleaning agents used for general sanitization), the following organisms should be considered when determining challenge organisms:

- Gram-positive coccus
- Gram-positive sporulating rod
- Gram-negative rod
- Mold
- Yeast

Local isolates and reference organisms shall be selected for use with sanitizer validation studies. The facilities should determine the number and type of local isolates to be used based on process or environmental monitoring organism trending data of an appropriate time period.

Hold Time Development

The durations between cleaning, sterilization/sanitization, use of processing equipment/components, and cleaning following use shall be subjected to time limits to control the microbial risks associated with equipment usage. A documented risk assessment shall be completed to aid in establishment of the time limits.

All time limits shall be validated as part of the equipment cleaning validation program. Local procedures shall specify storage requirements, time limitations, and actions to be taken if time limits are exceeded.

Dirty Hold Time

The dirty hold time is required to address the maximum time allowed for chemical residues to remain on equipment (potential dry) and for microbial contamination (potential grow). The maximum time that dirty equipment can be held following production and *prior* to cleaning shall be validated.

Clean Hold Time

The time that clean equipment can be held without requiring additional recleaning *prior* to use shall be validated.

Following cleaning, equipment that is to be reused shall be stored in a manner to protect it against microbial contamination. In so much as possible to limit bioburden proliferation, equipment shall be held in a dry or drained state.

Additional Hold Times/Cleaning Frequencies

Dependent on the production process, the following additional timeframes may be considered:

- Maximum campaign length (duration and/or number of batches between cleaning operations).
- Frequency of teardown.
- Duration between sterilization, decontamination or sanitization, and use (sterile hold time).

Continuous Process Verification

Continuous process verification (cleaning verification) requirements necessary for maintenance of the validated state of the cleaning process shall be defined in local procedures, CPV plan, or validation document based upon risk assessment. Trending of cleaning cycle performance shall be assessed as part of routine monitoring. Facilities shall have procedures in place identifying the investigation requirements if adverse or anomalous results are identified, but the results are within validated parameters. CPV plan and actions should be managed as per dedicated PQP. Facilities shall perform an evaluation of cleaning validation process CPV periodically.

Failure Investigations

Any cleaning validation, or routine monitoring testing failure, requires an investigation. The investigation shall be performed and documented as per dedicated PQP.

16

Validation of Filling Equipment

This is to establish consistency in the validation of new or modified filling equipment or processes. This procedure lists minimum validation requirements for new or modified filling equipment or processes.

The Project Engineer leading the project/work will be responsible for initiating the applicable change controls and obtaining approvals prior to the beginning of work [46]. Whenever operating parameters are to be established or changed, the Project Leader will validate new parameters per the applicable Standard Operating Procedure (SOP). The Project Leader will inspect the equipment to assure it meets the department's good manufacturing practices (GMP) standards [47]. When a machine is to be qualified, which has established parameters, the parameters at which the validation samples are produced will be verified to be at current parameter limits and will be recorded.

New installations of filling major equipment will require a protocol as a minimum requirement. This validation is used primarily for validating parameters and modifications to equipment that have been previously validated under protocol and the results were acceptable. If it can be demonstrated through objective evidence that the performance of a piece of equipment is equivalent or better than the previous equipment or the current equipment, previous validation in part or whole can be referenced [48]. If no adverse effect on the product or the manufacturing process is anticipated, comparative testing (two-sample t-tests, two-sample standard deviations, capability analyses, equivalency tests, etc.) or other appropriate testing may be used to qualify a change of very little risk. Rationale for this expectation and the testing method may be documented [49].

For equipment where both high and low parameters exist, and when both high and low parameters are modified, and they are modified in the same direction, e.g. hot stamp temperature (°C): 170 high 165 high, 160 low 155 low, only one side must be verified. If the parameters shift higher, the high parameter shall be tested. If the parameters shift lower, the low parameter shall be tested.

Sterile Processing of Pharmaceutical Products: Engineering Practice, Validation, and Compliance in Regulated Environments, First Edition. Sam A. Hout.

If a protocol was used to process validation samples and an SOP change is made based upon validation results, the protocol number must be documented on the SOP change request form in addition to the validation form.

For qualification/testing requirements that use capability testing, attribute-based (pass/fail) testing may be substituted during a post-performance qualification (PPQ). Examples of product defects include vertical print placement, barcode ANSI grading, solution volume, and air volume.

Modified parameters within validated parameter windows will not require a validation. Update the appropriate parameter addendum in filling to show the modified parameters.

Visually inspect high and low parameter units for missing/incomplete/illegible:

a. Volume
b. Drug dose statement
c. Ingredients of the label copy
d. Product code
e. Expiration date
f. Customer icons

Standard testing will consist of high and low parameter samples to be filled and sterilized. Evaluate all packs produced at high parameters and check for primary container leaks. When performing a process performance qualification (PPQ), testing for primary container leaks during PPQ may be supplemented by a postproduction review of leaker analysis data.

> Measure **30** units from the low parameters and **30** units from the high parameters for an ANSI grade. ANSI grading is an additional requirement for enhanced process optimization and barcode quality visibility. Product acceptability is determined through barcode scanability.

Check units for particulate matter in filled unit caused by the fill machine. Check units for leakers caused by the fill machine. Alongside the solution measurement system, a control system is in place to reject primary packs that are filled beyond the specification limits. Filled packs that are determined to be within the specification limits are allowed to be plugged and are evaluated through the method next:

1. Set up the machine and, with all nozzle stations in use, fill 50 packs at each nozzle station.
2. Record the scale instrument number. Record the stock number for the primary packs and the enhanced closures. For all nozzle station(s) impacted by the change, weigh each unit on a properly leveled scale. If less than all of the nozzle stations are anticipated to be impacted by the change, rationale will be documented.

3. Calculate average, standard deviation, and Ppk using specification limits.
4. Ppk must equal or exceed 1.33. Zero samples on each nozzle may be outside of the specification limits.

Note: When applicable, a 95% confidence level (see example next) may be used as the acceptance level. If the sample size is between 45 and 50, the applicable minimum Ppk value may be recalculated; otherwise, the lower sample size shall be used next.

AQL	Sample size	Minimum Ppk value
0.65	50	1.02
0.65	45	1.03
0.065	50	1.31
0.065	45	1.32
1.0	50	1.00
1.0	45	1.00

Changes that potentially affect flushing, cleaning, or sanitization may have some documented testing in the equipment-specific protocol/qualification (e.g. related alarm testing, minimum flush volumes functional testing, etc.). Changes affecting solution flush quantities or sequence will require a protocol. Flush volume quantities are procedurally controlled with the applicable filling line-specific. If meeting these requirements could be potentially affected (e.g. a change in the mechanical or electrical solution volume control), the applicable fill machine minimum flush volume requirements will be confirmed on each nozzle. The use of calibrated measurement devices will be documented.

To have dosing/dispensing valves that cycle on/off during sanitization so that the liquid will come in contact with the total possible inner surface area of the valves. Modifications that potentially affect this requirement will be confirmed.

For new products, new materials, or changes affecting all machines – parameters will be tested on all affected machines. For new machines – test worst-case conditions for all products that could be produced on the equipment. High parameters are high temperature, high pressure, and lower speed (or most dwell time). Low parameters are low temperature, low pressure, and higher speed (or least dwell time).

1. Perform parameter testing listed for parameter or sealing modifications
2. After sterilization, inspect **120** samples (or as established in acceptance criteria) for marks and stains on product such as caramelized product, dirt, grease, print transfer, watermarks.
3. Accept zero defects.

A new code may use the seal-to-seal distance for one of the groups listed:

1. fill volume, pack size, sterilization cycle, and displacement range all match
2. dimensions match one of those listed
3. sealer configuration differences can be shown to reduce the displacement volume the same seal-to-seal distance may be used.

The validation of new conveyors and changes affecting pack contact areas of product conveyors are considered such that all transition points for the conveyor (i.e. downstream and upstream) shall be tested prior to the release of the equipment. If the material, and the application of that material, has been qualified, this material and application have been proven with a high degree of confidence to be effective; traceability to the previous testing must be documented in the qualification package.

Validation will use capability studies to assure 95% confidence of acceptable quality limit (AQL) 1.0. Capability will test **30** values per nozzle with the Ppk equal to or greater than 1.33 for a 95% confidence of AQL 1.00. Zero samples on each nozzle may be outside of the specification limits.

This specifies testing required when equipment location is changed or power sources are disconnected and reconnected, without mechanical or software changes to the equipment.

Note: For fill lines that produce a combination of many different types of codes – i.e. materials containers, different container sizes, and different print colors. A matrix approach can be used, when applicable, to establish testing requirements provided objective evidence such as RBLCM data to substantiate the re-qualification strategy.

For solution of volume measurement using density, the process is used to establish empty container tare weights and solution density for new codes or changes to existing codes. Solution volume measurement using density must have been validated for the fill line under protocol prior to the use of this process for changes/additions to the validated fill line product code portfolio. Tare weights for individual stock numbers will be collected using capability study. A minimum of **30** data points will be required per stock number. It is recommended that parts from three production lots are utilized, if available. Weights will be recorded to the nearest milligram (mg) utilizing a calibrated scale. The average and standard deviation of the **30** data points will be calculated and included with the capability study. Established solution density values and individual tare weights for closures and packs are maintained. Density for individual solutions is collected using capability study. A minimum of **30** data points will be required for each specification.

Samples for density measurement may come from production samples taken directly from a mix tank containing the appropriate solution or from laboratory samples formulated per the applicable specification. It is recommended that three production tanks or samples are utilized if available.

The average and standard deviation of the **30** data points will be calculated and included with the capability study. Calculate the standard deviation for each tank individually, and then average the standard deviations of the tanks to calculate the overall standard deviation. (Example Tanks A, B, and C are each sampled 10 times for density determination.) Use all 30 data points to determine the average density. Determine the standard deviation for Tank A, then B, then C. Determine the overall standard deviation via the formula:

$$\frac{\text{Std Dev A}^2 + \text{Std Dev B}^2 + \text{Std Dev C}^2}{N}$$

N = total number of standard deviations.

Sample Plan Justification

Sample sizes

120 accept 0 defects yields a 95% confidence of meeting a 2.5 AQL.
300 accept 0 defects yields a 95% confidence of meeting a 1.0 AQL.
462 accept 0 defects yields a 95% confidence of meeting a 0.65 AQL.
4615 accept 0 defects yields a 95% confidence of meeting a 0.065 AQL.

For capability analyses:

AQL	Sample size	Minimum Ppk value	*K* value (reference only)
0.065	50	1.31	3.92
	45	1.32	3.96
	30	1.39	4.18
0.65	50	1.02	3.05
	45	1.03	3.09
	30	1.09	3.26
1.00	50	1.00	2.87
	45	1.00	2.90
	30	1.02	3.06

Note: A given sample size and *K* value yields 95% confidence of a corresponding AQL (e.g. a sample size of 50 and a *K* value of 3.92 yields a 95% confidence that the AQL is 0.065). Additionally, the *K* value is three times the Ppk value (i.e. *K* value = 3 * Ppk value).

Technical References

Dovich, R.A. (1992). *Quality Engineering Statistics*. Wisconsin: ASQC Quality Press.

Evans, J.R. and Lindsay, W.M. (1993). *The Management and Control of Quality*, 2e. Minnesota: West Publishing.

17

Manufacturing Process Validation

Process validation is the collection and evaluation of data from the process design stage through commercial production, which establishes scientific evidence that the process is capable of consistently delivering quality products [50].

Process validation involves a series of activities taking place over the life cycle of the product and process [51]:

- Stage 1: Process Design
- Stage 2: Process Qualification
- Stage 3: Continued Process Verification

The objectives of this section are to define requirements for process validation, understand the life cycle approach to process validation, and describe elements of Stages 1, 2, and 3 of process validation. There are different types of validations, namely critical systems validation, design validation, cleaning validation, sterilization validation [52], analytical instrument validation, software validation, hardware validation, test method validation, and analytical method validation. Nonetheless, this section is concerned with process validation [53].

A multidisciplinary team is needed to plan and execute the activities involved in process validation: quality, engineering, manufacturing, research and development, laboratory personnel for microbiology, chemistry, environmental monitoring, and sterility assurance for total process qualification. The life cycle approach to validation involves regulations. Process validation is a requirement of current good manufacturing practices (cGMP)

- Finished pharmaceutical products
  - o 21 CFR parts 210 and 211
  - o Directive 1572

Sterile Processing of Pharmaceutical Products: Engineering Practice, Validation, and Compliance in Regulated Environments, First Edition. Sam A. Hout.

- Medical devices
  - 21 CFR part 820
  - Directive 745

Stages 1–3 of the validation life cycle come from Food and Drug Administration (FDA) process validation guidance on therapeutic products. In addition, EU Annex 15, and Global Harmonization Task Force (International Council for Harmonization [ICH]) process validation guidance.

The validation life cycle consists of three stages:

> During stage 1, the commercial manufacturing process is defined based on knowledge gained through development and scale-up activities (Process Design). During stage 2, the process design is evaluated to determine if the process is capable of reproducible commercial manufacturing (Process Qualification). During stage 3, ongoing assurance is gained through routine production that the process remains in a validated state (Continued Process Verification). Continued evaluation and trending during stage 3 can identify potential process improvements. This may require additional process design and qualification activities, which will cause a return to stages 1 or 2.

The validation life cycle is a process flow that details the specific elements of stages 1, 2, and 3 of process validation.

Stage 1: Process designs start with product requirements definition. Quality attributes identification and product specification development will to follow complete process description and process flow. In parallel, risk assessment scope in terms of process Failure Modes and Effects Analysis (pFMEA) tabulation is started. User requirements specifications (URSs) are drafted to initiate stage 1 – studies for process characterization. A control/automation strategy is contemplated at this time. The output from these steps will lead to Design Review and Design Qualification.

Stage 2: Process qualifications are performed in steps starting with installation qualification (IQ). Once all equipment and systems are checked for appropriate connections, operational qualifications (OQ) protocols are executed to evaluate process parameters and all variables control limits. Following media fills or products are run to ensure that product characteristics and specification are reproduced for microbiological and chemistry formulation aspects, and checking for product quality. All requirements are tracked in a traceability matrix and a validation summary report is issued [54].

Stage 3: Continued process verification is dictated by a plan that would cover heightened monitoring if applicable or conduct routine monitoring making sure that all necessary investigation are documented and appropriate action is taken as needed.

To ensure compliance, new product development/introductions, the validation life cycle will always begin with stage 1, and then progress to stage 2, and stage 3 [55]. For legacy products that are in routine manufacture, the validation of life cycle begins in stage 3, as there is already a process that has been designed and validated. When stage 3 data indicates a signal, the validation life cycle may return to stage 1 and stage 2 activities.

The benefits of the life cycle approach are

- Safety
  - Patient safety assuring quality of product
  - Operator safety by reducing workplace hazards
  - Environmental hazards mitigation outside facilities
- Business
  - Increased throughput
  - Reduction in scrap and rework
  - Capital expenditure control
  - Reduced in-process and finished goods testing
  - Reduced business risk
- Quality
  - Increased and consistent quality
  - Reduced rejects
  - Fewer complaints related to process failures
  - Improved compliance to regulations and guidelines

Process validation elements in quality systems cover

- Production and process control
  - Manufacturing process validation: URSs, process design and qualifications, stage 1 and stage 2 guidance for therapeutics, and requirements for trace matrix.
  - Continued process verification: CPV plan and reaction plan
- Product development and life cycle management (PDLM)
  - PDLM inputs: I/O traceability guidance, and quality attributes identification
  - PDLM transfers: In-process and release specifications, transfer guides, quality attributes transfer list/checklist
- Product risk management
  - Risk assessment and reduction: pFMEA

The goal of stage 1 – process design – is to design a process suitable for routine commercial manufacturing that can consistently deliver a product that meets its quality attributes. Stage 1 activities include:

- Quality attributes identification/product specifications development
- Process description flow chart

- Risk assessment, and process parameters identification
- URS development
- Process characterization
- Controls strategy
- Design qualification review

A quality attribute is a property or characteristic that is bound by an appropriate limit, range, or distribution to ensure desired product quality. A critical quality attribute is a quality attribute, the variations of which outside the specified tolerance have a significant impact on product quality and patient safety.

Quality attributes are identified from product requirements and transferred to manufacturing through product specifications. Quality attributes identification:

- Defines key design characteristics, which translates product requirements at a meaningful level of detail for transfer to manufacturing
- Establishes a ranking system that can be used to set validation requirements and release criteria
- Provides traceability of quality attributes essential design outputs and product requirements

The quality attribute level of detail is more appropriate for transfer to manufacturing than the requirement level of detail. For example, the hanger hole must withstand the primary container system closure removal forces, for closures that are removed through pulling (requirement). The associated quality attributes are (i) primary film thickness, (ii) primary film tensile strength, (iii) hanger hole shape, and (iv) hanger hole vertical distance to seal.

As the manufacturing process is defined, it must be described in a process description, block diagram, or process flow diagram. The process description/flow should detail each unit operation and should be used to assist in the execution of risk assessments and in the development of the control strategy.

A process description shows process inputs, outputs, yields, in-process tests, and references of all documented related reports. A process flow details all major steps, sub-steps of the process, which must function for successful operation, and reflects on any interdependence between steps.

A pFMEA is a risk assessment tool that systematically

- Identifies potential failure modes, their causes, and the effect on product and process;
- Identifies process parameters that have the potential to affect each quality attribute;
- Documents prevention and detection controls for each failure.

A risk assessment should be initiated at the start of process design and updated with knowledge gained throughout stage 1 and throughout the life cycle.

A URS describes the quality and business needs for a process or an individual piece of equipment. Minimally, the URS must identify requirements related to quality attributes and process parameters. Other requirements may include:

1. Build requirements
2. Operational requirements
3. Control system requirements
4. Maintenance requirements
5. EH&S requirements
6. Calibration requirements
7. Documentation requirements

Process characterization or process development studies are performed to better understand the impact of process parameters on quality attributes. The information gained may be used to define parameter ranges and acceptance criteria for stage 2 activities.

Other stage 1 studies may include factory acceptance testing (FAT) or site acceptance testing (SAT) to evaluate equipment against the requirements in the URS prior to accepting equipment from the supplier.

A process control strategy must be developed based on knowledge and experience gained through stage 1 and through the life cycle. There are two elements of the control strategy:

1. Specific controls related to each potential failure, which may be documented in a risk assessment or control plan
2. A strategy for compiling and reviewing data from specific controls to ensure maintenance of the validated state, which is documented in a Continued Process Validation (CPV) plan.

CPV is an element of stage 3; however, the planning for a successful CPV program deigns during stage 1. A CPV plan does not need to be finalized until the start of routine manufacture.

The proposed process design against the URS must be performed and documented as part of Design qualification review, which covers:

1. Approved process flow
2. Approved risk assessment
3. Approved URS
4. Approved list of controls

Stage 2 – Process Qualification

During this stage, the process design is evaluated to determine if it is capable of reproducible commercial manufacturing. Stage 2 activities include:

- Installation qualification (IQ)
- Operational Qualification (OQ)
- Performance qualification (PQ)

Results of IQ, OQ, and PQ are documented in requirements traceability matrix and validation summary report [56]. IQ includes demonstrating by objective evidence that all key aspects of the process equipment and ancillary system installation adhere to the manufacturer's recommendations of the supplier of the equipment are suitably considered. IQ verifies that the equipment/systems:

1. Is correctly installed
2. Is connected to specified utilities
3. Is positioned in a suitable environment
4. Has correct supporting documentation

Operation qualification acceptance criteria must be scientifically sound or statistically justified and based on the risk level of the quality attributes documented in the product specifications. Prior to closure of OQ, the following approval are required:

- SOPs
- Product specifications [57]
- Cleaning procedures
- Operator training
- Preventive maintenance requirements

PQ includes demonstrating by objective evidence that the process under anticipated conditions consistently produces a product which meets all predetermined requirements, within the scope of validation. PQ must be performed under normal operating conditions:

1. Parameters are within validated ranges
2. Process is in the intended operating environment
3. Operations are using personnel expected to perform these tasks
4. Operating procedures are approved

PQ should occur across a sufficient number of runs (batches) to include any potential source of routine variation such as raw materials, shifts, tank, batch size, or equipment selections [58]. The number of PQ batches must be adequate to provide evidence that the process consistently produces a product that meets all predetermined requirements. Acceptance criteria must be scientifically sound or statistically justified and based on the risk level of the quality attributes documented in the product specifications. PQ must have a higher level of sampling, additional testing, and greater scrutiny of process performance than routine commercial production.

A separate PQ is not required for each piece of equipment or subprocesses if performed in conjunction with a PQ of the entire process. PQ executed for the entire process meets the intent of the FDA Process Performance Qualification (PPQ) definition for therapeutic products. However, PQ and PPQ may be performed separately as a business decision.

The requirements trace matrix (RTM) demonstrates traceability between the requirements in the URS to the protocols and tests where the requirements are challenged to demonstrate test coverage. The RTM must be finalized and approved along with the validation summary report (or included as part of the validation summary report) to ensure that all requirements of validation were met, and the objective evidence is referenced and available.

Stage 3 – Continued Process Verification (CPV) goal is continual assurance that the process remains in the validated state during commercial manufacture. Stage 3 activities include:

- Continued Process Verification plan
- Heightened monitoring where applicable
- Routine monitoring and continuous process improvements

By continuing to monitor the process, the impact of equipment fatigue and small changes over time can be assessed. The CPV plan is a tool to document the approach that will be used to collect and analyze data and demonstrate that the process remains in a validated state. It defines the routine monitoring mechanism for validated quality attributes and process parameters and relationships. Minimally, the CPV plan must cover all finished good quality attributes.

The initial CPV plan is drafted during stage 1, confirmed during stage 2, and finalized prior to routine production or heightened monitoring. The CPV plan is a living document and the control strategy must be reviewed and updated throughout the life cycle as process knowledge is gained. For new products and processes with limited historical data and subject-matter expert's (SME) knowledge, continued process verification may include a period of increased frequency of review and sampling following completion of stage 2 (heightened monitoring). If heightened monitoring is completed, it should be documented under a protocol. For new processes, rationale for not completing heightened monitoring must be documented.

Signals during routine monitoring may identify deficiencies in the process or opportunities to improve the process. Product or process changes may require a return to stage 1 and stage 2 of process validation as part of change management.

With a robust CPV program, revalidation on regularly defined frequency may not be needed. Revalidation is only needed when the CPV data indicates a return to stage 1 or stage 2 is necessary, or when required by local or international standards.

In conclusion, the validation life cycle involving process validation is the collection and evaluation of data, from the process design stage through commercial production, which establishes scientific evidence that the process is capable of consistently delivering quality products. Process validation involves a series of activities taking place over the life cycle of the product and process:

- Stage 1: Process Design
- Stage 2: Process Qualification
- Stage 3: Continued Process Verification

Appendix A

Installation Test Plans

For systems controlled by programmable logic controller (PLCs) and monitored on the building management system (BMS), verification testing will be performed such that the equipment can be controlled and monitored at the local operator interphase terminal (OIT) in addition to the BMS human machine interphase (HMI).

Test case	Description
Receipt verification	Verifies if the component ordered and associated documentation was received and complies with the approved submittals and project purchase order requirements
Documentation verification	Verification that the system has been assembled and approved. This verification will serve as the evidence of construction quality as defined in project specifications
Drawing verification	All controlled drawings (e.g. piping and instrumentation drawing [P&IDs]) are walked down and marked up to verify against field conditions
Component verification	Verification that all equipment components, system elements, and instruments of the system are installed in accordance with the manufacturers and user requirements. This verification should include materials of construction. Information from the receipt verification (RV) system can be used and referenced as part of this verification
Weld documentation	Verifies that all required welding documentation for piping/components are on file and are consistent with the installer's/client's standard operating procedure (SOP(s)) and/or specifications
Material of construction verification	Verifies the materials of construction of equipment components, system elements, and instruments of the system are as specified

Sterile Processing of Pharmaceutical Products: Engineering Practice, Validation, and Compliance in Regulated Environments, First Edition. Sam A. Hout.

Test case	Description
Materials compatibility verification	Verifies the materials of construction are nonreactive with the fluid or media used
Surface finish verification	Verifies the texture of the surface of equipment, piping, and components, to ensure their roughness does not exceed the specified surface finish
Filter installation verification	Verifies if the appropriate types of filters (housings and elements) are installed in the system
Safety device installation verification	Verifies correct type, set point, and tolerance of safety devices. This also will verify safety device release path/orientation
System lubricant verification	Verifies type of lubricant is compatible for use as it pertains to the product or component, as applicable
Calibration verification	Verifies that instruments are calibrated to a National Institute of Standards and Technology (NIST) traceable standard, as required to support system operations
Utility verification	Verification that all utilities that support the system (e.g. electricity, instrument air, etc.) are available and meet the manufacturer's specifications (e.g. voltage, frequency, fuse size, pressure, etc.)
Circuit protection verification	Verification that circuit breaker settings, OL settings, fuse ratings, etc. are as specified
Spare parts list verification	Verifies that a vendor-provided spare part's list is available for each major component or equipment. Critical parts identified. Long lead time parts identified
Preventive maintenance verification	Verification that preventive maintenance procedures are available and scheduled following manufacturer and user requirements. Procedures and schedules that are not in a computerized maintenance management system can be used as support documentation
Cleaning/flushing verification	Verifies that systems and/or equipment have been flushed and/or cleaned, as applicable
EP and/or passivation certification	Verifies if the EP and/or passivation certification documentation is available
Process drain and floor drain verification	Verifies that drains are available and will collect system drainage as required. Check for air break gaps 2× pipe diameter
Hydrostatic and/or pressure test verification	Verifies proper test parameters (pressure, duration, test medium), and test boundaries
Room architectural materials/finish verification	Verifies if room architectural materials/finish meets the specification

Test case	Description
Room cleaning agent verification	Verifies the effectiveness of the room cleaning agent
Duct leak test	Verifies the airtightness of ductwork for forced air of heating, ventilating, and air conditioning (HVAC)
Duct cleaning	Verifies duct cleaning was performed and documentation is available
High-efficiency particulate air (HEPA) filter certification verification	Verifies HEPA Filter Certification documents are available and meet the specification. HEPA filter installation grid numbered
Lighting verification	Verifies that adequate lighting exists for the intended operations and maintenance
Floor slope verification	Verifies floor slope that meets the specification
Hardware installation verification	Verifies documentation and drawings for automation hardware that is available, and that the hardware is installed properly
Software installation verification	Verifies that the system software is installed as designed
Wiring termination verification	Verifies that wires are labeled and terminated as depicted on design drawings
Control system input/output test (I/O)	Verifies the control system I/O points to/from for each device are addressed, connected and function per the specification
Logbook verification	Verifies that a logbook exists for the system; this may be electronic
Authority having jurisdiction inspections	Verification that mandated inspections are completed

Appendix B

Operational Tests Plans

Programmable Logic Controllers (PLCs)/Building Management System (BMS) are operational and process can be controlled from interphase terminals (Operator Interface Terminal [OIT])/Human Machine Interface [HMI].

Test case	Description
Operator interface tests	Verifies that the operator interface functions (e.g. menu options, screens, switches, buttons, etc.) of the system function in accordance with the manufacturer's specifications. Documentation from vendor start-up and/or work instructions can be used for this test
System security verification	Verify that the security measures for the system function as specified by the manufacturer. System security verification may include password, lock, and key switch testing
21 CFR 11 verification	Verifies that the applicable control system complies with the requirements of 21 CFR Part 11
Software backup, archiving, and version verification	Documentation of software version, that a backup copy of the software associated with the system is securely stored and identified. Additionally, disaster recovery and backup procedures exist to restore the system's software
Radio frequency (RF) and electromagnetic (EM) interference verification	Verifies that the automation system controllers are not affected by RF or EM
Alarm and interlock test	Verifies that all alarms and interlocks function as specified when an alarm or interlock condition is introduced and that the system's responds per specifications. Audio/visual
Standard operating procedure (SOP) verification	Verifies that operational SOPs are accurate while performing functional tests. The SOPs will be redlined, as required

Sterile Processing of Pharmaceutical Products: Engineering Practice, Validation, and Compliance in Regulated Environments, First Edition. Sam A. Hout.

Test case	Description
Training verification	Verifies that all applicable training activities have been completed to support the commissioning activities. The signature/date log will also serve as the mechanism to document this verification
Life safety device test	Verifies the operability of atmosphere monitoring equipment upon an alert or alarm condition
Start-up and shutdown verification	Verifies that the control system functions as designed during start-up and shutdown sequences
Safety feature functionality testing	Verifies that automatic safety features are triggered and operate correctly; emergency stops are one example of a safety feature
Emergency stop test	Verifies if equipment or system deactivates when emergency stop is activated
Stop/Abort/Hold/Restart/Run for BMS	Verifies if system performs the following functions when commanded by the control system: Stop, Abort, Hold, Restart, and Run
Valve operation (automatic)	Verifies that valves operate smoothly and as designed when commanded by the control system. Verify failure safe mode (Open/close)
Valve operation (manual)	Verifies that valves operate smoothly and as designed
Vessel spray ball verification	Verifies for spray coverage (Riboflavin) test
Agitator test	Verifies the agitator works within the specified range
Sequence of operation test	Verification that the system's operations function in the specified order
Rotating equipment vibration testing	Verifies that vibration readings for applicable equipment are within the required tolerance
System performance verification	Verifies that the system operates as designed over a wide range of operating limits and delivers the required performance as specified in the applicable User Requirements Specification (URS). Simulation of values and variables may be required to demonstrate that the logic works
Temperature, pressure, and flow control loop performance	Verifies that any applicable temperature, pressure, and/or flow control loops function as designed
Power failure and response test	Verifies that the system responds as designed under abnormal circumstances
Water and air system balancing as applicable	Verifies that the balancing report is available for review

Test case	Description
Air change rate verification	Measure velocity at the face of high-efficiency particulate air (HEPA). Compare to standard
Room temperature and humidity testing	Verifies that room atmospheric conditions measurements are as designed
Room differential pressure verification	Verifies the measured differential pressure meets specification
Airflow visualization verification	Verifies that airflow patterns meet specification. Document smoke studies, including video
Dew point verification	Verifies the dew point temperature of the air meets specification
Airlock recovery verification	Verifies the recovery time of parameters after opening and closing of the airlock door
Environmental monitoring for classified spaces	Verifies the parameters are within required limits based on space classification
Compressed gas system quality tests	Verifies that the specified tests are within acceptable limits. Pressure reading of PI is steady
Boiler feedwater test	Verifies water meets manufacturer's specifications
Boiler exhaust gas test	Verifies that exhaust gases meet environmental limits
Water system quality tests	Verifies specified tests are within acceptable limits
Water system chemical tests	Verifies that the applicable system chemistry is controlled in the required tolerances
Test equipment calibration verification	Verifies equipment used for a required test is within calibration and National Institute of Standards and Technology (NIST) certificates are available

Appendix C

WFI Turbulence Flow Requirements

	Minimum flow to hit 20 000 Reynolds				
	Ambient (20 °C)		Hot (85 °C)		For 5 ft/s
Pipe diameter (in.)	Flow (gpm)	Velocity (ft/s)	Flow (gpm)	Velocity (ft/s)	Flow (gpm)
½	2.35	7.00	0.81	2.40	1.68
¾	3.93	4.18	1.35	1.43	4.71
1	5.51	2.98	1.89	1.02	9.26
1½	8.68	1.89	2.98	0.65	22.97
2	11.85	1.38	4.07	0.48	42.80
3	18.19	0.90	6.25	0.31	100.82

Water properties	Ambient (20 °C)	Hot (85 °C)	Units
Density	8.33	8.08	lbm/gal
Dynamic viscosity	2.42	0.81	lbm/ft h

Sterile Processing of Pharmaceutical Products: Engineering Practice, Validation, and Compliance in Regulated Environments, First Edition. Sam A. Hout.

Appendix D

Water For Injection (WFI) – Design Requirements

User requirements specification (URS) for water for injection (WFI) distribution system considers the normal range of equipment operation. The purpose is to specify the user requirements for equipment that is to be used to support the production of finished pharmaceuticals in a current good manufacturing practice (cGMP) environment.

URS serves as the input to subsequent project risk assessment activities, design objectives, control strategy, and acceptance criteria for testing and qualification of the manufacturing system.

1. Materials of construction of WFI loops will be SS 316L
2. All piping weld connections must be horoscope tested at 100% of all welds. Visual inspection of welds to determine finish quality must be random and cover at least 10% of the welds.

Compliance Determination

Regulatory body	Number	Title
FDA	21 CFR parts 11	Electronic Records and Signatures
FDA	21 CFR part 210	cGMP in Manufacturing, Processing, Packaging or Holding of Drugs, General
FDA	21 CFR part 211	Current Good Manufacturing Practice for Finished Pharmaceuticals
FDA	21 CFR part 820	Quality System Regulation

Sterile Processing of Pharmaceutical Products: Engineering Practice, Validation, and Compliance in Regulated Environments, First Edition. Sam A. Hout.

Engineering Standards/Guidelines

Regulatory body	Standard/Guide
FDA	High-Purity Water System Inspection Guidance (7/93)
USP	<85> Bacterial Endotoxins Test, <643> Total Organic Carbon <645> Water Conductivity, and <1231> Water for Pharmaceutical Purposes v.40
ASME	BPE 2019
ISPE	Approaches to Commissioning and Qualification of Pharmaceutical Water and Steam Systems, Commissioning and Qualification, Applied Risk Management for Commissioning and Qualification, Water and Steam Systems
ISO	22519 2019
GAMP 5	Good Automated Manufacturing Practices
ASTM E2500	Standard Guide for Specification, Design, and Verification of Pharmaceutical and Biopharmaceutical Manufacturing Systems and Equipment

Key Abbreviations and Terms

Abbreviation/Term	Definition
CPP	Critical process parameter
CQA	Critical quality attribute
DW	Distilled water
FDA	Food and Drug Administration
FDS	Functional design specification
OSHA	Occupational Safety Health Administration
P&ID	Piping and instrument diagram
TOC	Total organic carbon
URS	User requirements specification
USP	United States Pharmacopeia
WFI	Water for injection

Process Description

WFI is generated utilizing pretreatment system and distillation units and is then stored in silos (storage tanks). Water is recirculated via primary pump skids through primary loops. Secondary pump skids, pump WFI through secondary recirculating loops to use points. Pump skids provide enough flow to assure adequate volume for operations, as well as maintenance of minimum Reynolds' number to assure turbulent flow on both supply and return legs of the loops. The skids also utilize heat exchangers to accommodate periodic sanitization. Silos and tanks will each be equipped with a vent pipeline with two 0.22-μm filters allowing pressure equalization when the water level changes. Typically, WFI water temperature is kept at 80 °C in aseptic operations, but the temperature is usually reduced using heat exchangers to meet point of use requirements for operator handling and product requirements conditions [59].

Specifications

Life Cycle Requirements

Equipment suppliers of the silos, tanks, equipment, piping, instrumentation, etc. are expected to provide for manufacturing cycle:

- Written communication of any deviations from user specifications.
- Written communication of changes to the design specification.
- Manufacturing/completion project schedule.

Product and Process User Requirements

Product Description

WFI Distribution System distributes WFI throughout the facility. WFI Distribution System transports WFI primarily to the mix tanks to become compounded solutions when active pharmaceutical ingredient (API) and excipients are added.

WFI Critical Quality Attributes (CQA)

Process step reference	CQA	Specification
Distribution	Micro	≤10 cfu/100 ml (absence of USP indicator pathogenic organisms)
Distribution	Endotoxin	<0.25 EU/ml
Distribution	TOC	≤500 ppb
Distribution	Conductivity	≤1.3 μS/cm @ 25°C

Critical Design Aspects (CDA)

Process step reference	CDA	Associated CQA	Specification
Distribution	Materials of construction	Micro, endotoxin, TOC, conductivity	316/316L or USP Class VI Elastomers
Distribution	Surface finish	Micro, endotoxin, TOC	≤30 Ra
Distribution	Slope	Micro, endotoxin, TOC	0.5%
Sanitization	Absence of dead legs	Micro, endotoxin, TOC	None with $L/D \geq 2.0$[a)]
Sanitization	Drain ability	Micro, endotoxin, TOC	Low point drains present at low points

a) Dead legs should be avoided wherever possible. If dead leg cannot be avoided due to design constraints or other means, examine the risk of dead leg section with $L/D \geq 2.0$ and determine risk acceptability on a case-by-case basis, including manual ability to drain that section.

Critical Component (CC)

Process step reference	CC	Associated CQA	Specification
Distribution	Vent filter	Micro, endotoxin, TOC	0.22 μm

Process Parameter Requirements

Process step reference	CPP	Operating range	Associated CQA
Distribution	Flow	Reynolds number ≥ 2500	Micro, endotoxin, TOC
Sanitization	Temperature	≥80 °C @ coldest or worst-case point	Micro, endotoxin, TOC
Sanitization	Time	TBD in validation (capability NLT 8 min)	Micro, endotoxin, TOC

Critical Instrument (CI)

Process step reference	CI	Associated CPP	Associated CQA
Sanitization	Temperature transmitter	Temperature	Micro, endotoxin, TOC
Sanitization	Programmable logic controller (PLC) clock	Time	Micro, endotoxin, TOC
Distribution	TOC analyzer	N/A	TOC
Distribution	Conductivity probe	N/A	Conductivity
Distribution	Flowmeter	Flow	Micro, endotoxin, TOC
Distribution	Pressure transducer	Pressure	N/A

Operating Ranges

WFI system shall be a recirculating loop-based design with any sections of piping length/diameter $\geq 2:1$ preapproved by the quality unit. WFI Distribution System shall have a validated process control system. WFI system shall be capable of monitoring, in real time, line pressure, temperature, TOC, flow, and conductivity at a minimum.

Non-pressurized vessels in uncontrolled environments shall be equipped with a vent filter for bidirectional flow of make-up air. All points-of-use shall contain a means of collecting a sample aseptically and without adversely affecting the environment. All loops shall contain at least one means of collecting a representative sample for each loop. Rinse assemblies will be automated to decrease moisture generation in controlled environments.

Each point of use/product contact valve shall be sanitary diaphragm e.g. zero static design, double block-and-bleed or mix-proof valves. Heat exchangers will be sized to heat water in each circuit to sanitization temperature in $\leq$30 minutes. Pumps will be controlled by VFD and pump motors must be VFD rated, high efficiency, totally enclosed fan cooled (TEFC). In addition, all equipment operating ranges must be clearly communicated and controlled within the system manufacturer's recommended safe operating range.

Recommended spare parts' list must be provided and spare parts are required to be on-site prior to system start-up. Most point-of-use valves shall be automated and controlled for operation. Manual points of use (POU) shall be placed throughout the system to allow for tap in for cleaning where applicable. Modulating valves shall be used to control flow into tanks to achieve target weights. The recirculating loops in the distribution system shall be capable of being hot water sanitized in a USP 1231-compliant manner. The system will heat WFI at user-defined temperatures for process requirements.

Material of Construction

Materials of construction requirement for all components have solution contact:

Components include pipes, fittings, instruments, valves, tanks, pumps, heat exchangers, etc.

Metallics

All metallic materials of construction must be 316L stainless steel and have a maximum internal surface finish of 0.6 μm/25 μin. Ra.

Plastics

All plastic materials shall be polyvinylidene fluoride (PVDF) or polypropylene (PP).

Elastomers

All elastomers and alternate materials shall be capable of withstanding system temperature and chemical exposure and be constructed of ethylene propylene diene monomer (EPDM), FKM Viton, teflon/polytetrafluoroethylene (PTFE), or platinum silicon.

Glass

Tempered glass, such as sight glass, shall be homogeneous, scratch-proof, impervious, inert, and nontoxic.

Material traceability will be provided with every metallic component that will have solution contact. Each pipe component should be marked with an indelible identifier that can be traced back to the original material certificate. Identifiers can be heat, lot, or a reference number linked to the weld record. Instruments, valves, and hardware are to be marked or tagged. Fittings will be identified in the weld map with traceability to the materials of construction and internal surface finish. Examples of acceptable material certificates are as follows:

- Certified material test report
- Certificate of conformance
- Cut sheet

A surface finish verification certification shall be provided for each component that will have solution contact, along with a certificate of calibration of the profilometer. The surface of each metallic component that comes into contact with the product must be passivated in accordance with standard operating procedure (SOP). Passivation must have official documentation and the passivation numbers shall each be linked to the corresponding components.

Exterior surfaces of components that reside in a controlled environment are preferred to be 316L, 316, or 304 series stainless.

Welding Requirements

Piping shall be orbital welded. Welding meets ASME BPE 2019 for Bioprocessing Equipment. Purge gas is ≥99.99% pure argon. Pressure vessels shall be welded in accordance to appropriate regulatory agency requirements and ASME. Welders and weld inspectors must be qualified for weld procedures. Welder and inspector welding program credentials must be furnished to project manager prior to performing any welding job.

Welds on solution piping must be passivated. The passivation must be documented and included with the welding documentation package. These welds may be spot-passivated if reachable. Non-accessible welds must be bath or circuit passivated. Weld repairs shall be permissible during initial fabrication if caused by a lack of penetration, lack of fusion, or unconsumed tack welds. Defective welds that cannot be repaired shall be cut out. Welding procedures shall include:

- Cleaning procedure
- Weld repairs
- Visual inspection of the external of all welds
- Visual inspection ≥10% of internal portion of welds. The percent inspection shall increase with each failed weld
- Visual inspection of all manual welds
- Inspection documentation
- Unique weld numbers
- Date performed
- Welder identification

- Description of materials being joined
- Heat numbers

Verification of all welds will be 100% verified through borescope videos. Weld documentation/weld map (drawing of the area/components), including:

a. Weld location identification and reference numbers
b. Welder signature and date at end of weld job
c. Weld inspector signature and date at end of weld job
   (1) Weld record (table with the following columns):
      (a) Weld map reference number
      (b) Weld type – manual/orbital/etc.
      (c) Component #1 material and reference to the traceability document (if solution contact)
      (d) Component #2 material and reference to the traceability document (if solution contact)
      (e) Rod/wire material and reference to the traceability document (if solution contact)
      (f) Performed by/date
      (g) Inspected by/date

Note: Welder and Inspector welding program credentials must be on file. This documentation must be completed immediately following any welding activity (daily, at a minimum) and shared with the project manager on a weekly basis for review. Welding symbols should be in accordance with AWS A2.4 or other equivalent ASME standards.

Construction Requirements

Insulation will be utilized for piping for anything within 7 ft of the working floor or platform (following OSHA standards). Equipment insulation must be non-particulate generating and shall be jacketed per OSHA requirements. WFI storage vessels will have vent lines and be protected by vent filters (electrically heat jacketed where applicable). WFI Silos and Tanks Vent filter and Housings shall be sized to accommodate the design flow rate. Storage vessels shall be kept wetted via spray balls.

Distribution system piping shall be appropriately supported and sloped at a minimum of 0.5%. Slope measurements should be made between pipe hangers/supports and at each change of direction. All system low points shall have a low point drain. All piping shall be orbital welded. Orbital welded connections are required for product contact piping. All welds will meet SOP requirements and surface finish requirements.

All piping will be clearly labeled for content and direction of flow. All POU, silos, loops, subloops, and buffer tanks shall be clearly labeled and identifiable. Each vessel shall have a manway with a minimum opening diameter of 22 in. Passivation procedure and reports of all 316/316L components being passivated and absence of passivation chemicals post-passivation. Manufacturer's drawings shall be provided for all components, piping excluded. Hydrostatic/leak test reports shall be provided for all components. All attempts will be made that all pump head will be directly after the pump. If this cannot be accomplished, appropriate engineering controls must be in place to ensure adequate turbulent flow and performance of the system. The vendor shall ensure that loop can operate with either one of a redundant pump system.

Lubrication Requirements

All lubricants used must be detailed in the design documentation. Include lubricants used in the mechanical aspects of the equipment as well as the lubricants used in the product assembly. Where direct lubrication is required, the design and construction must be such that the lubrication cannot leak or drain into the product or into the product contact areas. Where pneumatic machine functions require the use of lubricated compressed air, only USP and FDA-approved lubricants are acceptable: e.g. White Mineral Oil, USP.

Lubricants with USDA and FDA approval and "AA" classification shall be used where incidental contact may occur.

Electrical Requirements

Electrical systems must be designed to applicable Electrical and Controls Specification. Where applicable, "as built" instrumentation wiring diagrams shall also be provided. Electrical panel penetrations should be made from bottom of panel when possible. Reference NFPA 79, OSHA, and User Standards for additional Electrical Requirements.

All electrical panels need to have externally mounted disconnects for Arc Flash compliance. Panels in controlled environments to be Nat'l Electrical Manufacturer's Association (NEMA) 4×, all others a minimum of 3R rating. All electrical panel connections will be IR mapped and tested periodically to identify hot spots and should be corrected to prevent ignition and fire hazard. High-voltage cabinets should be Underwriters Laboratory (UL) listed.

Safety Requirements

All pinch points and hazardous areas must be guarded per OSHA standards using required materials of construction. E-stop buttons must be located close to operator work areas and must release all potential energy immediately upon engagement. Restarting the equipment shall require a reset and disengaging of the E-stop. All-access doors must have interlocks as described in Electrical and Controls Specification. Access doors must be interlocked with magnetic mechanisms to not allow entry to the machine while it is running.

The design must permit the electrical, pneumatic, and hydraulic devices to be locked out during maintenance and repair activities. All lockout/tagout points must be easily accessible. The control system must be designed to operate in a fail-safe manner and meet OSHA safety requirements. The safety circuit shall be at least category 3. Jog capability is acceptable only with a single momentary switch. Sharp corners and edges must be eliminated. Safety circuits must not be able to be bypassed. Electrical panels with >50 V are to be designed with incident energy below 1.2 cal/cm^2. Fixed-mount human machine interfaces (HMIs) are to be set at a height of 54″ from the floor to the center of the screen. Adjustable mounts are to cover the range of 54″–68″ to center of the screen. All system components and connections (pumps, heat exchangers, instruments, POU valves, tank connections etc.) must be easily, safely, and ergonomically assessable.

All heat exchangers shall be, on the shell side, protected by an appropriate pressure relief device where required. Distribution piping in contact with hot WFI and readily accessible by personnel shall be insulated for safety. Each storage vessel shall be equipped with a certified rupture disc.

All equipment shall be appropriately guarded. Required EHS checklists will be completed at the appropriate time in the project life cycle. All equipment with potential electric discharge must be grounded and bonded to code.

Operational Requirements

Process Constraints and Limitations

WFI velocity in the piping system shall be limited to no more than 10 ft/s.

Process Control System

Vessel volume, liquid additions, temperature monitoring, quality monitoring capability, and sanitization shall be interfaced with and controlled by an automation system. Vessel systems shall be capable of being monitored and/or controlled from a local HMI within controlled space interfaced to an automation system.

Electrical components shall be mounted in NEMA-rated enclosures suitable for the local environment and service. Control panel assembly and electrical motors and systems shall be UL approved.

- Loops have low, high supply pressure alarms.
- Loops have low, high, return pressure alarms.

- Tanks have low, high, level alarms.
- Loops have Hi TOC and Conductivity alarms.

The rupture disks in the WFI system will generate an alarm if triggered. All instruments shall be suitable for the local environment and service. There shall be no passwords that reside in programming code and no protected source code. All electrical panels that contain controllers must have externally mounted communications ports with 120-V AC receptacle for programming and servicing.

Operator Interface

Display measurements in inches. Display time using 24-hour clock. Display speed in gallons per minute.

Control system must be able to generate accurate and complete copies of records in both human readable and electronic form for inspection, review, and copying. Control system must be able to discern invalid or altered records or entries. Control system must be able to protect records to enable their accurate and ready retrieval throughout the records retention period. Control system must be able to ensure that system access is limited to authorized individuals where:

- System access is restricted based on user responsibilities and application functions.
- System user access levels are documented.
- Password expiration rules exist for the system.

Control system must have authority checks to ensure that only authorized individuals can:

- Access and use the system
- Alter a record
- Perform the specified operation.

Control system must provide audit trails for all objects and users that must:

- Be secure
- Be computer generated
- Be time and date stamped (e.g. server time).

Audit trails must independently record the date/time of operator entries and actions that:

- Create electronic records
- Modify electronic records
- Maintain electronic records
- Delete electronic records.

Audit trail must record changes and must not obscure previously recorded information. Control system must ensure the audit trail is secure from being modified/deleted and cannot be disabled. Control system must ensure that the audit trail documentation must be retained for a period at least as long 90 days and must be available for review and printing.

Power Loss and Recovery

On power restoration, the system shall not restart without operator or communication-link input. No damage to machine will occur as a result of going to the safe state. Integrator will supply instructions for recovery from catastrophic control system failures. It is generally accepted that the system shall protect in the following priority: Personnel, Equipment, and then Product. Machine will be manually restarted based on operator inputs (operator initiates power recovery sequence).

Cleaning and Sanitizing

All components that will contact solution will be initially cleaned by the passivation requirement. Further cleaning will be accomplished when the system comes online though sanitization requirement.

Maintenance Requirements

Recommended maintenance procedures will be available prior to completion of Factory Acceptance Test (FAT). Original equipment manufacturer part numbers are required for all spare parts. Spare parts shall be described as either "critical" or "general." Drawings will be provided in a format agreed to by user engineering for all machined parts.

The Operations Manual will include a section on troubleshooting and repair and preventive maintenance.

Training and Documentation Requirements

Material cut sheets or specs are required for all WFI connections. Training documents will be supplied for Operation and Maintenance procedures. Operator training will be available during commissioning, potentially on multiple shifts. Maintenance-specific training will be available during commissioning, potentially on multiple shifts. A draft Operations Manual will be available prior to FAT and the final version at FAT. Operations and Maintenance Manuals and training will be provided prior to system go-live. Start-up and tuning instructions (water flow quantities, correct instrument readings, etc.) shall be provided in these manuals. Design documents, parameters, data sheets, technical specifications, P&IDs, and drawing files will be provided and will include the requirements for preventive maintenance.

Appendix E

Solution Transfer System (STS) – Design Requirements

Solution transmission system (STS) user requirements specification (URS) covers product URSs, process URSs, and general URSs. STS is limited to the distribution system and does not include equipment or systems designed to formulate or fill solutions.

Term	Definition
ANSI	American National Standards Institute
ASME	American Society of Mechanical Engineers
BMS	Building management system
BOM	Bill of material
CIP	Clean-in-place
CPP	Critical process parameters
CQA	Critical quality attributes
DW	Distilled water
EBR	Electronic batch record
ENDO	Test for enumerating coliforms in water
NEMA	National Electrical Manufacturer Association
NFPA	National Fire Protection Association
SCADA	Supervisory Control and Data Acquisition
STS	Solution transmission system
TOC	Total organic carbon
UL	Underwriters laboratory
UPS	Uninterrupted power supply
URS	User requirement specification
VAC	Voltage alternating current

Sterile Processing of Pharmaceutical Products: Engineering Practice, Validation, and Compliance in Regulated Environments, First Edition. Sam A. Hout.

Term	Definition
VDC	Voltage direct current
WFI	Water for injection
Zero Static	Point of use (Zero Static) critical diaphragm valve allows process fluids to be transferred, sampled, drained, or diverted without impact on recirculating distribution WFI loop

System Description

Solutions are formulated in Mix Complex. Solutions are discharged from mix tanks through a valve array and pumped through sanitary filter housings to filling complex. Process piping consists of a recirculating loop for each of the fill lines. The supply pumps provide motive force to assure adequate volume and pressure for operations and maintaining minimum velocities to assure turbulent flow on both supply and return legs of the loops.

STS fluid system shall be a recirculating loop-based design with zero dead legs.

On each filling line, the system shall be designed to provide solution delivery from one tank while another tank is being cleaned. System shall be capable of continuously monitoring supply and return pressure, temperature, conductivity, flow, and TOC at a minimum. System shall provide valve arrays to allow multiple, isolated flow paths to accommodate recirculation, filling, cleaning, and draining. Distribution system shall be capable of being cleaned via CIP. All loops shall contain at least one means of collecting a representative sample prior to and after filtration. Distribution system shall be designed to provide specific flows and pressures as required for each product on a line-by-line basis. System shall be designed for continuous operation. Equipment insulation shall be non-particulate generating and shall be jacketed. Product contact materials shall be 316L with a surface finish of 25 RA or better and approved elastomers. All product contact surfaces shall be passivated using approved procedure. Pumps, valves, piping, and other system components shall be sized to accommodate design flow rates and pressures. Distribution system piping shall be appropriately supported and sloped at a minimum of 1/8″ per foot. Slope measurements shall be made between pipe hangers/supports and at each change of direction. All system low points shall have a low point drain. Orbital welded connections are required for product contact piping. All welds will meet governing standard operating procedure (SOP) requirements. Piping inside the clean rooms shall be minimized. All piping shall be clearly labeled with environmentally suitable indelible stickers with system (STS), line number, and direction of flow. Distribution pumps shall be mounted on base plates, with sanitary flexible coupling connections to minimize the effect of transient shock waves (water hammer). Distribution system routing valves shall be sanitary diaphragm, double block-and-bleed, or mix-proof valves. All system components and connections (pumps, heat exchangers, instruments, valves, connections, etc.) shall be easily, safely, and ergonomically accessible.

Pumps shall be controlled by VFD and pump motors shall be VFD rated, high-efficiency, totally enclosed fan-cooled (TEFC) motor.

All equipment operating ranges shall be clearly communicated and controlled within the system manufacturer's recommended safe operating range. Recommended spare parts' list shall be provided and spare parts are required to be on-site prior to system commissioning activities, specifically critical parts. All solution routing valves shall be automated and controlled for operation, with position feedback.

Pressure modulating valves shall be used to control distribution system pressure, with closed-loop control, and an adjustable set point from a human machine interface (HMI). New power distribution gear shall be enrolled in the site arc flash model and an approved arc flash hazard label affixed. All distribution system monitoring, control, and routing devices shall be interfaced with and controlled by an automation system. Distribution system shall be capable of being monitored and/or controlled from a local HMI within controlled space interfaced to an automation system. Electrical components shall be mounted in NEMA-rated enclosures suitable for the local environment and service. Electrical motors and field electrical components shall be UL listed.

- Loops shall have low, high, supply pressure alarms.
- Loops shall have low, high, return pressure alarms.
- Loops shall have low, high, temperature alarms.

Wired HMI's shall be utilized for critical and noncritical functions and deployed on existing EBR workstations. Pneumatics, Programmable Logic Controller (PLC's), and greater than 120 VAC Power shall be segregated into separate enclosures. PLC and Remote I/O panels shall be connected to a UPS. PLC programming shall place the process in a safe state upon detection of loss of line power. Solution routing valve fails to open and fails to close alarms shall index the process to a safe state. Motor fails to run alarms shall index the process to a safe state.

Distribution system shall be capable of being cleaned via CIP. The skidded CIP system URS shall be a separate document. All equipment safety and guarding shall be per ANSI B11. All pressurized systems shall be protected by an ASME-rated relief device. Emergency power shutoff shall provide immediate pump shutdown in the event of a line break or rupture. When the emergency stop (ESTOP) button is pushed, all sequences shall stop, and valves shall return to their safe position.

ESTOPs shall be easily accessible. Operator reset is required to resume normal operation.

Sample locations must be accessible to allow easy application of aseptic sampling techniques.

Design documents and drawing files shall be provided and shall include the requirements for preventive maintenance. Operations and Maintenance Manuals and training shall be provided.

Bulk Powder Handling System

Bulk powders in super sacks are connected to discharge stations above tanks. Discharge stations will dispense powders by loss of weight using gravity through a chute to a specified tank via a transfer pipe. The discharge stations will be cleaned in position using dedicated wash hoods.

Bulk powders in flexible intermediate bulk containers (FIBC) are moved to the discharge station where barcodes unique to the station and FIBC are scanned into batch processing system. Batch processing system will display a message stating that the FIBC is ready for loading into the discharge station. An FIBC will be placed in front of the discharge station and connected to a hoist. The hoist will be used to move the FIBC into position for unloading. Once in position, FIBC is connected to the valve then the operator acknowledges that connection is complete in the batch processing system. This acknowledgment may be automated through the control system. When the batch processing step calls for bulk powder, the valve will open, and powder will transfer the appropriate weight from the FIBC to the mix tank. When multiple FIBCs are required for a batch, the operator will be prompted to load additional FIBCs as needed and will follow the same process described above. Discharge station will be capable of being cleaned in place.

RACI Analysis – Project Responsibilities

Task deliverable	Engineering	Quality validation	Manufacturing maintenance	EH&S	Process designer	Construction manager	Process piping installer	Control systems integrator
URS	A, R	R	R	C	C	C	C	C
FRS	A	C	C	I	C	I	I	R
SDS	A	I	C	I	C	I	I	R
HDS	A	I	C	I	C	I	I	R
FAT	R	A	C	I	C	R	R	R
SAT	R	A	C	I	C	R	R	R
IQ/OQ	R	A	C	I	I	I	I	R

RACI Analysis – Project Responsibilities

Task deliverable	Engi-neering	Quality vali-dation	Manu-facturing maintenance	EH&S	Process designer	Cons-truction manager	Process piping installer	Control systems integrator
PQ	R	A	C	C	C	C	C	C
Change control	R	A	I	I	I	I	I	I
Construction document	R	C	C	I	C	A	I	I
Submittal approval	R	I	I	I	C	A	C	R
P&ID	A	C	C	I	R	C	I	I
Instrument sheet	A	C	C	I	R	C	C	C
General arrangement	A	C	C	I	R	C	I	I
BOM	A	C	C	I	R	C	C	C
Mechanical routing	A	I	C	I	R	R	R	I
Electrical routing	A	I	C	I	R	R	I	I
MEP	R	C	C	I	A, R	R	R	I
Material control	R	C	C	I	R	A, R	R	I
Mechanical install	R	C	C	I	C	A, R	R	I
Electrical install	R	I	C	I	R	A, R	R	I
Power distribution	R	I	C	I	R	A, R	I	I
Control panels	R	I	C	I	R	R	I	A, R
IT infrastructure	R	I	C	I	R	I	I	A, R
HMI	R	C	C	I	R	R	I	A, R
PLC	R	I	C	I	R	I	I	A, R
Purchasing	R	I	I	I	R	A, R	I	I
Demolition	R	C	C	I	R	A, R	R	I
Asset induction	A	C	R	I	I	C	C	I
As-built CAD DWG	A	I	I	I	R	C	C	R
Turnover package	A	C	C	I	R	R	C	R

RACI Analysis – Project Responsibilities

Task deliverable	Engineering	Quality validation	Manufacturing maintenance	EH&S	Process designer	Construction manager	Process piping installer	Control systems integrator
Training	A, R	C	C	C	I	I	I	R
Preventive maintenance	R	C	A, R	I	R	C	I	C
Safety	R	R	R	A	R	R	R	R
Permit	R	I	I	A	R	R	R	I
Arc flash	R	I	R	I	A	I	I	I
Security	R	C	C	A	R	I	I	C
Fire protection	A	I	C	C	R	C	C	C
Cost control	A	C	C	I	R	R	R	R
GAMP 5 compliance	A, R	I	I	I	I	I	I	I

R = Responsible
Those who do the work to complete the task. There is at least one role with a participation type of responsible, although others can be delegated to assist in the work required.
A = Accountable
The one ultimately answerable for the correct and thorough completion of the deliverable or task, the one who ensures the prerequisites of the task are met and who delegates the work to those responsible. An accountable must sign off (approve) work that responsible provides. There must be only one accountable specified for each task or deliverable.
C = Consulted
Those whose opinions are sought, typically subject-matter experts (SME); and with whom there is two-way communication.
I = Informed
Those who are kept up to date on progress, often only on completion of the task or deliverable; and with whom there is just one-way communication.

Projects Risks and Mitigation Strategies

Risk	Mitigation strategy
Lost time injury	• Compliance to site EH&S policies and procedures. Owner safety professional oversight. Owner safety professional oversight • Owner-approved master project plan provides for trade labor staffing at levels to minimize overtime and worker fatigue • Reinforce safety as top priority
Budget exceeded	• Produce accurate, reviewed, signed, and owner-approved URS, FRS, SDS, HDS, FAT, and SAT protocol documents prior to fabrication or construction, traceable to owner-approved drawings • Implement project controls aligned with minimizing engineering change orders • Vendors shall draw upon contingency funds with owner consent prior to change order discussions

Projects Risks and Mitigation Strategies

Risk	Mitigation strategy
Scope expansion	• Produce owner-approved engineering drawings traceable to URS, FRS, SDS, HDS, FAT, and SAT documents • Maintain compliance to standards • Owner-approved engineering drawings and submittals shall be the basis of execution and construction change order management
In-service date delayed	• Independent owner project management leadership in support of control systems integration, tracking progress against owner-approved project plan • Promptness in furnishing submittals for owner review and approval • Promptness in owner approval of protocol documents • Site management alignment with key site professional and trade labor personnel plans in support of project deliverables within timeline
Performance failure	• Review and approval of vendor material submittals by owner SME • Owner SME inspection of materials prior to installation • Owner SME inspection of tasks upon completion • Executed FAT and SAT protocols by Owner SME
Technology transfer failure	• Identify user team leads for project • Management time allocation for project • Identify technology training requirements • Align engineering, quality, maintenance, and operations personnel plans • Induct asset and integrate in financial system • Maintenance plans shall include preventive procedures and tasks list

Glossary

Design qualification (DQ): The documented verification that the proposed design of the facilities, systems, and equipment is suitable for the intended purpose.

Installation qualification (IQ): The documented verification that the facilities, systems, and equipment as installed or modified comply with the approved design and manufacture's recommendations.

Operations qualification (OQ): The documented verification that the facilities, systems, and equipment as installed or modified can perform as intended throughout the anticipated operation ranges.

Performance qualification (PQ): The documented verification that the facilities, systems, and equipment as connected together can perform effectively and reproducibly, based on the approved process method and product specification.

Process validation: The documented evidence that the process operated within established parameters can perform effectively and reproducibly to produce a medicinal product meeting its predetermined specifications and quality attributes.

FDA

Guideline on General Principles of Process Validation: Validation – Establishing documented evidence that provides a high degree of assurance that a specific process will consistently produce a product meeting its predetermined specifications and quality attributes.

Guidance for Industry: Quality Systems Approach to Pharmaceutical Current Good Manufacturing Practice Regulations: With proper design and reliable mechanisms to transfer process knowledge from development to commercial production, a manufacturer should be able to validate the manufacturing process. In a quality system, process validation provides initial proof, through commercial batch manufacture, that the design of the process produces the intended product quality.

Sterile Processing of Pharmaceutical Products: Engineering Practice, Validation, and Compliance in Regulated Environments, First Edition. Sam A. Hout.

EU GMP Guide Annex 15

Chapter Process Validation § 20: The requirements and principles outlined in this chapter are applicable to the manufacture of pharmaceutical dosage forms. They cover initial validation of new processes, subsequent validation of modified processes, and revalidation.

Site Acceptance Testing (SAT): Documented testing similar in nature to that of the Factory Acceptance Test; however, testing is conducted in the facility (site) on a much scaled-down version. SAT is conducted to ensure that no damage has occurred to the equipment in transit and that the unit functions properly. Determination of units for product OQ is specified by product specifications only.

Nomenclature

ANVISA	National Health Surveillance Agency – Brazil
API	Active pharmaceutical ingredient
BFS	Blow-fill-seal
CAPA	Corrective Action Preventive Action
CAR T	CAR T-cell therapy – T immune system cells taken from patient blood and changed in lab. The special receptor chimeric antigen receptor (CAR) gene, which binds to a protein on the patient cancer cell, is added in the lab. CAR T cells are grown in the lab and then given to patients by infusion
CEBR	Center for Biologics Evaluation and Research – FDA
CEDR	Center for Drug Evaluation and Research – FDA
CFU	Colony-forming unit (microbes)
CIP	Clean-in-place system for cleaning and sanitization
COFEPRIS	Federal Commission for Protection against Sanitary Risk – Mexico
DOE	Design of experiments
EMA	European Medicines Agency
EN ISO	International standard adopted by European Union
EU	European Union
FDA	Food and Drug Administration
FMEA	Failure Modes and Effects Analysis
FMECA	Failure Modes, Effects, and Criticality Analysis
GMP	Good Manufacturing Practice (GLP – Laboratory; GEP – Engineering; GXP – generic)
HACCP	Hazard Analysis Critical Control Points
HEPA	High-efficiency particulate air – filter
ICH	International Council for Harmonization
ISO	International Organization for Standardization
ISPE	International Society of Pharmaceutical Engineers
IV	Intravenous injection (Infusion)

Sterile Processing of Pharmaceutical Products: Engineering Practice, Validation, and Compliance in Regulated Environments, First Edition. Sam A. Hout.

LAL	Limulus amebocyte lysate testing (Bacterial endotoxin testing)
MES	Manufacturing Execution Systems
MHRA	Medicines and Healthcare Products Regulatory Agency (UK)
Ø	Diameter
PAI	Pre-approval inspection (by FDA)
PFS	Prefilled syringe
PIC/S	Pharmaceutical Inspection Cooperation Scheme
PLC	Programmable logic controller
QA/QC	Quality Assurance/Quality Control
SIP	Steam-in-place system for sterilization
SOP	Standard operating procedure
UV	Ultraviolet light
WFI	Water for injection (pharmaceutical high grade, highly purified water) (intravenous injection water)
WHO	World Health Organization

References

Standards and reference documents	Title
ISPE Baseline Guide	Volume 4, Water and Steam Systems
ISPE Baseline Guide	Commissioning and Qualification
ASME BPE 2019	Bioprocessing Equipment
USP <1231>	Water for Pharmaceutical Purposes
Title 21 Code of Federal Regulations (21 CFR)	Parts 210, 211 and 820
ICH Q9	Quality Risk Management
EU GMP Annex 15, PIC/S Annex 15	Qualification and Validation
ASTM E2500	Standard Guide for Specification, Design, and Verification of Pharmaceutical and Biopharmaceutical Manufacturing Systems and Equipment

References

For manufacturing of API:

- ICH Q7 Good Manufacturing Practice Guidance for Active Pharmaceutical Ingredients (APIs), 1 April 2003
- FDA's Guide to inspection of Bulk Pharmaceutical Chemicals, 1985

Sterile Processing of Pharmaceutical Products: Engineering Practice, Validation, and Compliance in Regulated Environments, First Edition. Sam A. Hout.

For manufacturing of drug product:

- The Code of Federal regulations, 21 CFR Parts 210 and 211
- Pharmaceutical Inspection Convention, PIC-GMP
- EU GMP Guidelines

US FDA, Guidance for Industry – Process Validation: General Principles and Practices, www.fda.gov

ICH Q8 (R2) Pharmaceutical Development, ICH Q9 Quality Risk Management, ICH Q10 Pharmaceutical Quality Systems, www.ich.org

ASTM E2500-07, Standard Guide for Specifications, Design, and Verification Pharmaceutical and Biopharmaceutical Manufacturing Systems and Equipment, ASTM International, www.astm.org

US FDA, Pharmaceutical cGMP for the twenty-first century – A Risk-Based Approach, www.fda.gov

21 CFR Part 211, Subpart C, Buildings and Facilities, www.fda.gov

ISPE Guide: Science and Risk-Based Approach for Delivery of Facilities, Systems, and Equipment, International Society of pharmaceutical Engineering (ISPE), June 2011, www.ispe.org

ISPE Good Practice Guide: Applied Risk Management for Commissioning and Qualification, International Society of pharmaceutical Engineering (ISPE), First Edition, October 2011, www.ispe.org

References

1 *Effective Implementation of Audit Programs*, Miguel Montalvo.
2 *Confronting Variability: A Framework for Risk Assessment*, Diane Petitti, Richard Prince.
3 *Visual Inspection and Particulate Control*, Scott Aldrich, Roy Cherris, John Shabushnig.
4 *Torbeck's Statistical Cookbook for Scientists and Engineers*, Lynn Torbeck.
5 *Cleanroom Microbiology*, Tim Sandle, R. VijayaKumar.
6 *Microbiological Monitoring of Pharmaceutical Process Water*, Michael Jahnke.
7 *Lessons of Failure: When Things Go Wrong In Pharmaceutical Manufacturing*, Maik Jornitz, Russell Madsen.
8 *Media Fill Validation Environmental Monitoring During Aseptic Processing*, Michael Jahnke.
9 *Technology and Knowledge Transfer: Keys to Successful Implementation and Management*, Mark Gibson, Siegfried Schmitt.
10 Pflug, I. (2007). *Microbiology, and Engineering of Sterilization Processes*, 12e.

11 *Fungi: A Handbook for Life Science Manufacturers and Researchers*, Jeanne Moldenhauer.

12 Sandle, T. and Vesper, J. *GMP in Practice: Regulatory Expectations for the Pharmaceutical Industry*, 5e.

13 Global Sterile Manufacturing Regulatory Guidance Comparison.

14 *Validation by Design: The Statistical Handbook for Pharmaceutical Process Validation*, Lynn Torbeck.

15 Baker, D., Hawana, J., and Matsumura, T.(ed. B. Jentges). *Pharmaceutical Legislation of the European Union, Japan and the United States of America - An Overview*.

16 *Hosting a Compliance Inspection*, Janet Gough.

17 *Introduction to Environmental Monitoring in Pharmaceutical Areas*, Michael Jahnke.

18 *Lifecycle Risk Management for Healthcare Products: From Research Through Disposal*, Edwin Bills, Stan Mastrangelo.

19 PDA Technical Series: Endotoxin Analysis and Risk Management.

20 *Laboratory Design: Establishing the Facility and Management Structure*, Scott Sutton.

21 *Pharmaceutical Contamination Control: Practical Strategies for Compliance*, Nigel Halls.

22 *Systems Based Inspection for Pharmaceutical Manufacturers*, Jeanne Moldenhauer.

23 *Ethylene Oxide Sterilization Validation and Routine Operations Handbook*, Anne Booth.

24 *Radiation Sterilization: Validation and Routine Operations Handbook*, Anne Booth.

25 LeBlanc, D. *Cleaning Validation: Practical Compliance Solutions for Pharmaceutical Manufacturing*, vol. 1.

26 *Sterility Testing of Pharmaceutical Products*, Tim Sandle.

27 *Root Cause Investigations for CAPA: Clear and Simple*, James Vesper.

28 *Environmental Monitoring*, Jeanne Moldenhauer.

29 *FDA Warning Letters: Analysis and Guidance*, Jeanne Moldenhauer.

30 *Microbial Risk Assessment in Pharmaceutical Clean Rooms (single user digital version)*, Bengt Ljungqvist, Berit Reinmueller.

31 *Square Root of (N) Sampling Plans: Procedures and Tables for Inspection of Quality Attributes*, Joyce Torbeck, Lynn Torbeck.

32 *Pharmaceutical Outsourcing: Quality Management and Project Delivery*, Trevor Deeks, Karen Ginsbury, Susan Schniepp.

33 *Microbial Control and Identification: Strategies Methods Applications*, Mary Griffin, Dona Reber.

34 *Pharmaceutical Quality Control Microbiology: A Guidebook to the Basics*, Scott Sutton.

35 *Pharmaceutical Quality*, Richard Prince.

36 *Validating Enterprise Systems: A Practical Guide*, David Stokes.

37 Prince, R. (ed.). *Microbiology in Pharmaceutical Manufacturing*, 2e, vols. 1 and 2.

38 *Trend and Out-of-Trend Analysis for Pharmaceutical Quality and Manufacturing Using Minitab*, Lynn Torbeck.

39 *The Bacterial Endotoxins Test: A Practical Guide*, Karen McCullough.

40 *Risk-Based Software Validation*, Janet Gough, David Nettleton.

41 *Software as a Service (SaaS): Risk-Based Validation with Time-Saving Templates*, Janet Gough, David Nettleton.

42 *Practical Aseptic Processing Fill and Finish*, vols. 1 and 2, Jack Lysfjord.

43 Container/Closure Integrity Assessment A Compilation of Papers from the PDA Journal of Pharmaceutical Science and Technology.

44 *Validation of Analytical Methods for Biopharmaceuticals: A Guide to Risk-Based Validation and Implementation Strategies*, Stephan Krause.

45 Madsen, R. and Moldenhauer, J. (eds.). *Contamination Control in Healthcare Product Manufacturing*, vol. 1.

46 *Steam Sterilization: A Practitioner's Guide*, Jeanne Moldenhauer.

47 *Thermal Validation in Moist Heat Sterilization*, Jeanne Moldenhauer.

48 *Rapid Sterility Testing*, Jeanne Moldenhauer.

49 *Computerized Systems in the Modern Laboratory: A Practical Guide*, Joseph Liscouski.

50 *Phase Appropriate GMP for Biological Processes: Pre-Clinical to Commercial Production*, Trevor Deeks.

51 *SOPs Clear and Simple: For Healthcare Manufacturers*, Brian Matye, Jeanne Moldenhauer, Susan Schniepp.

52 *Risk Assessment and Management for Healthcare Manufacturing: Practical Tips and Case Studies*, Tim Sandle.

53 *Quality by Design: Putting Theory Into Practice*, Siegfried Schmitt.

54 *Microbial Risk and Investigations*, Karen McCullough, Jeanne Moldenhauer.

55 *Risk-Based Compliance Handbook*, Siegfried Schmitt.

56 *The Internal Quality Audit*, Janet Gough, Monica Grimaldi.

57 *Validation Master Plan: The Streetwise Downtown Guide*, Trevor Deeks.

58 *Why Life Science Manufacturers Do What They Do in Development, Formulation, Production and Quality: A History*, Lynn Torbeck.

59 *Contamination Prevention for Nonsterile Pharmaceutical Manufacturing*, Andrew Dick

Further Reading

Abbott, C.F., Cockton, J., and Jones, W. (1956). Resistance of crystalline substances to gas sterilization. *J. Pharm. Pharmacol.* 8: 709–720.

Ad hoc Committee on Infection Control in the Handling of Endoscopic Equipment (1980). Guidelines for preparation of laparoscopic instrumentation. *AORN J.* 32: 65–66, 70, 74, 76.

Adler, S., Scherrer, M., and Daschner, F.D. (1998). Costs of low-temperature plasma sterilization compared with other sterilization methods. *J. Hosp. Infect.* 40: 125–134.

Agalloco, J.P., Akers, J.E., and Madsen, R.E. (1998). Moist heat sterilization–myths and realities. *PDA J. Pharm. Sci. Technol.* 52: 346–350.

Agerton, T., Valway, S., Gore, B. et al. (1997). Transmission of a highly drug-resistant strain (strain W1) of *Mycobacterium tuberculosis*. Community outbreak and nosocomial transmission via a contaminated bronchoscope. *JAMA* 278: 1073–1077.

Agolini, G., Russo, A., and Clementi, M. (1999). Effect of phenolic and chlorine disinfectants on hepatitis C virus binding and infectivity. *Am. J. Infect. Control* 27: 236–239.

Akamatsu, T., Tabata, K., Hironga, M. et al. (1996). Transmission of *Helicobacter pylori* infection via flexible fiberoptic endoscopy. *Am. J. Infect. Control* 24: 396–401.

Alasri, A., Roques, C., Michel, G. et al. (1992). Bactericidal properties of peracetic acid and hydrogen peroxide, alone and in combination, and chlorine and formaldehyde against bacterial water strains. *Can. J. Microbiol.* 38: 635–642.

Alfa, M.J. (1997). Flexible endoscope reprocessing. *Infect. Control Steril. Technol.* 3: 26–36.

Alfa, M.J. (1999). Importance of lumen flow in liquid chemical sterilization. *Am. J. Infect. Control* 27: 373–375.

Alfa, M.J. and Jackson, M. (2001). A new hydrogen peroxide-based medical-device detergent with germicidal properties: comparison with enzymatic cleaners. *Am. J. Infect. Control* 29: 168–177.

Sterile Processing of Pharmaceutical Products: Engineering Practice, Validation, and Compliance in Regulated Environments, First Edition. Sam A. Hout.

Alfa, M.J. and Nemes, R. (2003). Inadequacy of manual cleaning for reprocessing single-use, triple-lumen sphinctertomes: simulated-use testing comparing manual with automated cleaning methods. *Am. J. Infect. Control* 31: 193–207.

Alfa, M.J. and Nemes, R. (2004). Manual versus automated methods for cleaning reusable accessory devices used for minimally invasine surgical procedures. *J. Hosp. Infect.* 58: 50–58.

Alfa, M.J. and Nemes, R. (2004). Reprocessing of lumened instruments. In: *Disinfection, Sterilization and Antisepsis: Principles, Practices, Challenges, and New Research* (ed. W.A. Rutala), 189–199. Washington, DC: Association for Professionals in Infection Control and Epidemiology.

Alfa, M.J. and Sitter, D.L. (1994). In-hospital evaluation of orthophthalaldehyde as a high level disinfectant for flexible endoscopes. *J. Hosp. Infect.* 26: 15–26.

Alfa, M.J., DeGagne, P., Olson, N., and Puchalski, T. (1996). Comparison of ion plasma, vaporized hydrogen peroxide and 100% ethylene oxide sterilizers to the 12/88 ethylene oxide gas sterilizer. *Infect. Control Hosp. Epidemiol.* 17: 92–100.

Alfa, M.J., DeGagne, P., and Olson, N. (1997). Bacterial killing ability of 10% ethylene oxide plus 90% hydrochlorofluorocarbon sterilizing gas. *Infect. Control Hosp. Epidemiol.* 18: 641–645.

Alfa, M.J., Olson, N., Degagne, P., and Hizon, R. (1998). New low temperature sterilization technologies: microbicidal activity and clinical efficacy. In: *Disinfection, Sterilization, and Antisepsis in Healthcare* (ed. W.A. Rutala), 67–78. Champlain, NY: Polyscience Publications.

Alfa, M.J., DeGagne, P., Olson, N., and Hizon, R. (1998). Comparison of liquid chemical sterilization with peracetic acid and ethylene oxide sterilization for long narrow lumens. *Am. J. Infect. Control* 26: 469–477.

Alfa, M.J., Degagne, P., and Olson, N. (1999). Worst-case soiling levels for patient-used flexible endoscopes before and after cleaning. *Am. J. Infect. Control* 27: 392–401.

Alfa, M.J., Olson, N., DeGagne, P., and Jackson, M. (2002). Evaluation of rapid readout biological indicators for 132°C gravity and 132°C vacuum-assisted steam sterilization cycles using a new automated fluorescent reader. *Infect. Control Hosp. Epidemiol.* 23: 388–392.

Alfa, M.J., Nemes, R., Olson, N., and Mulaire, A. (2006). Manual methods are suboptimal compared with automated methods for cleaning of single-use biopsy forceps. *Infect. Control Hosp. Epidemiol.* 27: 841–846.

Alfano, E.M., Cole, E.C., and Rutala, W.A. (1988). Quantitative evaluation of bacteria washed from stainless steel penicylinders during AOAC use-dilution method. *J. Assoc. Off. Anal. Chem.* 71: 868–871.

Ali, Y., Dolan, M.J., Fendler, E.J., and Larson, E.L. (2001). Alcohols. In: *Disinfection, Sterilization, and Preservation* (ed. S.S. Block), 229–254. Philadelphia, PA: Lippincott Williams & Wilkins.

Al-Masaudi, S.B., Day, M.J., and Russell, A.D. (1988). Sensitivity of methicillin-resistant *Staphylococcus aureus* strains to some antibiotics, antiseptics and disinfectants. *J. Appl. Bacteriol.* 65: 329–337.

Alter, M.J., Tokars, J.I., Arduino, M.J., and Favero, M.S. (2004). Nosocomial infections with hemodialysis. In: *Infection Control and Hospital Epidemiology* (ed. C.G. Mayhall), 1139–1160. Philadelphia, PA: Lippincott Williams & Wilkins.

Alvarado, C.J. and Reichelderfer, M. (2000). APIC guideline for infection prevention and control in flexible endoscopy. Association for Professionals in Infection Control. *Am. J. Infect. Control* 28: 138–155.

Alvarado, C.J., Stolz, S.M., and Maki, D.G. (1991). Nosocomial infections from contaminated endoscopes: a flawed automated endoscope washer. An investigation using molecular epidemiology. *Am. J. Med.* 91: 272S–280S.

Amato, R.L. and Curtis, J.M. (2002). The practical application of ozone in dialysis. *Nephrol. News Issues*: 27–29.

American Academy of Ophthalmology (1988). *Updated Recommendations for Ophthalmic Practice in Relation to the Human Immunodeficiency Virus*. San Francisco, CA: American Academy of Ophthalmology.

American Conference of Governmental Industrial Hygienists (ACGIH) (2001). *Threshold Limit Values for Chemical Substances and Physical Agents and Biological Exposure Indices*. Cincinnati, OH: ACGIH.

American Dental Association (1996). Infection control recommendations for the dental office and the dental laboratory. *JADA* 127: 672–680.

American Institute of Architects (2001). *Guidelines for Design and Construction of Hospital and Health Care Facilities*. Washington, DC: The American Institute of Architects Press.

American Society for Gastrointestinal Endoscopy (1996). Position statement: reprocessing of flexible gastrointestinal endoscopes. *Gastrointest. Endosc.* 43: 541–546.

American Society for Healthcare Central Service Professionals (2001). *Training Manual for Health Care Central Service Technicians* (ed. Association AH), 1–271. Chicago, IL: The Jossey-Bass/American Hospital Association Press Series.

American Society for Hospital Central Service Personnel (2001). *Training Manual for Central Service Technicians*, 1–271. Chicago, IL: American Hospital Association.

Amis, S., Ruddy, M., Kibbler, C.C. et al. (2000). Assessment of condoms as probe covers for transvaginal sonography. *J. Clin. Ultrasound* 28: 295–298.

Anderson, R.L., Holland, B.W., Carr, J.K. et al. (1990). Effect of disinfectants on pseudomonads colonized on the interior surface of PVC pipes. *Am. J. Public Health* 80: 17–21.

Anderson, R.L., Vess, R.W., Carr, J.H. et al. (1991). Investigations of intrinsic *Pseudomonas cepacia* contamination in commercially manufactured povidone-iodine. *Infect. Control Hosp. Epidemiol.* 12: 297–302.

Anderson, R.L., Carr, J.H., Bond, W.W., and Favero, M.S. (1997). Susceptibility of vancomycin-resistant enterococci to environmental disinfectants. *Infect. Control Hosp. Epidemiol.* 18: 195–199.

Andres, M.T., Tejerina, J.M., and Fierro, J.F. (1995). Reliability of biologic indicators in a mail-return sterilization-monitoring service: a review of 3 years. *Quintessence Int.* 26: 865–870.

Angelillo, I.F., Bianco, A., Nobile, C.G., and Pavia, M. (1998). Evaluation of the efficacy of glutaraldehyde and peroxygen for disinfection of dental instruments. *Lett. Appl. Microbiol.* 27: 292–296.

Anonymous (1984). Recommendations for decontaminating manikins used in cardiopulmonary resuscitation training, 1983 update. *Infect. Control* 5: 399–401.

Anonymous (1992). *Glutaraldehyde*. New York: Occupational Health Services, Inc.

Anonymous (1993). *Memorandum of Understanding Between the Food and Drug Administration*. Public Health Service, and the Environmental Protection Agency.

Anonymous (1994). Ethylene oxide sterilization: How hospitals can adapt to the changes. *Health Devices* 23: 485–492.

Anonymous (1999). Dental unit waterlines: approaching the Year 2000. ADA Council on Scientific Affairs. *JADA* 130: 1653–1664.

Anonymous (2001). *Hydrogen Peroxide, ACS Reagent*. Vol. 2001: Sigma Product Information Sheet [This link is no longer active: http://www.sigma.sial.com/sigma/proddata/h0904.htm] (accessed 21 August 2021).

Ansari, S.A., Sattar, S.A., Springthorpe, V.S. et al. (1988). Rotavirus survival on human hands and transfer of infectious virus to animate and nonporous inanimate surfaces. *J. Clin. Microbiol.* 26: 1513–1518.

Ansari, S.A., Spingthorpe, S., and Sattar, S.A. (1991). Survival and vehicular spread of human rotaviruses: possible relation to seasonality of outbreaks. *Rev. Infect. Dis.* 13: 448–461.

ANSI/AAMI ST46:2002 (2002). *Steam Sterilization and Sterility Assurance in Health Care Facilities*. Arlington, VA: Association for the Advancement of Medical Instrumentation.

ANSI/AAMI ST79 (2006). *Comprehensive Guide to Steam Sterilization and Sterility Assurance in Health Care Facilities*. Association for the Advancement of Medical Instrumentation.

Ascenzi, J.M. (1991). Standardization of tuberculocidal testing of disinfectants. *J. Hosp. Infect.* 18: 256–263.

Ascenzi, J.M., Ezzell, R.J., and Wendt, T.M. (1987). A more accurate method for measurement of tuberculocidal activity of disinfectants. *Appl. Environ. Microbiol.* 53: 2189–2192.

Association for peri-Operative Registered Nurses (2002). Recommended practices for cleaning and caring for surgical instruments and powered equipment. *AORN J.* 75: 727–741.

Association for peri-Operative Registered Nurses (2006). Recommended pracice for sterilization in the perioperative practice setting. *AORN J.* 83: 700–722.

Association for the Advancement of Medical Instrumentation (1988). *Good Hospital Practice: Steam Sterilization and Sterility Assurance.* Arlington, VA: AAMI.

Association for the Advancement of Medical Instrumentation (1993). *Good Hospital Practice: Steam Sterilization and Sterility Assurance.* Arlington, VA: AAMI.

Association for the Advancement of Medical Instrumentation (1995). *Safe Use and Handling of Glutaraldehyde-Based Products in Healthcare Facilities.* Arlington, VA: AAMI.

Association for the Advancement of Medical Instrumentation (1996). *Flash Sterilization: Steam Sterilization of Patient Care Items for Immediate Use.* Arlington, VA: AAMI.

Association for the Advancement of Medical Instrumentation (1999). *Ethylene Oxide Sterilization in Health Care Facilities: Safety and Effectiveness.* Arlington, VA: AAMI.

Association for the Advancement of Medical Instrumentation (2003). *Reuse of Hemodialyzers*, 1–32. Arlington VA: Association for the Advancement of Medical Instrumentation. 2002/2003:ANSI/AAMI RD47:2002 & RD47:2002/A1:2003.

Association for the Advancement of Medical Instrumentation (2003). Technical Information Report on process challenge devices/test packs for use in health care facilities.

Association of Operating Room Nurses (1987). AORN standards and recommended practices for perioperative nursing. Section III:14.1-III:14.11. Denver, CO: AORN.

Association of Operating Room Nurses (2000). *Recommended Practices for Use and Care of Endoscopes. 2000 Standards, Recommended Practices, and Guidelines*, 243–247. Denver, CO: AORN.

Association of Operating Room Nurses (2000). *Recommended Practices for Sterilization in Perioperative Practice Settings. 2000 Standards, Recommended Practices, and Guidelines*, 347–358. Denver, CO: AORN.

Association of peri-Operative Registered Nurses (2005). Recommended practices for high-level disinfection. *AORN J.* 81: 402–412.

Association of peri-Operative Registered Nurses (2006). Recommended practices for sterilization in the perioperative practice setting. *AORN J.* 83: 700–722.

Atlas, R.M. (1999). *Legionella*: from environmental habitats to disease pathology, detection and control. *Environ. Microbiol.* 1: 283–293.

Aton, E.A., Murray, P., Fraser, V. et al. (1994). Safety of reusing cardiac electrophysiology catheters. *Am. J. Cardiol.* 74: 1173–1175.

Avitall, B., Khan, M., Krum, D. et al. (1993). Repeated use of ablation catheters: a prospective study. *J. Am. Coll. Cardiol.* 22: 1367–1372.

Ayliffe, G.A.J., Collins, D.M., and Lowbury, E.J.L. (1966). Cleaning and disinfection of hospital floors. *Brit. Med. J.* 2: 442–445.

Ayliffe, G.A., Collins, B.J., and Lowbury, E.J. (1966). Cleaning and disinfection of hospital floors. *BMJ* 5511: 442–445.

Ayliffe, G.A., Collins, B.J., Lowbury, E.J. et al. (1967). Ward floors and other surfaces as reservoirs of hospital infection. *J. Hyg. (Lond).* 65: 515–536.

Ayliffe, G.A., Babb, J.R., and Bradley, C.R. (1986). Disinfection of endoscopes. *J. Hosp. Infect.* 7: 296–299.

Babb, J.R. and Bradley, C.R. (1995). Endoscope decontamination: where do we go from here? *J. Hosp. Infect.* 30: 543–551.

Babb, R.R. and Paaso, B.T. (1995). Glutaraldehyde proctitis. *West. J. Med.* 163: 477–478.

Babb, J.R., Bradley, C.R., and Ayliffe, G.A.J. (1980). Sporicidal activity of glutaraldehydes and hypochlorites and other factors influencing their selection for the treatment of medical equipment. *J. Hosp. Infect.* 1: 63–75.

Babb, J.R., Bradley, C.R., Deverill, C.E. et al. (1981). Recent advances in the cleaning and disinfection of fibrescopes. *J. Hosp. Infect.* 2: 329–340.

Baird, R.M. (1999). Sterility assurance: concepts, methods and problems. In: *Principles and Practice of Disinfection, Preservation and Sterilization* (eds. A.D. Russell, W.B. Hugo and G.A.J. Ayliffe), 787–799. Oxford, England: Blackwell Scientific Publications.

Baker, J.L., Kelen, G.D., Sivertson, K.T., and Quinn, T.C. (1987). Unsuspected human immunodeficiency virus in critically ill emergency patients. *JAMA* 257: 2609–2611.

Bar, W., Marquez de Bar, G., Naumann, A., and Rusch-Gerdes, S. (2001). Contamination of bronchoscopes with *Mycobacterium tuberculosis* and successful sterilization by low-temperature hydrogen peroxide plasma sterilization. *Am. J. Infect. Control* 29: 306–311.

Barbee, S.L., Weber, D.J., Sobsey, M.D., and Rutala, W.A. (1999). Inactivation of *Cryptosporidium parvum* oocyst infectivity by disinfection and sterilization processes. *Gastrointest. Endosc.* 49: 605–611.

Bargmann, L.S., Bargmann, B.C., Collier, J.P. et al. (1999). Current sterilization and packaging methods for polyethylene. *Clin. Orthop.*: 49–58.

Barker, J., Vipond, I.B., and Bloomfield, S.F. (2004). Effects of cleaning and disinfection in reducing the spread of *Norovirus contamination* via environmental surfaces. *J. Hosp. Infect.* 58: 42–49.

Barone, M.A., Faisel, A.J., Andrews, L. et al. (1997). Adaptation and validation of a portable steam sterilizer for processing intrauterine device insertion instruments and supplies in low-resource settings. *Am. J. Infect. Control* 25: 350–356.

Barrett, T. (2001). Flash sterilization: what are the risks? In: *Disinfection, Sterilization and Antisepsis: Principles and Practices in Healthcare Facilities* (ed. W.A. Rutala), 70–76. Washington, DC: Association for Professional in Infection Control and Epidemiology.

Bates, C.J. and Pearse, R. (2005). Use of hydrogen peroxide vapour for environmental control during a Serratia outbreak in a neonatal intensive care unit. *J. Hosp. Infect.* 61: 364–366.

Baxter, R.L., Baxter, H.C., Campbell, G.A. et al. (2006). Quantitaive analysis of residual protein contamination on reprocessed surgical instruments. *J. Hosp. Infect.* 63: 439–444.

Bean, H.S. (1967). Types and characteristics of disinfectants. *J. Appl. Bacteriol.* 30: 6–16.

Beauchamp, R.O., St Clair, M.B., Fennell, T.R. et al. (1992). A critical review of the toxicology of glutaraldehyde. *Crit. Rev. Toxicol.* 22: 143–174.

Beck-Sague, C.M. and Jarvis, W.R. (1989). Epidemic bloodstream infections associated with pressure transducers: a persistent problem. *Infect. Control Hosp. Epidemiol.* 10: 54–59.

Beekmann, S.E., Vlahov, D., Koziol, D.E. et al. (1994). Temporal association between implementation of universal precautions and a sustained, progressive decrease in percutaneous exposures to blood. *Clin. Infect. Dis.* 18: 562–569.

Benson, W.G. (1984). Exposure to glutaraldehyde. *J. Soc. Occup. Med.* 34: 63–64.

Bergan, T. and Lystad, A. (1971). Antitubercular action of disinfectants. *J. Appl. Bacteriol.* 34: 751–756.

Berkelman, R.L., Lewin, S., Allen, J.R. et al. (1981). Pseudobacteremia attributed to contamination of povidone-iodine with *Pseudomonas cepacia*. *Ann. Intern. Med.* 95: 32–36.

Berkelman, R.L., Holland, B.W., and Anderson, R.L. (1982). Increased bactericidal activity of dilute preparations of povidone-iodine solutions. *J. Clin. Microbiol.* 15: 635–639.

Bernhang, A.M. (1987). Clostridium pyoarthrosis following arthroscopy. *Arthroscopy* 3: 56–58.

Berrington, A.W. and Pedler, S.J. (1998). Investigation of gaseous ozone for MRSA decontamination of hospital side-rooms. *J. Hosp. Infect.* 40: 61–65.

Best, M. (1994). *Development of a Combined Carrier Test for Disinfectant Efficacy.* Ottawa: University of Ottawa.

Best, M., Sattar, S.A., Springthorpe, V.S., and Kennedy, M.E. (1988). Comparative mycobactericidal efficacy of chemical disinfectants in suspension and carrier tests. *Appl. Environ. Microbiol.* 54: 2856–2858.

Best, M., Sattar, S.A., Springthorpe, V.S., and Kennedy, M.E. (1990). Efficacies of selected disinfectants against *Mycobacterium tuberculosis*. *J. Clin. Microbiol.* 28: 2234–2239.

Best, M., Kennedy, M.E., and Coates, F. (1990). Efficacy of a variety of disinfectants against *Listeria* spp. *Appl. Environ. Microbiol.* 56: 377–380.

Best, M., Springthorpe, V.S., and Sattar, S.A. (1994). Feasibility of a combined carrier test for disinfectants: studies with a mixture of five types of microorganisms. *Am. J. Infect. Control* 22: 152–162.

Bhalla, A., Pultz, N.J., Gries, D.M. et al. (2004). Acquisition of nosocomial pathogens on hands after contact with environmental surfaces near hospitalized patients. *Infect. Control Hosp. Epidemiol.* 25: 164–167.

Bhattachatyya, M. and Kepnes, L.J. (2004). The effectiveness of immersion disinfection for flexible fiberoptic laryngoscopes. *Otolaryngol. Head Neck* 130: 681–685.

Bilotta, J.J. and Waye, J.D. (1989). Hydrogen peroxide enteritis: the "snow white" sign. *Gastrointest. Endosc.* 35: 428–430.

Bland, L.A. and Favero, M.S. (1990). *Microbial Contamination Control Strategies for Hemodialysis System*, Plant, Technology & Safety Management Series: Infection Control Issues in PTSM. Oakbrook Terrace, IL.

Blob, R. and Kampf, G. (2004). Test models to determine cleaning efficacy with different types of bioburden and its clinical correlation. *J. Hosp. Infect.* 56 (Suppl): S44–S48.

Block, S.S. (2001). *Disinfection, Sterilization, and Preservation*. Philadelphia, PA: Lippincott Williams & Wilkins.

Block, S.S. (2001). Peroxygen compounds. In: *Disinfection, Sterilization, and Preservation* (ed. S.S. Block), 185–204. Philadelphia, PA: Lippincott Williams & Wilkins.

Block, S.S. (2001). Definition of terms. In: *Disinfection, Sterilization, and Preservation* (ed. S.S. Block), 19–28. Philadelphia, PA: Lippincott Williams & Wilkins.

Block, C. (2004). The effect of Perasafe® and sodium dichloroisocyanurate (NaDCC) against spores of *Clostridium difficile* and *Bacillus atrophaeus* on stainless steel and polyvinyl chloride surfaces. *J. Hosp. Infect.* 57: 144–148.

Bloomfield, S.F. and Miller, E.A. (1989). A comparison of hypochlorite and phenolic disinfectants for disinfection of clean and soiled surfaces and blood spillages. *J. Hosp. Infect.* 13: 231–239.

Bloomfield, S.F. and Uso, E.E. (1985). The antibacterial properties of sodium hypochlorite and sodium dichloroisocyanurate as hospital disinfectants. *J. Hosp. Infect.* 6: 20–30.

Bond, W.W. (1987). Virus transmission via fiberoptic endoscope: recommended disinfection. *JAMA* 257: 843–844.

Bond, W.W. (1991). Disinfection and endoscopy: microbial considerations. *J. Gastroenterol. Hepatol.* 6: 31–36.

Bond, W.W. (1993). Biological indicators for a liquid chemical sterilizer. *Infect. Control Hosp. Epidemiol.* 14: 565.

Bond, W.W. (1993). Biological indicators for a liquid chemical sterilizer: a solution to the instrument reprocessing problem? *Infect. Control Hosp. Epidemiol.* 14: 309–312.

Bond, W.W. (1998). Endoscope reprocessing: problems and solutions. In: *Disinfection, Sterilization, and Antisepsis in Healthcare* (ed. W.A. Rutala), 151–163. Champlain, NY: Polyscience Publications.

Bond, W.W. and Hedrick, E.R. (1992). Microbiological culturing of environmental and medical-device surfaces. In: *Clinical Microbiology Procedures Handbook, Section 11, Epidemiologic and Infection Control Microbiology* (eds. H.D. Isenberg and M.J.R. Gilchrist), 11.10.1–11.10.9. Washington, DC: American Society for Microbiology.

Bond, W.W., Favero, M.S., Petersen, N.J., and Ebert, J.W. (1983). Inactivation of hepatitis B virus by intermediate-to-high-level disinfectant chemicals. *J. Clin. Microbiol.* 18: 535–538.

Bonten, M.J.M., Hayden, M.J., Nathan, C. et al. (1996). Epidemiology of colonisation of patients and environment with vancomycin-resistant enterococci. *Lancet* 348: 1615–1619.

Crystal Booth. *Method Development and Validation for the Pharmaceutical Microbiologist.*

Borick, P.M., Dondershine, F.H., and Chandler, V.L. (1964). Alkalinized glutaraldehyde, a new antimicrobial agent. *J. Pharm. Sci.* 53: 1273–1275.

Borneff, M., Ruppert, J., Okpara, J. et al. (1995). Efficacy testing of low-temperature plasma sterilization (LTP) with test object models simulating practice conditions. *Zentr. Steril.* 3: 361–371.

Borneff-Lipp, M., Okpara, J., Bodendorf, M., and Sonntag, H.G. (1997). Validation of low-temperature-plasma (LPT) sterilization systems: comparison of two technical versions, the Sterrad 100, 1.8 and the 100S. *Hyg. Mikrobiol.* 3: 21–28.

Boucher, R.M. (1974). Potentiated acid 1,5 pentanedial solution–a new chemical sterilizing and disinfecting agent. *Am. J. Hosp. Pharm.* 31: 546–557.

Boyce, J.M., Potter-Bynoe, G., Chenevert, C., and King, T. (1997). Environmental contamination due to methicillin-resistant *Staphylococcus aureus*: possible infection control implications. *Infect. Control Hosp. Epidemiol.* 18: 622–627.

Boyce, J.M., Havill, N.L., Otter, J.A. et al. (2006). Impact of hydrogen peroxide vapor room bio-decontamination on environmental contamination and nosocomial transmission of *Clostridium difficile*. *Soc. Healthc. Epidemiol. Am.* Abstract 155: 109.

Bradley, C.R. and Babb, J.R. (1995). Endoscope decontamination: automated vs. manual. *J. Hosp. Infect.* 30: 537–542.

Bradley, C.R. and Fraise, A.P. (1996). Heat and chemical resistance of enterococci. *J. Hosp. Infect.* 34: 191–196.

Bradley, C.R., Babb, J.R., and Ayliffe, G.A. (1995). Evaluation of the Steris System 1 peracetic acid endoscope processor. *J. Hosp. Infect.* 29: 143–151.

Brady, M.J., Lisay, C.M., Yurkovetskiy, A.V., and Sawan, S.P. (2003). Persistent silver disinfectant for the environmental control of pathogenic bacteria. *Am. J. Infect. Control* 31: 208–214.

Brazier, J.G. (1998). The diagnosis of *Clostridium difficile*-associated disease. *J. Antimicrob. Chemother.* 41 (Suppl): 29–40.

Brazis, A.R., Leslie, J.E., Kabler, P.W., and Woodward, R.L. (1958). The inactivation of spores of *Bacillus globigii* and *Bacillus anthracis* by free available chlorine. *Appl. Microbiol.* 6: 338–342.

Bright, K.R., Gerba, C.P., and Rusin, P.A. (2002). Rapid reduction of *Staphylococcus aureus* populations on stainless stell surfaces by zeolite ceramic coatings containing silver and zinc ions. *J. Hosp. Infect.* 52: 307–309.

British Society of Gastroenterology (1998). Cleaning and disinfection of equipment for gastrointestinal endoscopy. Report of a working party of the British Society of Gastroenterology Endoscope Committee. *Gut* 42: 585–593.

British Thoracic Society (2001). British Thoracic Society guidelines on diagnostic flexible bronchoscopy. *Thorax* 56: 1–21.

Broadley, S.J., Furr, J.R., Jenkins, P.A., and Russell, A.D. (1993). Antimycobacterial activity of 'Virkon'. *J. Hosp. Infect.* 23: 189–197.

Bronowicki, J.P., Venard, V., Botte, C. et al. (1997). Patient-to-patient transmission of hepatitis C virus during colonoscopy. *N. Engl. J. Med.* 337: 237–240.

Brown, D.G., Skylis, T.P., and Fekety, F.R. (1983). Comparison of chemical sterilant/disinfectant solutions against spores of *Clostridium difficile. Abstract. Am. Soc. Microbiol.* Q39: 267.

Brown, M.L., Aldrich, H.C., and Gauthier, J.J. (1995). Relationship between glycocalyx and povidone-iodine resistance in *Pseudomonas aeruginosa* (ATCC 27853) biofilms. *Appl. Environ. Microbiol.* 61: 187–193.

Brown, S.A., Merritt, K., Woods, T.O. et al. (2002). Effects of different disinfection and sterilization methods on tensile strength of materials used for single-use devices. *Biomed. Instrum. Technol.*: 23–27.

Bruckner, N.I., Gordon, M.D., and Howell, R.G. (1989). Odorless aromatic dialdehyde disinfecting and sterilizing composition. US Patent 4, 851, 449. July, 1989.

Brumfitt, W., Dixson, S., and Hamilton-Miller, J.M. (1985). Resistance to antiseptics in methicillin and gentamicin resistant *Staphylococcus aureus. Lancet* 1: 1442–1443.

Bryce, E.A., Roberts, F.J., Clements, B., and MacLean, S. (1997). When the biological indicator is positive: investigating autoclave failures. *Infect. Control Hosp. Epidemiol.* 18: 654–656.

Bryce, E.A., Chia, E., Logelin, G., and Smith, J.A. (1997). An evaluation of the AbTox Plazlyte Sterilization System. *Infect. Control Hosp. Epidemiol.* 18: 646–653.

Brystrom, A. and Sundqvist, G. (1983). Bacteriologic evaluation of the effect of 0.5 percent sodium hypochlorite in endodontic therapy. *Oral Surg. Oral Med. Oral Pathol.* 55: 307–312.

Bucx, M.J., Veldman, D.J., Beenhakker, M.M., and Koster, R. (2000). The effect of steam sterilization at 134 degrees C on light intensity provided by fibrelight *Macintoch laryngoscopes. Anaesthesia* 55: 185–186.

Bueumer, R., Bloomfield, S.F., Exner, M. et al. (1999). The need for a home hygiene policy and guidelines on home hygiene. *Ann. Ig.* 11: 11–26.

Burgess, D.J. and Reich, R.R. (1993). Industrial ethylene oxide sterilization. In: *Sterilization Technology: A Practical Guide for Manufacturers and Users of Health Care Product* (eds. R.F. Morrissey and G.B. Phillips), 120–151. New York: Van Nostrand Reinhold.

Burns, S., Edwards, M., Jennings, J. et al. (1996). Impact of variation in reprocessing invasive fiberoptic scopes on patient outcomes. *Infect. Control Hosp. Epidemiol.* 17 (Suppl): P42.

Burtin, P., Ruget, O., Petit, R., and Boyer, J. (1993). Glutaraldehyde-induced proctitis after endorectal ultrasound examination: a higher risk of incidence than expected? *Gastrointest. Endosc.* 39: 859–860.

Butcher, W. and Ulaeto, D. (2005). Contact inactivation of orthopoxviruses by household disinfectants. *J. Appl. Microbiol.* 99: 279–284.

Butt, W.E., Bradley, D.V. Jr., Mayhew, R.B., and Schwartz, R.S. (1991). Evaluation of the shelf life of sterile instrument packs. *Oral Surg. Oral Med. Oral Pathol.* 72: 650–654.

Buxton, A.E., Anderson, R.L., Werdegar, D., and Atlas, E. (1978). Nosocomial respiratory tract infection and colonization with *Acinetobacter calcoaceticus*. Epidemiologic characteristics. *Am. J. Med.* 65: 507–513.

Byers, K.E., Durbin, L.J., Simonton, B.M. et al. (1998). Disinfection of hospital rooms contaminated with vancomycin-resistant *Enterococcus faecium*. *Infect. Control Hosp. Epidemiol.* 19: 261–264.

Cabrera-Martinez, R.M., Setlow, B., and Setlow, P. (2002). Studies on the mechanisms of the sporicidal action of ortho-phthalaldehyde. *J. Appl. Microbiol.* 92: 675–680.

Cardenas-Camarena, L. (1998). Ethylene oxide burns from improperly sterilized mammary implants. *Ann. Plast. Surg.* 41: 361–369.

Cardo, D.M. and Sehulster, L.M. (1999). Central sterile supply. In: *Infection Control and Hospital Epidemiology* (ed. C.G. Mayhall), 1023–1030. Philadelphia, PA: Lippincott Williams & Wilkins.

Carling, P.C., Briggs, J.L., Perkins, J., and Highlander, D. (2006). Improved cleaning of patient rooms using a new targeting method. *Clin. Infect. Dis.* 42: 385–388.

Carsauw, H. and Debacker, N. (2002). Recall of patients after use of inactive batch of Cidex disinfection solution in Belgian hospitals. *5th International Conference of the Hospital Infection Society*, Edinburgh (15–18 September 2002). Hospital Infections Society.

Carson, J.A. and Favero, M.S. (1984). Comparative resistance of nontuberculous mycobacteria to iodophor germicides. *Abstract. Annu. Meet. Am. Soc. Microbiol.* Q101: 221.

Casemore, D.P., Blewett, D.A., and Wright, S.E. (1989). Cleaning and disinfection of equipment for gastrointestinal flexible endoscopy: interim recommendations of a Working Party of the British Society of Gastroenterology. *Gut* 30: 1156–1157.

Castelli, M., Qizilbash, A., and Seaton, T. (1986). Post-colonoscopy proctitis. *Am. J. Gastroenterol.* 81: 887.

Castillo, A., Lucia, L.M., Kemp, G.K., and Acuff, G.R. (1999). Reduction of *Escherichia coli* O157:H7 and *Salmonella typhimurium* on beef carcass surfaces using acidified sodium chlorite. *J. Food Prot.* 62: 580–584.

Cavagnolo, R.Z. (1985). Inactivation of herpesvirus on CPR manikins utilizing a currently recommended disinfecting procedure. *Infect. Control* 6: 456–458.

Cefai, C., Richards, J., Gould, F.K., and McPeake, P. (1990). An outbreak of respiratory tract infection resulting from incomplete disinfection of ventilatory equipment. *J. Hosp. Infect.* 15: 177–182.

Centers for Disease Control (1981). False-positive results of spore tests in ethylene oxide sterilizers – Wisconsin. *MMWR* 30: 238–240.

Centers for Disease Control (1982). Acquired immune deficiency syndrome (AIDS): precautions for clinical and laboratory staffs. *MMWR* 31: 577–580.

Centers for Disease Control (1983). Formaldehyde exposures in a gross anatomy laboratory – Colorado. *MMWR* 52: 698–700.

Centers for Disease Control (1985). Recommendations for preventing possible transmission of human T-lymphotropic virus type III/lymphadenopathy-associated virus from tears. *MMWR* 34: 533–534.

Centers for Disease Control (1986). Bacteremia associated with reuse of disposable hollow-fiber hemodialyzers. *MMWR* 35: 417–418.

Centers for Disease Control (1986). Occupational exposures to formaldehyde in dialysis units. *MMWR* 35: 399–301.

Centers for Disease Control (1987). Recommendations for prevention of HIV transmission in health-care settings. *MMWR* 36: S3–S18.

Centers for Disease Control (1987). Symptoms of irritation associated with exposure to glutaraldehyde. *MMWR* 36: 190–191.

Centers for Disease Control (1989). Guidelines for prevention of transmission of human immunodeficiency virus and hepatitis B virus to health-care and public-safety workers. *MMWR* 38: 1–37.

Centers for Disease Control (1991). Federal regulatory action against sporicidin cold sterilizing solution. *MMWR* 40: 880–881.

Centers for Disease Control (1993). Recommended infection-control practices for dentistry, 1993. *MMWR* 41: 1–12.

Centers for Disease Control (1993). Use of bleach for disinfection of drug injection equipment. *MMWR* 42: 418–419.

Centers for Disease Control (1998). Ambulatory and inpatient procedures in the United States, 1996. Atlanta, GA, pp. 1–39.

Centers for Disease Control (2000). Biological and chemical terrorism: strategic plan for preparedness and response. *MMWR* 49 (RR-4): 1–14.

Centers for Disease Control (2001). Recommendations for preventing transmission of infections among chronic hemodialysis patients. *MMWR* 50: 1–43.

Centers for Disease Control (2003). Guidelines for environmental infection control in health-care facilities, 2003. *MMWR* 52 (RR-10): 1–44.

Centers for Disease Control and Prevention (1994). Preventing the spread of vancomycin resistance – report from the Hospital Infection Control Practices Advisory Committee. *Fed. Regist.*: 25758–25763.

Centers for Disease Control and Prevention (1998). Corneal decompensation after intraocular ophthalmic surgery-Missouri, 1998. *MMWR* 47: 306–309.

Centers for Disease Control and Prevention (1999). Bronchoscopy-related infections and pseudoinfections – New York, 1996 and 1998. *MMWR* 48: 557–560.

Centers for Disease Control and Prevention (2003). Update: severe acute respiratory syndrome – United States, May 14, 2003. *MMWR* 52: 436–438.

Centers for Disease Control and Prevention (2006). *Pseudomonas aeruginosa* infections associated with transrectal ultrasound-guided prostate biopsies–Georgia, 2005. *MMWR CDC Surveill. Summ.* 55: 776–777.

Chadwick, P.R. and Oppenheim, B.A. (1994). Vancomycin-resistant enterococci and bedpan washer machines. *Lancet* 344: 685.

Chang, S.L. (1971). Modern concept of disinfection. *J. Sanit. Eng. Div. Proc. Am. Soc. Civ. Eng.*: 689–705.

Chan-Myers, H. (2000). Sporicidal activity of ortho-phthalaldehyde as a function of temperature (abstract). *Infect. Control Hosp. Epidemiol.* 21: 101.

Chan-Myers, H. and Roberts, C. (2000). Effect of temperature and organic soil concentration on biocidal activity of ortho-phthalaldehyde solution (abstract). *2000 Education Meeting of the Association for Professional in Infection Control and Epidemiology*, Minneapolis, MN, p. 31.

Chan-Myers, H., McAlister, D., and Antonoplos, P. (1997). Natural bioburden levels detected on rigid lumened medical devices before and after cleaning. *Am. J. Infect. Control* 25: 471–476.

Chan-Yeung, M., McMurren, T., Catonio-Begley, F., and Lam, S. (1993). Occupational asthma in a technologist exposed to glutaraldehyde. *J. Allergy Clin. Immunol.* 91: 974–978.

Chanzy, B., Duc-Bin, D.L., Rousset, B. et al. (1999). Effectiveness of a manual disinfection procedure in eliminating hepatitis C virus from experimentally contaminated endoscopes. *Gastrointest. Endosc.* 50: 147–151.

Chapin, M., Yatabe, J., and Cherry, J.D. (1983). An outbreak of rotavirus gastroenteritis on a pediatric unit. *Am. J. Infect. Control* 11: 88–91.

Chataigner, D., Garnier, R., Sans, S., and Efthymiou, M.L. (1991). Acute accidental poisoning with hospital disinfectant. 45 cases of which 13 with fatal outcome. *Presse Med.* 20: 741–743.

Chaufour, X., Deva, A.K., Vickery, K. et al. (1999). Evaluation of disinfection and sterilization of reusable angioscopes with the duck hepatitis B model. *J. Vasc. Surg.* 30: 277–282.

Cheung, H.Y. and Brown, M.R. (1982). Evaluation of glycine as an inactivator of glutaraldehyde. *J. Pharm. Pharmacol.* 34: 211–214.

Cheung, R.J., Ortiz, D., and DiMarino, A.J. Jr. (1999). GI endoscopic reprocessing practices in the United States. *Gastrointest. Endosc.* 50: 362–368.

Chitnis, V., Chitnis, S., Patil, S., and Chitnis, D. (2004). Practical limitations of disinfection of body fluid spills with 10,000 ppm sodium hypochlorite (NaOCl). *Am. J. Infect. Control* 32: 306–308.

Chobin, N., Furr, D., and Nuyttens, A. (2004). Wet packs and plastic accessory cases. *Infect. Control Today* 24: 28–30.

Chronister, C.L. (1997). Structural damage to Schiotz tonometers after disinfection with solutions. *Optom. Vis. Sci.* 74: 164–166.

Chronister, C.L. and Russo, P. (1990). Effects of disinfecting solutions on tonometer tips. *Optom. Vis. Sci.* 67: 818–821.

Chu, N.S. and Favero, M. (2000). The microbial flora of the gastrointestinal tract and the cleaning of flexible endoscopes. *Gastrointest. Endosc. Clin. N. Am.* 10: 233–244.

Chu, N.S., McAlister, D., and Antonoplos, P.A. (1998). Natural bioburden levels detected on flexible gastrointestinal endoscopes after clinical use and manual cleaning. *Gastrointest. Endosc.* 48: 137–142.

Chu, N.S., Chan-Myers, H., Ghazanfari, N., and Antonoplos, P. (1999). Levels of naturally occurring microorganisms on surgical instruments after clinical use and after washing. *Am. J. Infect. Control* 27: 315–319.

Chuanchuen, R., Karkhoff-Schweizer, R.R., and Schweizer, H.P. (2003). High-level triclosan resistance in *Pseudomonas aeruginosa* is solely a result of efflux. *Am. J. Infect. Control* 31: 124–127.

Cleaning and Cleaning Validation: A Biotechnology Perspective.

Coates, D. (1985). A comparison of sodium hypochlorite and sodium dichloroisocyanurate products. *J. Hosp. Infect.* 6: 31–40.

Coates, D. (1988). Comparison of sodium hypochlorite and sodium dichloroisocyanurate disinfectants: neutralization by serum. *J. Hosp. Infect.* 11: 60–67.

Coates, D. (1991). Disinfection of spills of body fluids: how effective is a level of 10,000 ppm available chlorine? *J. Hosp. Infect.* 18: 319–322.

Coates, D. (1992). Disinfectants and spills of body fluids. *Nurs. RSA* 7: 25–27.

Coates, D. (2001). An evaluation of the use of chlorine dioxide (Tristel One-Shot) in an automated washer/disinfector (Medivator) fitted with a chlorine dioxide generator for decontamination of flexible endoscopes. *J. Hosp. Infect.* 48: 55–65.

Coates, D. and Wilson, M. (1989). Use of sodium dichloroisocyanurate granules for spills of body fluids. *J. Hosp. Infect.* 13: 241–251.

Coates, D. and Wilson, M. (1992). Powders, composed of chlorine-releasing agent acrylic resin mixtures or based on peroxygen compounds, for spills of body fluids. *J. Hosp. Infect.* 21: 241–252.

Cokendolpher, J.C. and Haukos, J.F. (1996). *The Practical Application of Disinfection and Sterilization in Health Care Facilities*. Chicago, IL: American Hospital Association.

Cole, E.C. and Rutala, W.A. (1988). Bacterial numbers on penicylinders used in disinfectant testing: use of 24 hour adjusted broth cultures. *J. Assoc. Off. Anal. Chem.* 71: 9–11.

Cole, E.C., Rutala, W.A., and Carson, J.L. (1987). Evaluation of penicylinders used in disinfectant testing: bacterial attachment and surface texture. *J. Assoc. Off. Anal. Chem.* 70: 903–906.

Cole, E.C., Rutala, W.A., and Samsa, G.P. (1987). Standardization of bacterial numbers of penicylinders used in disinfectant testing: interlaboratory study. *J. Assoc. Off. Anal. Chem.* 70: 635–637.

Cole, E.C., Rutala, W.A., and Samsa, G.P. (1988). Disinfectant testing using a modified use-dilution method: collaborative study. *J. Assoc. Off. Anal. Chem.* 71: 1187–1194.

Cole, E.C., Rutala, W.A., and Alfano, E.M. (1988). Comparison of stainless steel penicylinders used in disinfectant testing. *J. Assoc. Off. Anal. Chem.* 71: 288–289.

Cole, E.C., Rutala, W.A., Carson, J.L., and Alfano, E.M. (1989). Pseudomonas pellicle in disinfectant testing: electron microscopy, pellicle removal, and effect on test results. *Appl. Environ. Microbiol.* 55: 511–513.

Cole, E.C., Rutala, W.A., Nessen, L. et al. (1990). Effect of methodology, dilution, and exposure time on the tuberculocidal activity of glutaraldehyde-based disinfectants. *Appl. Environ. Microbiol.* 56: 1813–1817.

Collins, F.M. (1986). Kinetics of the tuberculocidal response by alkaline glutaraldehyde in solution and on an inert surface. *J. Appl. Bacteriol.* 61: 87–93.

Collins, F.M. (1986). Bactericidal activity of alkaline glutaraldehyde solution against a number of atypical mycobacterial species. *J. Appl. Bacteriol.* 61: 247–251.

Collins, F.M. (1987). Use of membrane filters for measurement of mycobactericidal activity of alkaline glutaraldehyde solution. *Appl. Environ. Microbiol.* 53: 737–739.

Collins, F.M. and Montalbine, V. (1976). Mycobactericidal activity of glutaraldehyde solutions. *J. Clin. Microbiol.* 4: 408–412.

Connor, C.G., Hopkins, S.L., and Salisbury, R.D. (1991). Effectivity of contact lens disinfection systems against *Acanthamoeba culbertsoni*. *Optom. Vis. Sci.* 68: 138–141.

Conviser, C.A. and CW (1997). Ethylene oxide sterilization: sterilant alternatives. In: *Sterilization Technology for the Health Care Facility* (eds. M. Reichert and J.H. Young), 189–199. Gaithersburg, MD: Aspen Publication.

Cooke, R.P., Whymant-Morris, A., Umasankar, R.S., and Goddard, S.V. (1998). Bacteria-free water for automatic washer-disinfectors: an impossible dream? *J. Hosp. Infect.* 39: 63–65.

Cooke, R.P.D., Goddard, S.V., Chatterley, R. et al. (2001). Monitoring glutaraldehyde dilution in automated washer/disinfectors. *J. Hosp. Infect.* 48: 242–246.

Cooke, R.P.D., Goddard, S.V., Whymant-Morris, A. et al. (2003). An evaluation of Cidex OPA (0.55% ortho-phthalaldehyde) as an alternative to 2% glutaraldehyde for high-level disinfection of endoscopes. *J. Hosp. Infect.* 54: 226–231.

Corrado, O.J., Osman, J., and Davies, R.J. (1986). Asthma and rhinitis after exposure to glutaraldehyde in endoscopy units. *Hum. Toxicol.* 5: 325–328.

Corson, S.L., Block, S., Mintz, C. et al. (1979). Sterilization of laparoscopes. Is soaking sufficient? *J. Reprod. Med.* 23: 49–56.

Corson, S.L., Dole, M., Kraus, R. et al. (1979). Studies in sterilization of the laparoscope: II. *J. Reprod. Med.* 23: 57–59.

Costerton, J.S., Steward, P.S., and Greenberg, E.P. (1999). Bacterial biofilms: a common cause of persistent infections. *Science* 284: 1318–1322.

Coulter, W.A., Chew-Graham, C.A., Cheung, S.W., and Burke, F.J.T. (2001). Autoclave performance and operator knowledge of autoclave use in primary care: a survey of UK practices. *J. Hosp. Infect.* 48: 180–185.

Coulthard, C.E. and Sykes, G. (1936). The germicidal effect of alcohol with special reference to its action on bacterial spores. *Pharm. J.* 137: 79–81.

Courtright, P., Lewallen, S., Holland, S.P., and Wendt, T.M. (1995). Corneal decompensation after cataract surgery. An outbreak investigation in Asia. *Ophthalmology* 102: 1461–1465.

Cozad, A. and Jones, R.D. (2003). Disinfection and the prevention of infectious disease. *Am. J. Infect. Control* 31: 243–254.

Craven, D.E., Moody, B., Connolly, M.G. et al. (1981). Pseudobacteremia caused by povidone-iodine solution contaminated with *Pseudomonas cepacia*. *N. Engl. J. Med.* 305: 621–623.

Craven, E.R., Butler, S.L., McCulley, J.P., and Luby, J.P. (1987). Applanation tonometer tip sterilization for adenovirus type 8. *Ophthalmology* 94: 1538–1540.

Cronmiller, J.R., Nelson, D.K., Jackson, D.K., and Kim, C.H. (1999). Efficacy of conventional endoscopic disinfection and sterilization methods against *Helicobacter pylori* contamination. *Helicobacter* 4: 198–203.

Crow, S. (1992). Peracetic acid sterilization: a timely development for a busy healthcare industry. *Infect. Control Hosp. Epidemiol.* 13: 111–113.

Crow, S. (1993). Steam sterilizers: an evolution in design. *Infect. Control Hosp. Epidemiol.* 14: 488–490.

Crow, S., Metcalf, R.W., Beck, W.C., and Birnbaum, D. (1983). Disinfection or sterilization? Four views on arthroscopes. *AORN J.* 37: 854–859, 862-868.

Crystal, H.A., Schaumburg, H.H., Grober, E. et al. (1988). Cognitive impairment and sensory loss associated with chronic low-level ethylene oxide exposure. *Neurology* 38: 567–569.

Cumberland, N.S. and Botting, F.G. (1991). Formaldehyde vapour cabinets. *J. Hosp. Infect.* 19: 67–70.

D'Angelo, G.L. and Ogilvie-Harris, D.J. (1988). Septic arthritis following arthroscopy, with cost/benefit analysis of antibiotic prophylaxis. *Arthroscopy* 4: 10–14.

Dagley, S., Dawes, E.A., and Morrison, G.A. (1950). Inhibition of growth of *Aerobacter aerogenes*: the mode of action of phenols, alcohols, acetone and ethyl acetate. *J. Bacteriol.* 60: 369–378.

Dailey, J.R., Parnes, R.E., and Aminlari, A. (1993). Glutaraldehyde keratopathy. *Am. J. Ophthalmol.* 115: 256–258.

Dancer, S.J. (2004). How do we assess hospital cleaning? A proposal for microbiological standards for surface hygiene in hospitals. *J. Hosp. Infect.* 56: 10–15.

Danforth, D., Nicolle, L.E., Hume, K. et al. (1987). Nosocomial infections on nursing units with floors cleaned with a disinfectant compared with detergent. *J. Hosp. Infect.* 10: 229–235.

Daschner, F. (1994). Steris System 1 in Germany. *Infect. Control Hosp. Epidemiol.* 15: 294, 296.

Daschner, F. (1997). The hospital and pollution: role of the hospital epidemiologist in protecting the environment. In: *Prevention and Control of Nosocomial Infections* (ed. R.P. Wenzel), 595–605. Baltimore, MD: Williams and Wilkins.

Daschner, F. and Schuster, A. (2004). Disinfection and the prevention of infectious disease-no adverse effects? *Am. J. Infect. Control* 32: 224–225.

Daschner, F., Rabbenstein, G., and Langmaack, H. (1980). Surface decontamination in the control of hospital infections: comparison of different methods (author's transl). *Dtsch. Med. Wochenschr.* 105: 325–329.

Dauendorffer, J.N., Laurain, C., Weber, M., and Dailloux, M. (2000). Evaluation of the bactericidal efficiency of a 2% alkaline glutaraldehyde solution on *Mycobacterium xenopi*. *J. Hosp. Infect.* 46: 73–76.

Dempsey, K.M., Chiew, R.F., McKenzie, J.A., and Mitchell, D.H. (2000). Evaluation of the cleaning and disinfection efficacy of the DEKO-190; award-based automated washer/disinfector. *J. Hosp. Infect.* 46: 50–54.

Denton, M., Wilcox, M.H., Parnell, P. et al. (2004). Role of environmental cleaning in controlling an outbreak of *Acinetobacter baumanni* on a neurosurgical intensive care unit. *J. Hosp. Infect.* 56: 106–110.

Department of Health and Human Services. Food and Drug Administration (1992). *Dental Handpiece Sterilization*. Rockville, MD: Food and Drug Administration.

DesCoteaux, J.G., Poulin, E.C., Julien, M., and Guidoin, R. (1995). Residual organic debris on processed surgical instruments. *AORN J.* 62: 23–30.

Dettenkofer, M., Wenzler, S., Amthor, S. et al. (2004). Does disinfection of environmental surfaces influence nosocomial infection rates? A systematic review. *Am. J. Infect. Control* 32: 84–89.

Deva, A.K., Vickery, K., Zou, J. et al. (1996). Establishment of an in-use testing method for evaluating disinfection of surgical instruments using the duck hepatitis B model. *J. Hosp. Infect.* 33: 119–130.

Dharan, S., Mourouga, P., Copin, P. et al. (1999). Routine disinfection of patients' environmental surfaces. Myth or reality? *J. Hosp. Infect.* 42: 113–117.

Di Prima, T., De Pasquale, R., and Nigro, M. (1988). Contact dermatitis from glutaraldehyde. *Contact Derm.* 19: 219–220.

Doan, H.M., Keith, L., and Shennan, A.T. (1979). Phenol and neonatal jaundice. *Pediatrics* 64: 324–325.

Dolce, P., Gourdeau, M., April, N., and Bernard, P.M. (1995). Outbreak of glutaraldehyde-induced proctocolitis. *Am. J. Infect. Control* 23: 34–39.

Dolman, P.J. and Dobrogowski, M.J. (1989). Contact lens disinfection by ultraviolet light. *Am. J. Ophthalmol.* 108: 665–669.

Donlan, R.M. (2001). Biofilms: a source of infection? In: *Disinfection, Sterilization and Antisepsis: Principles and Practices in Healthcare Facilities* (ed. W.A. Rutala), 219–226. Washington, DC: Association for Professional in Infection Control and Epidemiology.

Donlan, R.M. and Costerton, J.W. (2002). Biofilms: survival mechanisms of clinically relevant mirocorganisms. *Clin. Microbiol. Rev.* 15: 167–193.

Douglas, C., Burke, B., Kessler, D.L. et al. (1990). Microwave: practical cost-effective method for sterilizing urinary catheters in the home. *Urology* 35: 219–222.

Doultree, J.C., Druce, J.D., Birch, C.J. et al. (1999). Inactivation of feline calicivirus, a Norwalk virus surrogate. *J. Hosp. Infect.* 41: 51–57.

Doyle, J.E. and Ernst, R.R. (1967). Resistance of *Bacillus subtilis* var. niger spores occluded in water-insoluble crystals to three sterilization agents. *Appl. Microbiol.* 15: 726–730.

Druce, J.D., Russell, J.S., Birch, C.J. et al. (2003). A decontamination and sterilization protocol employed during reuse of cardiac electrophysiology catheters inactivates human immunodeficiency virus. *Infect. Control Hosp. Epidemiol.* 24: 184–190.

Duc, D.L., Ribiollet, A., Dode, X. et al. (2001). Evaluation of the microbicidal efficacy of Steris System I for digestive endoscopes using GERMANDE and ASTM validation protocols. *J. Hosp. Infect.* 48: 135–141.

Duffy, R.E., Brown, S.E., Caldwell, K.L. et al. (2000). An epidemic of corneal destruction caused by plasma gas sterilization. *Arch. Ophthalmol.* 118: 1167–1176.

Dunkelberg, H. and Fleitmann-Glende, F. (2006). Measurement of the microbial barrier effectiveness of sterilization containers in terms of the log reduction value for prevention of nosocomial infections. *Am. J. Infect. Control* 34: 285–289.

Dunne, W.M. (2002). Bacterial adhesion: seen any good biofilms lately? *Clin. Microbiol. Rev.* 15: 155–166.

Dunnigan, A., Roberts, C., McNamara, M. et al. (1987). Success of re-use of cardiac electrode catheters. *Am. J. Cardiol.* 60: 807–810.

Durante, L., Zulty, J.C., Israel, E. et al. (1992). Investigation of an outbreak of bloody diarrhea: association with endoscopic cleaning solution and demonstration of lesions in an animal model. *Am. J. Med.* 92: 476–480.

Dwyer, D.M., Klein, E.G., Istre, G.R. et al. (1987). Salmonella newport infections transmitted by fiberoptic colonoscopy. *Gastrointest. Endosc.* 33: 84–87.

Dyas, A. and Das, B.C. (1985). The activity of glutaraldehyde against *Clostridium difficile*. *J. Hosp. Infect.* 6: 41–45.

Dychdala, G.R. (2001). Chlorine and chlorine compounds. In: *Disinfection, Sterilization, and Preservation* (ed. S.S. Block), 135–157. Philadelphia, PA: Lippincott Williams & Wilkins.

Edens, A.L. (2001). Occupational Safety and Health Administration: regulations affecting healthcare facilities. In: *Disinfection, Sterilization and Antisepsis: Principles and Practices in Healthcare Facilities* (ed. W.A. Rutala), 49–58. Washington, DC: Association for Professionals in Infection Control and Epidemiology, Inc.

Education Design (1998). *Best Practices for the Prevention of Surgical Site Infection*. Denver, CO: Education Design.

Ehrenkranz, N.J., Bolyard, E.A., Wiener, M., and Cleary, T.J. (1980). Antibiotic-sensitive *Serratia marcescens* infections complicating cardiopulmonary operations: contaminated disinfectant as a reservoir. *Lancet* 2: 1289–1292.

Eickhoff, T.C. (1962). An outbreak of surgical wound infections due to *Clostridium perfringens*. *Surg. Gynecol. Obstet.* 114: 102–108.

Elson, C.O., Hattori, K., and Blackstone, M.O. (1975). Polymicrobial sepsis following endoscopic retrograde cholangiopancreatography. *Gastroenterology* 69: 507–510.

Embil, J.M., Zhanel, G.G., Plourde, J., and Hoban, D. (2002). Scissors: a potential source of nosocomial infection. *Infect. Control Hosp. Epidemiol.* 23: 147–151.

Emmons, C.W. (1933). Fungicidal action of some common dsinfectants on two dermatophytes. *Arch. Dermatol. Syphil.* 28: 15–21.

Engelhardt, J.P., Grun, L., and Dahl, H.J. (1984). Factors affecting sterilization in glass bead sterilizers. *J. Endod.* 10: 465–470.

Engelhart, S.K.L., Glasmacher, A., Fischnaller, E. et al. (2002). *Pseudomonas aeruginosa* outbreak in a haematology-oncology unit associated with contaminated surface cleaning equipment. *J. Hosp. Infect.* 52: 93–98.

Engley, F.B. Jr. and Dey, B.P. (1970). A universal neutralizing medium for antimicrobial chemicals. *Chem. Specialists Manuf. Assoc. Proc.*: 100–106.

Environmental Protection Agency (1991). R.E.D. Facts sodium and calcium hypochlorite salts.

Environmental Protection Agency (1993). Protection of stratospheric ozone; Proposed Rule. 40 CFR Part 82. Fed. Regist.

Environmental Protection Agency (2003). Pesticides: Regulating Pesticides [This link is no longer active: http://www.epa.gov/oppad001/chemregindex.htm] (accessed 21 August 2021).

Epstein, B.J., Lattimer, J.M., Matsen, J.M., and Garibaldi, R.A. (1983). False positive spore strip sterility tests with steam sterilization. *Am. J. Infect. Control* 11: 71–73.

Ernst, R.R. and Doyle, J.E. (1968). Sterilization with gaseous ethylene oxide: a review of chemical and physical factors. *Biotech. Bioeng.* 10.

Estrin, W.J. and Becker, C.E. (1987). Evidence of neurologic dysfunction related to long-term ethylene oxide exposure. *Arch. Neurol.* 44: 1283–1286.

Estrin, W.J., Bowler, R.M., Lash, A., and Becker, C.E. (1990). Neurotoxicological evaluation of hospital sterilizer workers exposed to ethylene oxide. *J. Toxicol. Clin. Toxicol.* 28: 1–20.

Evans, M.R., Meldrum, R., Lane, W. et al. (2002). An outbreak of viral gastroenteritis following environmental contamination at a concert hall. *Epidemiol. Infect.* 129: 355–360.

Exner, M., Vacata, V., Hornei, B. et al. (2004). Household cleaning and surface disinfection: new insights and strategies. *J. Hosp. Infect.* 56 (Suppl): S70–S75.

Fahey, B.J., Koziol, D.E., Banks, S.M., and Henderson, D.K. (1991). Frequency of nonparenteral occupational exposures to blood and body fluids before and after universal precautions training. *Am. J. Med.* 90: 145–153.

Fantry, G.T., Zheng, Q.X., and James, S.P. (1995). Conventional cleaning and disinfection techniques eliminate the risk of endoscopic transmission of *Helicobacter pylori*. *Am. J. Gastroenterol.* 90: 227–232.

Farina, A., Fievet, M.H., Plassart, F. et al. (1999). Residual glutaraldehyde levels in fiberoptic endoscopes: measurement and implications for patient toxicity. *J. Hosp. Infect.* 43: 293–297.

Favero, M.S. (1982). Iodine–champagne in a tin cup. *Infect. Control* 3: 30–32.

Favero, M.S. (2001). Sterility assurance: concepts for patient safety. In: *Disinfection, Sterilization and Antisepsis: Principles and Practices in Healthcare Facilities* (ed. W.A. Rutala), 110–119. Washington, DC: Association for Professional in Infection Control and Epidemiology.

Favero, M.S. (2004). Naturally occurring microrganisms and their resistance to physical and chemical agents. In: *Disinfection, Sterilization and Antisepsis: Principles, Practices, Challenges, and New Research* (ed. W.A. Rutala), 1–14.

Washington, DC: Association for Professionals in Infection Control and Epidemiology.

Favero, M.S. (2004). Naturally occurring microorganisms and their resistance to physical and chemical agents. In: *Disinfection, Sterilization and Antisepsis: Principles, Practices, Challenges, and New Research* (ed. W.A. Rutala), 1–15. Washington, DC: Association for Professionals in Infection Control and Epidemiology.

Favero, M.S. and Bond, W.W. (1991). Chemical disinfection of medical and surgical materials. In: *Disinfection, Sterilization, and Preservation* (ed. S.S. Block), 617–641. Philadelphia, PA: Lea & Febiger.

Favero, M.S. and Bond, W.W. (1993). The use of liquid chemical germicides. In: *Sterilization Technology: A Practical Guide for Manufacturers and Users of Health Care Products* (eds. R.F. Morrissey and G.B. Phillips), 309–334. New York: Van Nostrand Reinhold.

Favero, M.S. and Bond, W.W. (2001). Chemical disinfection of medical and surgical materials. In: *Disinfection, Sterilization, and Preservation* (ed. S.S. Block), 881–917. Philadelphia, PA: Lippincott Williams & Wilkins.

Favero, M.S. and Groschel, D.H.M. (1987). *Chemical Germicides in the Health Care Field: Current Status and Evaluation of Efficacy and Research Needs*. Washington, DC: American Society for Microbiology.

Favero, M.S., Petersen, N.J., Carson, L.A. et al. (1975). Gram-negative water bacteria in hemodialysis systems. *Health Lab. Sci.* 12: 321–334.

Favero, M.S., Alter, M.J., Tokars, J.I., and Bland, L.A. (1998). Dialysis-associated disease and their control. In: *Hospital Infections* (eds. J.V. Bennett and P.S. Brachman), 357–380. Boston, MA: Little, Brown and Company.

Favero, M.J., Tokars, J.I., Arduino, M.J., and Alter, M.J. (1999). Nosocomial infections associated with hemodialysis. In: *Infection Control and Hospital Epidemiology* (ed. C.G. Mayhall), 897–917. Philadelphia, PA: Lippincott Williams & Wilkins.

Fayer, R., Graczyk, T.K., Cranfield, M.R., and Trout, J.M. (1996). Gaseous disinfection of *Cryptosporidium parvum* oocysts. *Appl. Environ. Microbiol.* 62: 3908–3909.

Feigal, D.W., Gardner, S.N., and McClellan, M. (2003). Ensuring safe and effective medical devices. *N. Engl. J. Med.* 348: 191–192.

Feldman, L.A. and Hui, H.K. (1997). Compatibility of medical devices and materials with low-temperature hydrogen peroxide gas plasma. *Med. Dev. Diag. Indust.* 19: 57–62.

Fengier, T.W., Pahike, H., Bisson, S., and Michels, W. (2001). Are processed surgical instruments free of protein? *Zentr. Steril.* 9: 20–32.

Ferrier, A., Garin, D., and Crance, J.M. (2004). Rapid inactivation of vaccinia virus in suspension and dried on surfaces. *J. Hosp. Infect.* 57: 73–79.

Finelli, P.F., Morgan, T.F., Yaar, I., and Granger, C.V. (1983). Ethylene oxide-induced polyneuropathy. A clinical and electrophysiologic study. *Arch. Neurol.* 40: 419–421.

Finelli, L., Miller, J.T., Tokars, J.I. et al. (2005). National surveillance of dialysis-associated diseases in the United States, 2002. *Semin. Dial.* 18: 52–61.

Fisher, A.A. (1984). Ethylene oxide dermatitis. *Cutis* 34: 20, 22, 24.

Fisher, A.A. (1990). Allergic contact dermatitis of the hands from Sporicidin (glutaraldehyde-phenate) used to disinfect endoscopes. *Cutis* 45: 227–228.

Fleming, S.J., Foreman, K., Shanley, K. et al. (1991). Dialyser reprocessing with Renalin. *Am. J. Nephrol.* 11: 27–31.

Foliente, R.L.K.B., Aprecio, R.M., Bains, H.J. et al. (2001). Efficacy of high-level disinfectants for reprocessing gastrointestinal endoscopes in simulated-use testing. *Gastrointest. Endosc.* 53: 456–462.

Food and Drug Administration (2000). Content and format of premarket notification [510(k)] submissions for liquid chemical sterilants/high level disinfectants. [This link is no longer active: www.fda.gov/cdrh/ode/397] (accessed 21 August 2021).

Food and Drug Administration (2000). *Enforcement Priorities for Single-Use Devices Reprocessed by Third Parties and Hospitals.* Rockville, MD.

Food and Drug Administration (2001). Sterilants and high level disinfectants cleared by FDA in a 510(k) as of January 30, 2002 with general claims for processing reusable medical and dental devices. [This link is no longer active: www.fda.gov/cdrh/ode/germlab.html] (accessed 21 August 2021).

Food and Drug Administration (2005). FDA-cleared sterilants and high-level disinfectants with general claims for processing reusable medical and dental devices, May 13, 2005. [This link is no longer active: www.fda.gov/cdrh/ode/germlab.html] (accessed 21 August 2021).

Food and Drug Administration, Centers for Disease Control and Prevention (1999). *FDA and CDC Public Health Advisory: Infections from Endoscopes Inadequately Reprocessed by An Automated Endoscope Reprocessing System.* Rockville, MD: Food and Drug Administration.

Food and Drug Administration, Division of General and Restorative Devices (1993). *Guidance on Premarket Notification [510(k)] Submissions for Sterilizers Intended for Use in Health Care Facilities.* Rockville, MD: Food and Drug Administration, Division of General and Restorative Devices.

Fowler, J.F. Jr. (1989). Allergic contact dermatitis from glutaraldehyde exposure. *J. Occup. Med.* 31: 852–853.

Fowler, C. and McCracken, D. (1999). US probes: risk of cross infection and ways to reduce it–comparison of cleaning methods. *Radiology* 213: 299–300.

Fraise, A.P. (1999). Choosing disinfectants. *J. Hosp. Infect.* 43: 255–264.

Fraser, V.J., Jones, M., Murray, P.R. et al. (1992). Contamination of flexible fiberoptic bronchoscopes with *Mycobacterium chelonae* linked to an automated bronchoscope disinfection machine. *Am. Rev. Respir. Dis.* 145: 853–855.

Fraser, T.G., Reiner, S., Malcznski, M. et al. (2004). Multidrug-resistant *Pseudomonas aeruginosa* cholangiopancreatography: failure of routine endoscope cultures to prevent an outbreak. *Infect. Control Hosp. Epidemiol.* 25: 856–859.

Fraud, S., Maillard, J.-Y., and Russell, A.D. (2001). Comparison of the mycobactericidal activity of ortho-phthalaldehyde, glutaraldehyde, and other dialdehydes by a quantitative suspension test. *J. Hosp. Infect.* 48: 214–221.

Fraud, S., Hann, A.C., Maillard, J.-Y., and Russell, A.D. (2003). Effects of ortho-phthalaldehyde, glutaraldehyde and chlorhexidine diacetate on *Mycobacterium chelonae* and *Mycobacterium abscessus* strains with modified permeability. *J. Antimicrob. Chemother.* 51: 575–584.

French, R.J., Tabb, H.G., and Rutledge, L.J. (1970). Esophageal stenosis produced by ingestion of bleach: report of two cases. *South. Med. J.* 63: 1140–1144.

French, G.L., Otter, J.A., Shannon, K.P. et al. (2004). Tackling contamination of the hospital environment by methicillin-resistant *Staphylococcus aureus* (MRSA): a comparison between conventional terminal cleaning and hydrogen peroxide vapour decontamination. *J. Hosp. Infect.* 57: 31–37.

Frey, R. (2000). The structural and functional prerequisites for a central sterile supply department (CSSD). *Zentr. Steril.* 8: 128–140.

Fritz, S., Hust, M.H., Ochs, C. et al. (1993). Use of a latex cover sheath for transesophageal echocardiography (TEE) instead of regular disinfection of the echoscope? *Clin. Cardiol.* 16: 737–740.

Frobisher, M., Sommermeyer, L., and Blackwell, M.J. (1973). Studies on disinfection of clinical thermometers. I. Oral thermometers. *Appl. Microbiol.* 1: 187–194.

Fuselier, H.A. Jr. and Mason, C. (1997). Liquid sterilization versus high level disinfection in the urologic office. *Urology* 50: 337–340.

Gamble, M.R. (1977). Hazard: formaldehyde and hypochlorites. *Lab. Anim.* 11: 61.

Gannon, P.F., Bright, P., Campbell, M. et al. (1995). Occupational asthma due to glutaraldehyde and formaldehyde in endoscopy and X-ray departments. *Thorax* 50: 156–159.

Gapany-Gapanavicius, M., Yellin, A., Almog, S., and Tirosh, M. (1982). Pneumomediastinum. A complication of chlorine exposure from mixing household cleaning agents. *JAMA* 248: 349–350.

Garcia de Cabo, A., Martinez Larriba, P.L., Checa Pinilla, J., and Guerra Sanz, F. (1978). A new method of disinfection of the flexible fibrebronchoscope. *Thorax* 33: 270–272.

Garland, S.M. and de Crespigny, L. (1996). Prevention of infection in obstetrical and gynaecological ultrasound practice. *Aust. N. Z. J. Obstet. Gynaecol.* 36: 392–395.

Garner, J.S. (1996). Guideline for isolation precautions in hospitals. The Hospital Infection Control Practices Advisory Committee. *Infect. Control Hosp. Epidemiol.* 17: 53–80.

Garner, J.S. and Favero, M.S. (1986). CDC Guideline for handwashing and hospital environmental control, 1985. *Infect. Control* 7: 231–243.

Garner, J.S. and Favero, M.S. (1986). CDC guidelines for the prevention and control of nosocomial infections. Guideline for handwashing and hospital environmental control, 1985. Supersedes guideline for hospital environmental control published in 1981. *Am. J. Infect. Control* 14: 110–129.

Garner, J.S. and Simmons, B.P. (1983). Guideline for isolation precautions in hospitals. *Infect. Control* 4: 245–325.

Gaspar, M.C., Pelaez, B., Fernandez, C., and Fereres, J. (2002). Microbiological efficacy of Sterrad 100S and LTSF sterilisation systems compared to ethylene oxide. *Zentr. Steril.* 10: 91–99.

Gebel, J., Sonntag, H.-G., Werner, H.-P. et al. (2002). The higher disinfectant resistance of nosocomial isolates of *Klebsiella oxytoca*: how reliable are indicator organisms in disinfectant testing? *J. Hosp. Infect.* 50: 309–311.

Gehrke, C., Steinmann, J., and Goroncy-Bermes, P. (2004). Inactivation of feline calicivirus, a surrogate of norovirus (formerly Norwalk-like viruses), by different types of alcohol in vitro and in vivo. *J. Hosp. Infect.* 56: 49–55.

Gerba, C.P. and Rusin, P. (2001). Relationship between the use of antiseptics/disinfectants and the development of antimicrobial resistance. In: *Disinfection, Sterilization and Antisepsis: Principles and Practices in Healthcare Facilities* (ed. W.A. Rutala), 187–194. Washington, DC: Association for Professional in Infection Control and Epidemiology.

Gerberding, J.L., Littell, C., Tarkington, A. et al. (1991). Risk of exposure of surgical personnel to patients' blood during surgery at San Francisco General Hospital. *N. Engl. J. Med.* 324: 1788–1793.

Gerding, D.N., Peterson, L.R., and Vennes, J.A. (1982). Cleaning and disinfection of fiberoptic endoscopes: evaluation of glutaraldehyde exposure time and forced-air drying. *Gastroenterology* 83: 613–618.

Gilbert, P. and McBain, A.J. (2003). Potential impact of increased use of biocides in consumer products on prevalence of antibiotic resistance. *Clin. Microbiol. Rev.* 16: 189–208.

Gilbert, J.A. and Phillips, H.O. (1984). The effect of steam sterilization on plaster casting material. *Clin. Orthop. Relat. Res.*: 241–244.

Gillis, R.J. and Schmidt, W.C. (1983). Scanning electron microscopy of spores on inoculated product surfaces. *MD*: 46–49.

Goddard, P.A. and McCue, K.A. (2001). Phenolic compounds. In: *Disinfection, Sterilization, and Preservation* (ed. S.S. Block), 255–281. Philadelphia, PA: Lippincott Williams & Wilkins.

Goetz, A. and Yu, V.L. (1997). Copper-silver ionization: cautious optimism for *Legionella* disinfection and implications for environmental culturing. *Am. J. Infect. Control* 25: 449–451.

Goodman, H.S., Carpenter, R.D., and Cox, M.R. (1994). Sterilization of dental instruments and devices: an update. *Am. J. Infect. Control* 22: 90–94.

Gordon, M.D., Ezzell, R.J., Bruckner, N.I., and Ascenzi, J.M. (1994). Enhancement of mycobactericidal activity of glutaraldehyde with α,ß-unsaturated and aromatic aldehydes. *J. Indust. Microbiol.* 13: 77–82.

Gorham, R.A., Jacobs, P., and Roberts, C.G. (1998). Laboratory artifacts due to protein and salt crystals on the inactivation of *Bacillus stearothermophilus*. *J. Hosp. Infect.* 40: abstract P.9.2.2.

Gottardi, W. (1991). Iodine and iodine compounds. In: *Disinfection, Sterilization, and Preservation* (ed. S.S. Block), 152–166. Philadelphia, PA: Lea & Febiger.

Gottardi, W. (2001). Iodine and iodine compounds. In: *Disinfection, Sterilization, and Preservation* (ed. S.S. Block), 159–184. Philadelphia, PA: Lippincott Williams & Wilkins.

Janet Gough, Monica Grimaldi. *The External Quality Audit*.

Graham, D.Y. and Osato, M.S. (1999). Disinfection of biopsy forceps and culture of *Helicobacter pylori* from gastric mucosal biopsies. *Am. J. Gastroenterol.* 94: 1422–1423.

Graham, G.S. and Riley, R. (1998). Sterilization manufacturers: interactions with regulatory agencies. In: *Disinfection, Sterilization, and Antisepsis in Healthcare* (ed. W.A. Rutala), 41–48. Champlain, New York: Polyscience Publications.

Grant, D., Venneman, M., and Burns, R.M. (1982). Mycobactericidal activity of Alcide an experimental liquid sterilant. *Abstract. Annu. Meet. Am. Soc. Microbiol.* Q101: 226.

Green, J., Wright, P.A., Gallimore, C.I. et al. (1998). The role of environmental contamination with small round structured viruses in a hospital outbreak investigated by reverse-transcriptase polymerase chain reaction assay. *J. Hosp. Infect.* 39: 39–45.

Greene, V.W. (1992). Control of sterilization process. In: *Principles and Practice of Disinfection, Preservation and Sterilization* (eds. A.D. Russell, W.B. Hugo and G.A.J. Ayliffe), 605–624. Oxford, England: Blackwell Scientific Publications.

Greene, V.W. (1999). Reuse of disposable devices. In: *Infection Control and Hospital Epidemiology* (ed. C.G. Mayhall), 1201–1208. Philadelphia, PA: Lippincott Williams & Wilkins.

Gregory, A.W., Schaalje, G.B., Smart, J.D., and Robison, R.A. (1999). The mycobactericidal efficacy of ortho-phthalaldehyde and the comparative resistances of *Mycobacterium bovis, Mycobacterium terrae*, and *Mycobacterium chelonae*. *Infect. Control Hosp. Epidemiol.* 20: 324–330.

Greub, G. and Raoult, D. (2003). Biocides currently used for bronchoscope decontamination are poorly effective against free-living amoebae. *Infect. Control Hosp. Epidemiol.* 24: 784–786.

Mary Griffin, Dona Reber. *Microbial Identification: The Keys to a Successful Program*.

Griffith, C.J., Cooper, R.A., Gilmore, J. et al. (2000). An evaluation of hospital cleaning regimes and standards. *J. Hosp. Infect.* 45: 19–28.

Griffiths, P.A., Babb, J.R., Bradley, C.R., and Fraise, A.P. (1997). Glutaraldehyde-resistant *Mycobacterium chelonae* from endoscope washer disinfectors. *J. Appl. Microbiol.* 82: 519–526.

Griffiths, P.A., Babb, J.R., and Fraise, A.P. (1999). Mycobactericidal activity of selected disinfectants using a quantitative suspension test. *J. Hosp. Infect.* 41: 111–121.

Groschel, D.H.M. (1983). Caveat emptor: do your disinfectants work? *Infect. Control* 4: 144.

Gross, D. (1995). Ethylene oxide sterilization and alternative methods. *Surg. Serv. Manage.* 1: 16–17.

Guinto, C.H., Bottone, E.J., Raffalli, J.T. et al. (2002). Evaluation of dedicated stethoscopes as a potential source of nosocomial pathogens. *Am. J. Infect. Control* 30: 499–502.

Gundogdu, H., Ocal, K., Caglikulekci, M. et al. (1998). High-level disinfection with 2% alkalinized glutaraldehyde solution for reuse of laparoscopic disposable plastic trocars. *J. Laparoendosc. Adv. Surg. Tech. Part A* 8: 47–52.

Gurevich, I., Holmes, J.E., and Cunha, B.A. (1982). Presumed autoclave failure due to false-positive spore strip tests. *Infect. Control* 3: 388–392.

Gurevich, I., Tafuro, P., Ristuccia, P. et al. (1983). Disinfection of respirator tubing: a comparison of chemical versus hot water machine-assisted processing. *J. Hosp. Infect.* 4: 199–208.

Gurevich, I., Qadri, S.M.H., and Cunha, B.A. (1992). False-positive results of spore tests from improper clip use with the Steris chemical sterilant system. *Infect. Control Hosp. Epidemiol.* 21: 42–43.

Gurevich, I., Dubin, R., and Cunha, B.A. (1996). Dental instrument and device sterilization and disinfection practices. *J. Hosp. Infect.* 32: 295–304.

Gurevich, I., Jacobsen, E., and Cunha, B.A. (1996). Pseudoautoclave failure caused by differences in spore test steam sensitivities. *Am. J. Infect. Control* 24: 402–404.

Gwaltney, J.M. Jr. and Hendley, J.O. (1982). Transmission of experimental rhinovirus infection by contaminated surfaces. *Am. J. Epidemiol.* 116: 828–833.

Haag, J.R. and Gieser, R.G. (1983). Effects of swimming pool water on the cornea. *JAMA* 249: 2507–2508.

Hall, K.K., Giannetta, E.T., Getchell-White, S.I. et al. (2003). Ultraviolet light disinfection of hospital water for preventing nosocomial *Legionella* infection: a 13-year follow-up. *Infect. Control Hosp. Epidemiol.* 24: 580–583.

Hamasuna, R., Nose, K., Sueyoshi, T. et al. (2004). High-level disinfection of cystoscopic equipment with ortho-phthalaldehyde solution. *J. Hosp. Infect.* 57: 346–348.

Hamouda, T. and Baker, J.R. Jr. (2000). Antimicrobial mechanism of action of surfactant lipid preparations in enteric Gram-negative bacilli. *J. Appl. Microbiol.* 89: 397–403.

Hamouda, T., Hayes, M.M., Cao, Z.H. et al. (1999). A novel surfactant nanoemulsion with broad-spectrum sporicidal activity against *Bacillus species. J. Infect. Dis.* 180: 1939–1949.

Hamouda, T., Myc, A., Donovan, B. et al. (2001). A novel surfactant nanoemulsion with a unique non-irritant topical antimicrobial activity against bacteria, enveloped viruses and fungi. *Microbiol. Res.* 156: 1–7.

Handsfield, H.H., Cummings, M.J., and Swenson, P.D. (1987). Prevalence of antibody to human immunodeficiency virus and hepatitis B surface antigen in blood samples submitted to a hospital laboratory. Implications for handling specimens. *JAMA* 258: 3395–3397.

Hansen, K.S. (1983). Occupational dermatoses in hospital cleaning women. *Contact Derm.* 9: 343–351.

Hansen, J.M. and Shaffer, H.L. (2001). Sterilization and preservation by radiation sterilization. In: *Disinfection, Sterilization, and Preservation* (ed. S.S. Block), 729–746. Philadelphia, PA: Lippincott Williams & Wilkins.

Hanson, P.J., Gor, D., Clarke, J.R. et al. (1989). Contamination of endoscopes used in AIDS patients. *Lancet* 2: 86–88.

Hanson, P.J., Gor, D., Jeffries, D.J., and Collins, J.V. (1989). Chemical inactivation of HIV on surfaces. *Br. Med. J.* 298: 862–864.

Hanson, P.J., Gor, D., Jeffries, D.J., and Collins, J.V. (1990). Elimination of high titre HIV from fibreoptic endoscopes. *Gut* 31: 657–659.

Hanson, P.J., Gor, D., Clarke, J.R. et al. (1991). Recovery of the human immunodeficiency virus from fibreoptic bronchoscopes. *Thorax* 46: 410–412.

Hanson, P.J., Jeffries, D.J., and Collins, J.V. (1991). Viral transmission and fibreoptic endoscopy. *J. Hosp. Infect.* 18: 136–140.

Hanson, P.J., Chadwick, M.V., Gaya, H., and Collins, J.V. (1992). A study of glutaraldehyde disinfection of fibreoptic bronchoscopes experimentally contaminated with *Mycobacterium tuberculosis. J. Hosp. Infect.* 22: 137–142.

Hanson, P.J., Bennett, J., Jeffries, D.J., and Collins, J.V. (1994). Enteroviruses, endoscopy and infection control: an applied study. *J. Hosp. Infect.* 27: 61–67.

Hardy, K.J., Oppenheim, B.A., Gossain, S. et al. (2006). A study of the relationship between environmental contamination with methicillin-resistant *Staphylococcus aureus* (MRSA) and patients' acquisition of MRSA. *Infect. Control Hosp. Epidemiol.* 27: 127–132.

Harris, M.G., Rechberger, J., Grant, T., and Holden, B.A. (1990). In-office microwave disinfection of soft contact lenses. *Optom. Vis. Sci.* 67: 129–132.

Hastreiter, R.J., Molinari, J.A., Falken, M.C. et al. (1991). Effectiveness of dental office instrument sterilization procedures. *J. Am. Dent. Assoc.* 122: 51–56.

Heffelfinger, J.D., Kool, J.L., Fridkin, S. et al. (2003). Risk of hospital-acquired legionnaires' disease in cities using monchloramine versus other water disinfectants. *Infect. Control Hosp. Epidemiol.* 24: 569–574.

Hegna, I.K. (1977). A comparative investigation of the bactericidal and fungicidal effects of three phenolic disinfectants. *J. Appl. Bacteriol.* 43: 177–181.

Hegna, I.K. (1977). An examination of the effect of three phenolic disinfectants on *Mycobacterium tuberculosis*. *J. Appl. Bacteriol.* 43: 183–187.

Heidemann, S.M. and Goetting, M.G. (1991). Treatment of acute hypoxemic respiratory failure caused by chlorine exposure. *Pediatr. Emerg. Care* 7: 87–88.

Helms, C.M., Massanari, R.M., Zeitler, R. et al. (1983). Legionnaires' disease associated with a hospital water system: a cluster of 24 nosocomial cases. *Ann. Intern. Med.* 99: 172–178.

Helms, C., Massanari, R., Wenzel, R. et al. (1987). Control of epidemic nosocomial legionellosis: a 5 year progress report on continuous hyperchlorination of a water distribution system. *Abstracts of 27th Interscience Conference of Antimicrobial Agents and Chemotherapy*, Volume 349, p. 158.

Hemminki, K., Mutanen, P., Saloniemi, I. et al. (1982). Spontaneous abortions in hospital staff engaged in sterilising instruments with chemical agents. *Br. Med. J.* 285: 1461–1463.

Henderson, D.A. (1999). The looming threat of bioterrorism. *Science* 283: 1279–1282.

Hernandez, A., Martro, E., Pizo, C. et al. (2003). In-use evaluation of Perasafe compared with Cidex in fibreoptic bronchoscope disinfection. *J. Hosp. Infect.* 54: 46–52.

Hernandez, A., Martro, E., Matas, L., and Ausina, V. (2003). In-vitro evaluation of Pearsafe compared with 2% alkaline glutaraldehyde against *Mycobacterium* spp. *J. Hosp. Infect.* 54: 52–56.

Herruzo-Cabrera, R., Vizcaino-Alcaide, M.J., and Fernandez-Acenero, M.J. (2004). The influence of laboratory adaptation on test strains, such as *Pseudomonas aeruginosa*, in the evaluation of the antimicrobial efficacy of ortho-phthalaldehyde. *J. Hosp. Infect.* 57: 217–222.

Herruzo-Cabrera, R., Vizcaino-Alcaide, M.J., and Rodriquez, J. (2006). Comparison of the microbicidal efficacy on germ carriers of several tertiary amine compounds with ortho-phthalaldehyde and Perasafe. *J. Hosp. Infect.* 63: 73–78.

Hess, J.A., Molinari, J.A., Gleason, M.J., and Radecki, C. (1991). Epidermal toxicity of disinfectants. *Am. J. Dent.* 4: 51–56.

Hession, S.M. (2003). Endoscopic disinfection by ortho-phthalaldehyde in a clinical setting: an evaluation of reprocessing time and costs compared with glutaraldehyde. *Gastroenterol. Nurs.* 26: 110–114.

Hignett, M. and Claman, P. (1995). High rates of perforation are found in endovaginal ultrasound probe covers before and after oocyte retrieval for in vitro fertilization-embryo transfer. *J. Assist. Reprod. Genet.* 12: 606–609.

Hobson, D.W. and Seal, L.A. (2000). Evaluation of a novel, rapid-acting, sterilizing solution at room temperature. *Am. J. Infect. Control* 28: 370–375.

Hoffman, P.N. and Layzell, S.K. (1993). Household bleach as disinfectant for use by injecting drug users. *Lancet* 342: 743.

Hoffman, P.N., Death, J.E., and Coates, D. (1981). The stability of sodium hypochlorite solutions. In: *Disinfectants: Their Use and Evaluation of Effectiveness* (eds. C.H. Collins, M.C. Allwood, S.F. Bloomfield and A. Fox), 77–83. London: Academic Press.

Hoffmann, K.K., Weber, D.J., and Rutala, W.A. (1989). Pseudoepidemic of *Rhodotorula rubra* in patients undergoing fiberoptic bronchoscopy. *Infect. Control Hosp. Epidemiol.* 10: 511–514.

Holler, C., Martiny, H., Christiansen, B. et al. (1993). The efficacy of low temperature plasma (LTP) sterilization, a new sterilization technique. *Zentralbl. Hyg. Umweltmed.* 194: 380–391.

Holton, J. and Shetty, N. (1997). In-use stability of Nu-Cidex. *J. Hosp. Infect.* 35: 245–248.

Honeybourne, D. and Neumann, C.S. (1997). An audit of bronchoscopy practice in the United Kingdom: a survey of adherence to national guidelines. *Thorax* 52: 709–713.

Hood, E., Stout, N., and Catto, B. (1997). Flash sterilization and neurosurgical site infections: guilt by association. *Am. J. Infect. Control* 25: 156.

Hota, B. (2004). Contamination, disinfection, and cross-contamination: are hospital surfaces reservoirs for nosocomial infection? *Clin. Infect. Dis.* 39: 1182–1189.

Lisa Hornback. *Combination Products: Implementation of cGMP Requirements.*

Hoy, R.H. (1981). Accidental systemic exposure to sodium hypochlorite (Clorox) during hemodialysis. *Am. J. Hosp. Pharm.* 38: 1512–1514.

Hughes, C.E., Gebhard, R.L., Peterson, L.R., and Gerding, D.N. (1986). Efficacy of routine fiberoptic endoscope cleaning and disinfection for killing *Clostridium difficile*. *Gastrointest. Endosc.* 32: 7–9.

Hulka, J.F., Wisler, M.G., and Bruch, C. (1977). A discussion: laparoscopic instrument sterilization. *Med. Instrum.* 11: 122–123.

Humphreys, H., McGrath, H., McCormick, P.A., and Walsh, C. (2002). Quality of final rinse water used in washer-disinfectors for endoscopes. *J. Hosp. Infect.* 51: 151–153.

Husni, L., Kale, E., Climer, C. et al. (1989). Evaluation of a new disinfectant for dialyzer reuse. *Am. J. Kidney Dis.* 14: 110–118.

Hutchisson, B. and LeBlanc, C. (2005). The truth and consequences of enzymatic detergents. *Gastroenterol. Nurs.* 28: 372–376.

Ingram, T.A. (1990). Response of the human eye to accidental exposure to sodium hypochlorite. *J Endodontics* 16: 235–238.

International Scientific Forum on Home Hygiene (2016). Current version of this document may differ from original. (accessed August 2016).

Isenberg, H.D., Giugliano, E.R., France, K., and Alperstein, P. (1988). Evaluation of three disinfectants after in-use stress. *J. Hosp. Infect.* 11: 278–285.

Ishino, Y., Ido, K., and Sugano, K. (2005). Contamination with hepatitis B virus DNA in gastrointestinal endoscope channels: risk of infection on reuse after on-site cleaning. *Endoscopy* 37: 548–551.

Isomoto, H., Urata, M., Kawazoe, K. et al. (2006). Endoscope disinfection using chlorine dioxide in an automated washer-disinfector. *J. Hosp. Infect.* 63: 298–305.

Jackson, F.W. and Ball, M.D. (1997). Correction of deficiencies in flexible fiberoptic sigmoidoscope cleaning and disinfection technique in family practice and internal medicine offices. *Arch. Fam. Med.* 6: 578–582.

Jackson, J., Leggett, J.E., Wilson, D.A., and Gilbert, D.N. (1996). *Mycobacterium gordonae* in fiberoptic bronchoscopes. *Am. J. Infect. Control* 24: 19–23.

Jacobs, P. (1998). Cleaning: principles, methods and benefits. In: *Disinfection, Sterilization, and Antisepsis in Healthcare* (ed. W.A. Rutala), 165–181. Champlain, NY: Polyscience Publications.

Jacobs, P.T. and Lin, S.M. (2001). Sterilization processes utilizing low-temperature plasma. In: *Disinfection, Sterilization, and Preservation* (ed. S.S. Block), 747–763. Philadelphia, PA: Lippincott Williams & Wilkins.

Jacobs, P.T. and Smith, D. (1998). The new Sterrad 100S sterilization system: Features and advantages. *Zentr. Steril.* 6: 86–94.

Jakobsson, S.W., Rajs, J., Jonsson, J.A., and Persson, H. (1991). Poisoning with sodium hypochlorite solution. Report of a fatal case, supplemented with an experimental and clinico-epidemiological study. *Am. J. Forensic Med. Pathol.* 12: 320–327.

Japp, N.F. (1997). Packaging: shelf life. In: *Sterilization Technology* (eds. M. Reichert and J.H. Young), 99–102. Gaithersburg, MD: Aspen Publication.

Jarvis, W.R. (2001). Hospital Infections Program, Centers for Disease Control and Prevention: on-site outbreak investigations, 1990–1999: how often are germicides or sterilants the source? In: *Disinfection, Sterilization and Antisepsis: Principles and Practices in Healthcare Facilities* (ed. W.A. Rutala), 41–48. Washington, DC: Association for Professional in Infection Control and Epidemiology.

Jatzwauk, L., Schone, H., and Pietsch, H. (2001). How to improve instrument disinfection by ultrasound. *J. Hosp. Infect.* 48 (Suppl): S80–S83.

Jay, W.M., Swift, T.R., and Hull, D.S. (1982). Possible relationship of ethylene oxide exposure to cataract formation. *Am. J. Ophthalmol.* 93: 727–732.

Jeanes, A., Rao, G., Osman, M., and Merrick, P. (2005). Eradication of persistent environmental MRSA. *J. Hosp. Infect.* 61: 85–86.

Jeng, D.K. and Woodworth, A.G. (1990). Chlorine dioxide gas sterilization of oxygenators in an industrial scale sterilizer: a successful model. *Artif. Organs* 14: 361–368.

Jernigan, J.A., Lowry, B.S., Hayden, F.G. et al. (1993). Adenovirus type 8 epidemic keratoconjunctivitis in an eye clinic: risk factors and control. *J. Infect. Dis.* 167: 1307–1313.

Jernigan, J.A., Siegman-Igra, Y., Guerrant, R.C., and Farr, B.M. (1998). A randomized crossover study of disposable thermometers for prevention of *Clostridium difficile* and other nosocomial infections. *Infect. Control Hosp. Epidemiol.* 19: 494–499.

Jette, L.P. and Lambert, N.G. (1988). Evaluation of two hot water washer disinfectors for medical instruments. *Infect. Control Hosp. Epidemiol.* 9: 194–199.

Jette, L.P., Ringuette, L., Ishak, M. et al. (1995). Evaluation of three glutaraldehyde-based disinfectants used in endoscopy. *J. Hosp. Infect.* 30: 295–303.

Jimenez, L. and Chiang, M. (2006). Virucidal activity of a quaternary ammonium compound disinfectant against feline calicivirus: a surrogate for norovirus. *Am. J. Infect. Control* 34: 269–273.

Johansen, C., Falholt, P., and Gram, L. (1997). Enzymatic removal and disinfection of bacterial biofilms. *Appl. Environ. Microbiol.* 63: 3724–3728.

Johnson, L.L., Shneider, D.A., Austin, M.D. et al. (1982). Two per cent glutaraldehyde: a disinfectant in arthroscopy and arthroscopic surgery. *J. Bone Joint Surg.* 64: 237–239.

Johnson, C.H., Rice, E.W., and Reasoner, D.J. (1997). Inactivation of *Helicobacter pylori* by chlorination. *Appl. Environ. Microbiol.* 63: 4969–4970.

Johnston, M.D., Lambert, R.J.W., Hanlon, G.W., and Denyer, S.P. (2002). A rapid method for assessing the suitability of quenching agents for individual biocides as well as combinations. *J. Appl. Microbiol.* 92: 784–789.

Joint Commission for the Accreditation of Healthcare Organizations (2003). *Comprehensive Accreditation Manual for Hospitals*. Chicago, IL: JCAHO.

Jonas, G., Mahoney, A., Murray, J., and Gertler, S. (1988). Chemical colitis due to endoscope cleaning solutions: a mimic of pseudomembranous colitis. *Gastroenterology* 95: 1403–1408.

Jones, R.D., Jampani, H.B., Newman, J.L., and Lee, A.S. (2000). Triclosan: a review of effectiveness and safety in health care settings. *Am. J. Infect. Control* 28: 184–196.

Jordan, S.L. (1995). The correct use of glutaraldehyde in the healthcare environment. *Gastroenterol. Nurs.* 18: 143–145.

Jordan, S.L.P., Russo, M.R., Blessing, R.L., and Grab, L.A. (1997). Glutaraldehyde safety: inactivation and disposal. *Abstract. Am. J. Infect. Control* 25: 154–155.

Joslyn, L. (2001). Sterilization by heat. In: *Disinfection, Sterilization, and Preservation* (ed. S.S. Block), 695–728. Philadelphia, PA: Lippincott Williams & Wilkins.

Joslyn, L. (2001). Gaseous chemical sterilization. In: *Disinfection, Sterilization, and Preservation* (ed. S.S. Block), 337–360. Philadelphia, PA: Lippincott Williams & Wilkins.

Judd, P.A., Tomlin, P.J., Whitby, J.L. et al. (1968). Disinfection of ventilators by ultrasonic nebulisation. *Lancet* 2: 1019–1020.

Kaatz, G.W., Gitlin, S.D., Schaberg, D.R. et al. (1988). Acquisition of *Clostridium difficile* from the hospital environment. *Am. J. Epidemiol.* 127: 1289–1294.

Kaczmarek, R.G., Moore, R.M. Jr., McCrohan, J. et al. (1992). Multi-state investigation of the actual disinfection/sterilization of endoscopes in health care facilities. *Am. J. Med.* 92: 257–261.

Kahn, G. (1970). Depigmentation caused by phenolic detergent germicides. *Arch. Dermatol.* 102: 177–187.

Kaneko, H., Mitsuma, T., Kotera, H. et al. (1993). Are routine cleaning methods sufficient to remove *Helicobacter pylori* from endoscopic equipment? *Endoscopy* 25: 435.

Kanemitsu, K., Kunishima, H., Imasaka, T. et al. (2003). Evaluation of a low-temperature steam and formaldehyde sterilizer. *J. Hosp. Infect.* 55: 47–52.

Kanemitsu, K., Imasaka, T., Ishikawa, S. et al. (2005). A comparative study of ethylene oxide gas, hydrogen peroxide gas plasma, and low-temperature steam formaldehyde sterilization. *Infect. Control Hosp. Epidemiol.* 26: 486–489.

Kaplan, J.C., Crawford, D.C., Durno, A.G., and Schooley, R.T. (1987). Inactivation of human immunodeficiency virus by Betadine. *Infect. Control* 8: 412–414.

Karapinar, M. and Gonul, S.A. (1992). Effects of sodium bicarbonate, vinegar, acetic and citric acids on growth and survival of *Yersinia enterocolitica*. *Int. J. Food Microbiol.* 16: 343–347.

Kariwa, H., Fujii, N., and Takashima, I. (2004). Inactivation of SARS coronavirus by means of povidone-iodine, physical conditions, and chemical reagents. *Jpn. J. Vet. Res.* 52: 105–112.

Kaulfers, P.M. and Laufs, R. (1985). Transmissible formaldehyde resistance in *Serratia marcescens*. *Zentralbl. Bakteriol. Mikrobiol. Hyg. [B]* 181: 309–319.

Kearns, A.M., Freeman, R., and Lightfoot, N.F. (1995). Nosocomial enterococci: resistance to heat and sodium hypochlorite. *J. Hosp. Infect.* 30: 193–199.

Kelen, G.D., Fritz, S., Qaqish, B. et al. (1988). Unrecognized human immunodeficiency virus infection in emergency department patients. *N. Engl. J. Med.* 318: 1645–1650.

Keswick, B.H., Pickering, L.K., DuPont, H.L., and Woodward, W.E. (1983). Survival and detection of rotaviruses on environmental surfaces in day care centers. *Appl. Environ. Microbiol.* 46: 813–816.

Kilvington, S. (1991). Moist-heat disinfection of *Acanthamoeba cysts*. *Rev. Infect. Dis.* 13: S418.

Kim, K.H., Fekety, R., Batts, D.H. et al. (1981). Isolation of *Clostridium difficile* from the environment and contacts of patients with antibiotic-associated colitis. *J. Infect. Dis.* 143: 42–50.

Kindle, G., Busse, A., Kampa, D. et al. (1996). Killing activity of microwaves in milk. *J. Hosp. Infect.* 33: 273–278.

Kinney, T.P., Kozarek, R.A., Raltz, S., and Attia, F. (2002). Contamination of single-use biopsy forceps: a prospective in vitro analysis. *Gastrointest. Endosc.* 56: 209–212.

Kirschke, D.L., Jones, T.F., Craig, A.S. et al. (2003). Pseudomonas aeruginosa and *Serratia marcescens* contamination associated with a manufacturing defect in bronchoscopes. *N. Engl. J. Med.* 348: 214–220.

Klapes, N.A. and Vesley, D. (1990). Vapor-phase hydrogen peroxide as a surface decontaminant and sterilant. *Appl. Environ. Microbiol.* 56: 503–506.

Klapes, N.A., Greene, V.W., and Langholz, A.C. (1987). Microbial contamination associated with routine aseptic practice. *J. Hosp. Infect.* 10: 299–304.

Klapes, N.A., Greene, V.W., Langholz, A.C., and Hunstiger, C. (1987). Effect of long-term storage on sterile status of devices in surgical packs. *Infect. Control* 8: 289–293.

Kleier, D.J. and Averbach, R.E. (1990). Glutaraldehyde nonbiologic monitors. *Infect. Control Hosp. Epidemiol.* 11: 439–441.

Kleier, D.J., Tucker, J.E., and Averbach, R.E. (1989). Clinical evaluation of glutaraldehyde nonbiologic monitors. *Quintessence Int.* 20: 271–277.

Klein, M. and DeForest, A. (1963). The inactivation of viruses by germicides. *Chem. Spec. Manuf. Assoc. Proc.* 49: 116–118.

Kleinegger, C.L., Yeager, D.L., Huling, J.K., and Drake, D.R. (2001). The effects of contamination on biological monitoring. *Infect. Control Hosp. Epidemiol.* 22: 391–392.

van Klingeren, B. and Pullen, W. (1993). Glutaraldehyde resistant mycobacteria from endoscope washers. *J. Hosp. Infect.* 25: 147–149.

Knapp, J.E. and Battisti, D.L. (2001). Chloride dioxide. In: *Disinfection, Sterilization, and Preservation* (ed. S.S. Block), 215–227. Philadelphia, PA: Lippincott Williams & Wilkins.

Kobayashi, H. and Tsuzuki, M. (1984). The effect of disinfectants and heat on hepatitis B virus. *J. Hosp. Infect.* 5: 93–94.

Kohn, W.G., Collins, A.S., Cleveland, J.L. et al. (2003). Guidelines for infection control in dental health-care settings-2003. *MMWR* 52 (RR-17): 1–67.

Koncur, P., Janes, J.E., and Ortiz, P.A. (1998). 20 second sterilization indicator tests equivalent to BIs. *Infect. Control Steril. Technol.* 26–28, 30, 32–34.

Koo, D., Bouvier, B., Wesley, M. et al. (1989). Epidemic keratoconjunctivitis in a university medical center ophthalmology clinic; need for re-evaluation of the design and disinfection of instruments. *Infect. Control Hosp. Epidemiol.* 10: 547–552.

Korich, D.G., Mead, J.R., Madore, M.S. et al. (1990). Effects of ozone, chlorine dioxide, chlorine, and monochloramine on *Cryptosporidium parvum* oocyst viability. *Appl. Environ. Microbiol.* 56: 1423–1428.

Kotilainen, H.R. and Gantz, N.M. (1987). An evaluation of three biological indicator systems in flash sterilization. *Infect. Control* 8: 311–316.

Kovacs, B.J., Chen, Y.K., Kettering, J.D. et al. (1999). High-level disinfection of gastrointestinal endoscopes: are current guidelines adequate? *Am. J. Gastroenterol.* 94: 1546–1550.

Kowalski, J.B. (1998). Sterilization of medical devices, pharmaceutical components, and barrier isolation systems with gaseous chlorine dioxide. In: *Sterilization of Medical Products* (eds. R.F. Morrissey and J.B. Kowalski), 313–323. Champlain, NY: Polyscience Publications.

Kralovic, R.C. (1993). Use of biological indicators designed for steam or ethylene oxide to monitor a liquid chemical sterilization process. *Infect. Control Hosp. Epidemiol.* 14: 313–319.

Kruse, A. and Rey, J.F. (2000). Guidelines on cleaning and disinfection in GI endoscopy. Update 1999. The European Society of Gastrointestinal Endoscopy. *Endoscopy* 32: 77–80.

Kruse, R.H., Green, T.D., Chambers, R.C., and Jones, M.W. (1963). Disinfection of aerosolized pathogenic fungi on laboratory surfaces. 1. Tissue phase. *Appl. Microbiol.* 11: 436–445.

Kruse, R.H., Green, T.D., Chambers, R.C., and Jones, M.W. (1964). Disinfection of aerosolized pathogenic fungi on laboratory surfaces. II Culture phase. *Appl. Microbiol.* 12: 155–160.

Kurtz, J.B., Lee, T.W., and Parsons, A.J. (1980). The action of alcohols on rotavirus, astrovirus and enterovirus. *J. Hosp. Infect.* 1: 321–325.

Kyi, M.S., Holton, J., and Ridgway, G.L. (1995). Assessment of the efficacy of a low temperature hydrogen peroxide gas plasma sterilization system. *J. Hosp. Infect.* 31: 275–284.

Lagergren, E.R. (1998). Recent advances in sterilization. *J. Infect. Control (Asia)* 1: 11–13.

Landau, G.D. and Saunders, W.H. (1964). The effect of chlorine bleach on the esophagus. *Arch. Otolaryngol.* 80: 174–176.

Landeen, L.K., Yahya, M.T., and Gerba, C.P. (1989). Efficacy of copper and silver ions and reduced levels of free chlorine in inactivation of *Legionella pneumophila*. *Appl. Environ. Microbiol.* 55: 3045–3050.

Langenberg, W., Rauws, E.A., Oudbier, J.H., and Tytgat, G.N. (1990). Patient-to-patient transmission of *Campylobacter pylori* infection by fiberoptic gastroduodenoscopy and biopsy. *J. Infect. Dis.* 161: 507–511.

Laskowski, L.F., Marr, J.J., Spernoga, J.F. et al. (1977). Fastidious mycobacteria grown from porcine prosthetic-heart-valve cultures. *N. Engl. J. Med.* 297: 101–102.

Latimer, J.M. and Matsen, J.M. (1977). Microwave oven irradiation as a method for bacterial decontamination in a clinical microbiology laboratory. *J. Clin. Microbiol.* 6: 340–342.

Lauer, J.L., Battles, D.R., and Vesley, D. (1982). Decontaminating infectious laboratory waste by autoclaving. *Appl. Environ. Microbiol.* 44: 690–694.

Lawrentschuk, N. and Chamberlain, M. (2005). Sterile disposable sheath sytsem for flexible cytoscopes. *Urology* 66: 1310–1313.

Leach, E.D. (1981). A new synergized glutaraldehyde-phenate sterilizing solution and concentrated disinfectant. *Infect. Control* 2: 26–30.

Leaper, S. (1984). Influence of temperature on the synergistic sporicidal effect of peracetic acid plus hydrogen peroxide in *Bacillus subtilis* SA22(NCA 72-52). *Food Microbiol.* 1: 199–203.

LeBlanc, D. *Cleaning Validation: Practical Compliance Solutions for Pharmaceutical Manufacturing*, vol. 2.

LeBlanc, D. *Cleaning Validation: Practical Compliance Solutions for Pharmaceutical Manufacturing*, vol. 3.

LeBlanc, D. *Cleaning Validation: Practical Compliance Solutions for Pharmaceutical Manufacturing*, vol. 4.

LeChevallier, M.W., Cawthon, C.D., and Lee, R.G. (1988). Inactivation of biofilm bacteria. *Appl. Environ. Microbiol.* 54: 2492–2499.

LeChevallier, M.W., Cawthon, C.D., and Lee, R.G. (1988). Factors promoting survival of bacteria in chlorinated water supplies. *Appl. Environ. Microbiol.* 54: 649–654.

Lee, D.H., Miles, R.J., and Perry, B.F. (1985). The mycoplasmacidal properties of sodium hypochlorite. *J. Hyg. (Lond).* 95: 243–253.

Lee, R.M., Kozarek, R.A., Sumida, S.E., and Raltz, S.L. (1998). Risk of contamination of sterile biopsy forceps in disinfected endoscopes. *Gastrointest. Endosc.* 47: 377–381.

Lee, C.H., Cheng, S.M., Humar, A. et al. (2000). Acute febrile reactions with hypotension temporally associated with the introduction of a concentrated bioenzyme preparation in the cleaning and sterilization process of endomyocardial bioptones. *Infect. Control Hosp. Epidemiol.* 21: 102.

Lee, J.H., Rhee, P.L., Kim, J.H. et al. (2004). Efficacy of electolyzed acid water in reprocessing patient-used flexible upper endoscopes: comparison with 2% alkaline glutaraldehyde. *J. Gastroenterol. Hepatol.* 19: 897–904.

Leggiadro, R.J. (2000). The threat of biological terrorism: a public health and infection control reality. *Infect. Control Hosp. Epidemiol.* 21: 53–56.

Leinster, P., Baum, J.M., and Baxter, P.J. (1993). An assessment of exposure to glutaraldehyde in hospitals: typical exposure levels and recommended control measures. *Br. J. Ind. Med.* 50: 107–111.

Lensing, H.H. and Oei, H.L. (1985). Investigations on the sporicidal and fungicidal activity of disinfectants. *Zentralbl. Bakteriol. Mikrobiol. Hyg. [B]* 181: 487–495.

Leonard, D.L. and Mills, S.E. (1997). Comparison of automated instrument cleaning: preliminary results. *Infect. Control Steril. Technol.* 20–23: 26–28.

Leong, D., Dorsey, G., and Klapp, M. (1987). Dilution of glutaraldehyde by automatic endoscope machine washers: the need for a quality control program. *Abstracts of the 14th Annual Educational Conference of Association for Practitioners in Infection Control*, Volume 108, p. 130.

Lettau, L.A., Bond, W.W., and McDougal, J.S. (1985). Hepatitis and diaphragm fitting. *JAMA* 254: 752.

Leung, J., Vallero, R., and Wilson, R. (2003). Surveillance cultures to monitor quality of gastrointestinal endoscope reprocessing. *Am. J. Gastroenterol.* 98.

Levenson, J.E. (1989). Corneal damage from improperly cleaned tonometer tips. *Arch. Ophthalmol.* 107: 1117.

Levine, D.S. (1988). Proctitis following colonoscopy. *Gastrointest. Endosc.* 34: 269–272.

Levy, S.B. (1998). The challenge of antibiotic resistance. *Sci. Am.* 278: 46–53.

Levy, R.V. (2001). Sterile filtration of liquids and gases. In: *Disinfection, Sterilization, and Preservation* (ed. S.S. Block), 795–822. Philadelphia, PA: Lippincott Williams & Wilkins.

Lewis, D.L. and Arens, M. (1995). Resistance of microorganisms to disinfection in dental and medical devices. *Nat. Med.* 1: 956–958.

Lewis, D.L. and Boe, R.K. (1992). Cross-infection risks associated with current procedures for using high-speed dental handpieces. *J. Clin. Microbiol.* 30: 401–406.

Lewis, D.L., Arens, M., Appleton, S.S. et al. (1992). Cross-contamination potential with dental equipment. *Lancet* 340: 1252–1254.

Lindbohm, M.L., Hemminki, K., Bonhomme, M.G. et al. (1991). Effects of paternal occupational exposure on spontaneous abortions. *Am. J. Public Health* 81: 1029–1033.

Lingel, N.J. and Coffey, B. (1992). Effects of disinfecting solutions recommended by the Centers for Disease Control on Goldmann tonometer biprisms. *J. Am. Optom. Assoc.* 63: 43–48.

Lipscomb, I.P., Sihota, A.K., Botham, M. et al. (2006). Rapid method for the sensitive detection of protein contamination on surgical instruments. *J. Hosp. Infect.* 62: 141–148.

Liu, Z., Stout, J.E., Tedesco, L. et al. (1994). Controlled evaluation of copper-silver ionization in eradicating *Legionella pneumophila* from a hospital water distribution system. *J. Infect. Dis.* 169: 919–922.

Lloyd-Evans, N., Springthorpe, V.S., and Sattar, S.A. (1986). Chemical disinfection of human rotavirus-contaminated inanimate surfaces. *J. Hyg. (Lond).* 97: 163–173.

Loffer, F.D. (1980). Disinfection vs. sterilization of gynecologic laparoscopy equipment. The experience of the Phoenix Surgicenter. *J. Reprod. Med.* 25: 263–266.

Loukili, N.H., Zink, E., Grandadam, S. et al. (2004). Effectiveness of detergent-disinfecting agents on *Escherichia coli* 54127 biofilm. *J. Hosp. Infect.* 57: 175–178.

Lowry, P.W. and Jarvis, W.R. (1991). Use of tap water and disinfection practices in outpatient settings. A survey of otolaryngologists. *Arch. Otolaryngol. Head Neck Surg.* 117: 886–888.

Lowry, P.W., Jarvis, W.R., Oberle, A.D. et al. (1988). *Mycobacterium chelonae* causing otitis media in an ear-nose-and-throat practice. *N. Engl. J. Med.* 319: 978–982.

Luebbert, P. (2000). Home care. In: *APIC Text of Infection Control and Epidemiology*, vol. 1 (ed. J.A. Pfeiffer), 44–47. Washington, DC: Association for Professionals in Infection control and epidemiology.

Lynch, D.A., Porter, C., Murphy, L., and Axon, A.T. (1992). Evaluation of four commercial automatic endoscope washing machines. *Endoscopy* 24: 766–770.

MacDougall, K.D. and Morris, C. (2006). Optimizing disinfectant application in healthcare facilities. *Infect. Control Today*: 62–67.

Madsen, R. and Moldenhauer, J. (eds.). *Contamination Control in Healthcare Product Manufacturing*, vol. 2.

Madsen, R. and Moldenhauer, J. (eds.). *Contamination Control in Healthcare Product Manufacturing*, vol. 3.

Madsen, R. and Moldenhauer, J. (eds.). *Contamination Control in Healthcare Product Manufacturing*, vol. 4.

Madsen, R. and Moldenhauer, J. (eds.). *Contamination Control in Healthcare Product Manufacturing*, vol. 5.

Maki, D.G. and Hassemer, C.A. (1987). Flash sterilization: carefully measured haste. *Infect. Control* 8: 307–310.

Maki, D.G., Alvarado, C.J., Hassemer, C.A., and Zilz, M.A. (1982). Relation of the inanimate hospital environment to endemic nosocomial infection. *N. Engl. J. Med.* 307: 1562–1566.

Malchesky, P.S. (1993). Biological indicators for a liquid chemical sterilizer. *Infect. Control Hosp. Epidemiol.* 14: 563–566.

Malchesky, P.S. (2001). Medical applications of peracetic acid. In: *Disinfection, Sterilization, and Preservation* (ed. S.S. Block), 979–996. Philadelphia, PA: Lippincott Williams & Wilkins.

Malik, R.E., Cooper, R.A., and Griffith, C.J. (2003). Use of audit tools to evaluate the efficacy of cleaning systems in hospitals. *Am. J. Infect. Control* 31: 181–187.

Mallison, G.F. and Standard, P.G. (1974). Safe storage times for sterile packs. *Hospitals* 48: 77–78, 80.

Maloney, J.M. and Kohut, R.D. (1987). Infection control, barrier protection and the treatment environment. *Dent. Hyg. (Chic).* 61: 310–313.

Mandel, G.L., Bennett, J.E., and Dolin, R. (2000). *Principles and Practices of Infectious Diseases*. New York: Livingstone.

Mangram, A.J., Horan, T.C., Pearson, M.L. et al. (1999). Guideline for prevention of surgical site infection, 1999. Hospital Infection Control Practices Advisory Committee. *Infect. Control Hosp. Epidemiol.* 20: 250–278.

Mannion, P.T. (1995). The use of peracetic acid for the reprocessing of flexible endoscopes and rigid cystoscopes and laparoscopes. *J. Hosp. Infect.* 29: 313–315.

Marchand, M., Delesvonx, R., and Claeys, C. (1958). The toxicity of ethylene oxide and a report on three fatal cases of poisoning. *Am. Arch. Indust. Health* 18: 60.

Marion, K., Freney, J., James, G. et al. (2006). Using an efficient biofilm detaching agent: an essential step for the improvement of endoscope reprocessing protocols. *J. Hosp. Infect.*, in press.

Marion-Ferey, K., Pasmore, M., Stoodley, P. et al. (2003). Biofilm removal from silicone tubing: an assessment of the efficacy of dialysis machine decontamination procedures using an in vitro model. *J. Hosp. Infect.* 53: 64–71.

Marks, P.J., Vipond, I.B., Regan, F.M. et al. (2003). A school outbreak of Norwalk-like virus: evidence for airborne transmission. *Epidemiol. Infect.* 131: 727–736.

Marler, L.M., Siders, J.A., Wolters, L.C. et al. (1992). Comparison of five cultural procedures for isolation of *Clostridium difficile* from stools. *J. Clin. Microbiol.* 30: 514–516.

Marshburn, P.B., Rutala, W.A., Wannamaker, N.S., and Hulka, J.F. (1991). Gas and steam sterilization of assembled versus disassembled laparoscopic equipment. Microbiologic studies. *J. Reprod. Med.* 36: 483–487.

Martin, L.S., McDougal, J.S., and Loskoski, S.L. (1985). Disinfection and inactivation of the human T lymphotropic virus type III/Lymphadenopathy-associated virus. *J. Infect. Dis.* 152: 400–403.

Martin, M.A., Reichelderfer, M., and 1991, and 1993 APIC Guidelines Committee (1994). APIC guidelines for infection prevention and control in flexible endoscopy. *Am. J. Infect. Control* 22: 19–38.

Martiny, H. and Floss, H. (2001). Residuals on medical devices following reprocessing. *J. Hosp. Infect.* 48 (Suppl): S88–S92.

Martiny, H., Floss, H., and Zuhlsdorf, B. (2004). The importance of cleaning for the overall results of processing endoscopes. *J. Hosp. Infect.* 56: S16–S22.

Masferrer, R. and Marquez, R. (1977). Comparison of two activated glutaraldehyde solutions: Cidex Solution and Sonacide. *Respir. Care* 22: 257–262.

Mast, S.T., Woolwine, J.D., and Gerdberding, J.L. (1993). Efficacy of gloves in reducing blood volumes transferred during simulated needlestick injury. *J. Infect. Dis.* 168: 1589–1592.

Mata-Portuguez, V.H., Perez, L.S., and Acosta-Gio, E. (2002). Sterilization of heat-resistant instruments with infrared radiation. *Infect. Control Hosp. Epidemiol.* 23.

Mayfield, J.L., Leet, T., Miller, J., and Mundy, L.M. (2000). Environmental control to reduce transmission of *Clostridium difficile*. *Clin. Infect. Dis.* 31: 995–1000.

Mayworm, D. (1984). Sterile shelf life and expiration dating. *J. Hosp. Supply Process Distrib.* 2: 32–35.

Mbithi, J.N., Springthorpe, V.S., and Sattar, S.A. (1990). Chemical disinfection of hepatitis A virus on environmental surfaces. *Appl. Environ. Microbiol.* 56: 3601–3604.

Mbithi, J.N., Springthorpe, V.S., Sattar, S.A., and Pacquette, M. (1993). Bactericidal, virucidal, and mycobactericidal activities of reused alkaline glutaraldehyde in an endoscopy unit. *J. Clin. Microbiol.* 31: 2988–2995.

McCarthy, G.M., Koval, J.J., John, M.A., and MacDonald, J.K. (1999). Infection control practices across Canada: do dentists follow the recommendations? *J. Can. Dent. Assoc.* 65: 506–511.

McCulloch, E.C. and Costigan, S. (1936). A comparison of the efficiency of phenol, liquor cresolis, formaldehyde, sodium hypochlorite and sodium hydroxide against *Eberthella typhi* at various temperatures. *J. Infect. Dis.* 59: 281–284.

McDonnell, G. and Pretzer, D. (2001). New and developing chemical antimicrobials. In: *Disinfection, Sterilization, and Preservation* (ed. S.S. Block), 431–443. Philadelphia, PA: Lippincott Williams & Wilkins.

McDonnell, G. and Russell, A.D. (1999). Antiseptics and disinfectants: activity, action, and resistance. *Clin. Microbiol. Rev.* 12: 147–179.

McFarland, L.V., Mulligan, M.E., Kwok, R.Y., and Stamm, W.E. (1989). Nosocomial acquisition of *Clostridium difficile* infection. *N. Engl. J. Med.* 320: 204–210.

McMurry, L.M., Oethinger, M., and Levy, S.B. (1998). Triclosan targets lipid synthesis. *Nature* 394: 531–532.

Medcomp (2000). Medcomp Frequently Asked Questions. [This link is no longer active: www.medcompnet.com/faq/faq/html].

Meenhorst, P.L., Reingold, A.L., Groothuis, D.G. et al. (1985). Water-related nosocomial pneumonia caused by *Legionella pneumophila* serogroups 1 and 10. *J. Infect. Dis.* 152: 356–364.

Mehta, A.C., Prakash, U.B.S., Garland, R. et al. (2006). Prevention of flexible bronchoscopy-associated infection. *Chest* 128: 1742–1755.

Melli, M.C., Giorgini, S., and Sertoli, A. (1986). Sensitization from contact with ethyl alcohol. *Contact Derm.* 14: 315.

Mentel, R. and Schmidt, J. (1973). Investigations on rhinovirus inactivation by hydrogen peroxide. *Acta Virol.* 17: 351–354.

Merianos, J.J. (2001). Surface-active agents. In: *Disinfection, Sterilization, and Preservation* (ed. S.S. Block), 283–320. Philadelphia, PA: Lippincott Williams & Wilkins.

Merighi, A., Contato, E., Scagliarini, R. et al. (1996). Quality improvement in gastrointestinal endoscopy: microbiologic surveillance of disinfection. *Gastrointest. Endosc.* 43: 457–462.

Merritt, K., Hitchins, V.M., and Brown, S.A. (2000). Safety and cleaning of medical materials and devices. *J. Biomed. Mater. Res.* 53: 131–136.

Mervine, J. and Temple, R. (1997). Using a microwave oven to disinfect intermittent-use catheters. *Rehabil. Nurs.* 22: 318–320.

Meyers, H., Brown-Elliott, B.A., Moore, D. et al. (2002). An outbreak of *Mycobacterium chelonae* infection following liposuction. *Clin. Infect. Dis.* 34: 1500–1507.

Michele, T.M., Cronin, W.A., Graham, N.M. et al. (1997). Transmission of *Mycobacterium tuberculosis* by a fiberoptic bronchoscope. Identification by DNA fingerprinting. *JAMA* 278: 1093–1095.

Middleton, A.M., Chadwick, M.V., and Gaya, H. (1997). Disinfection of bronchoscopes, contaminated in vitro with Mycobacterium tuberculosis, *Mycobacterium avium-intracellulare* and *Mycobacterium chelonae* in sputum, using stabilized, buffered peracetic acid solution ('Nu-Cidex'). *J. Hosp. Infect.* 37: 137–143.

Miles, R.S. (1991). What standards should we use for the disinfection of large equipment? *J. Hosp. Infect.* 18: 264–273.

Milki, A.A. and Fisch, J.D. (1998). Vaginal ultrasound probe cover leakage: implications for patient care. *Fertil. Steril.* 69: 409–411.

Miller, C.H. and Sheldrake, M.A. (1994). The ability of biological indicators to detect sterilization failures. *Am. J. Dent.* 7: 95–97.

Miller, M. (ed.). *Encyclopedia of Rapid Microbiol Methods*, vol. 4.

Michael Miller. *Encyclopedia of Rapid Microbiological Methods*, vols. 1, 2 and 3.

Miller, C.H., Riggen, S.D., Sheldrake, M.A., and Neeb, J.M. (1993). Presence of microorganisms in used ultrasonic cleaning solutions. *Am. J. Dent.* 6: 27–31.

Miner, N.A. and Ross, C. (1991). Clinical evaluation of ColdSpor, a glutaraldehyde-phenolic disinfectant. *Respir. Care* 36: 104–109.

Miner, N.A., McDowell, J.W., Willcockson, G.W. et al. (1977). Antimicrobial and other properties of a new stabilized alkaline glutaraldehyde disinfectant/sterilizer. *Am. J. Hosp. Pharm.* 34: 376–382.

Mitchell, D.H., Hicks, L.J., Chiew, R. et al. (1997). Pseudoepidemic of *Legionella pneumophila* serogroup 6 associated with contaminated bronchoscopes. *J. Hosp. Infect.* 37: 19–23.

Miuetzner, S., Schwille, R.C., Farley, A. et al. (1997). Efficacy of thermal treatment and copper-silver ionization for controlling *Legionella pneumophila* in high-volume hot water plumbing systems in hospitals. *Am. J. Infect. Control* 25: 452–457.

Miyaji, H., Kohli, Y., Azuma, T. et al. (1995). Endoscopic cross-infection with *Helicobacter pylori*. *Lancet* 345: 464.

Moken, M.C., McMurry, L.M., and Levy, S.B. (1997). Selection of multiple-antibiotic-resistant (mar) mutants of *Escherichia coli* by using the disinfectant pine oil: roles of the mar and acrAB loci. *Antimicrob. Agents Chemother.* 41: 2770–2772.

Jeanne Moldenhauer. *Environmental Monitoring: A Comprehensive Handbook*, vols. 1, 2 and 3.

Moldenhauer, J. *Recent Warning Letters Review for Preparation of an Aseptic Processing Inspection*, vol. 1.

Moldenhauer, J. *Recent Warning Letters Review for Preparation of a Non-Sterile Processing Inspection*, vol. 2.

Moldenhauer, J. (ed.). *Environmental Monitoring: A Comprehensive Handbook*, vol. 4.

Moldenhauer, J. (ed.). *Environmental Monitoring: A Comprehensive Handbook*, vol. 5.

Moldenhauer, J. (ed.). *Environmental Monitoring: A Comprehensive Handbook*, vol. 6.

Moldenhauer, J. (ed.). *Environmental Monitoring: A Comprehensive Handbook*, vol. 7.

Moldenhauer, J. (ed.). *Environmental Monitoring: A Comprehensive Handbook*, vol. 8.

Molinari, J.A., Gleason, M.J., Cottone, J.A., and Barrett, E.D. (1987). Comparison of dental surface disinfectants. *Gen. Dent.* 35: 171–175.

Moore, M.B. (1990). Acanthamoeba keratitis and contact lens wear: the patient is at fault. *Cornea* 9: S33–S35; discussion S39-40.

Moore, M.R., Pryor, M., Fields, B. et al. (2006). Introduction of monochloramine into a municipal water system: impact on colonization of buildings by *Legionella* spp. *Appl. Environ. Microbiol.* 72: 378–383.

Morton, H.E. (1950). The relationship of concentration and germicidal efficiency of ethyl alcohol. *Ann. N.Y. Acad. Sci.* 53: 191–196.

Morton, H.E. (1983). Alcohols. In: *Disinfection, Sterilization, and Preservation* (ed. S.S. Block), 225–239. Philadelphia, PA: Lea & Febiger.

Moses, F.M. and Lee, J. (2003). Surveillance cultures to monitor quality of gastrointestinal endoscope reprocessing. *Am. J. Gastroenterol.* 98: 77–81.

Mrvos, R., Dean, B.S., and Krenzelok, E.P. (1993). Home exposures to chlorine/chloramine gas: review of 216 cases. *South. Med. J.* 86: 654–657.

Muradali, D., Gold, W.L., Phillips, A., and Wilson, S. (1995). Can ultrasound probes and coupling gel be a source of nosocomial infection in patients undergoing sonography? An in vivo and in vitro study. *AJR Am. J. Roentgenol.* 164: 1521–1524.

Murdoch, H., Taylor, D., Dickinson, J. et al. (2006). Surface decontamination of surgical instruments: an ongoing dilemma. *J. Hosp. Infect.* 63: 432–438.

Murphy, C. (1998). Inactivated glutaraldehyde: lessons for infection control. *Am. J. Infect. Control* 26: 159–160.

Murray, P.R., Baron, E.J., Pfaller, M.A. et al. (2003). Manual of clinical microbiology. In: (eds. P.R. Murray, E.J. Baron, M.A. Pfaller, et al.). Washington, DC: American Society for Microbiology Press.

Murtough, S.M., Hiom, S.J., Palmer, M., and Russell, A.D. (2001). Biocide rotation in the healthcare setting: is there a case for policy implementation? *J. Hosp. Infect.* 48: 1–6.

Murtough, S.M., Hiom, S.J., Palmer, M., and Russell, A.D. (2002). A survey of rotational use of biocides in hospital pharmacy aseptic units. *J. Hosp. Infect.* 50: 228–231.

Muscarella, L.F. (1995). Sterilizing dental equipment. *Nat. Med.* 1: 1223–1225.

Muscarella, L.F. (1996). High-level disinfection or "sterilization" of endoscopes? *Infect. Control Hosp. Epidemiol.* 17: 183–187.

Muscarella, L.F. (1996). Advantages and limitations of automatic flexible endoscope reprocessors. *Am. J. Infect. Control* 24: 304–309.

Muscarella, L.F. (1998). Are all sterilization processes alike? *AORN J.* 67: 966–970, 973–976.

Muscarella, L.F. (2000). Automatic flexible endoscope reprocessors. *Gastrointest. Endosc. Clin. N. Am.* 10: 245–257.

Muscarella, L.F. (2000). Deja Vu…All over again? The importance of instrument drying. *Infect. Control Hosp. Epidemiol.* 21: 628–629.

Muscarella, L.F. (2001). Current instrument reprocessing practices: results of a national survey. *Gastrointest. Nurs.* 24: 253–260.

Muscarella, L.F. (2002). Application of environmental sampling to flexible endoscope reprocessing: the importance of monitoring the rinse water. *Infect. Control Hosp. Epidemiol.* 23: 285–289.

Muscarella, L.F. (2002). Leading a horse to water: are crucial lessons in endoscopy and outbreak investigations being learned? *Infect. Control Hosp. Epidemiol.* 23: 358–360.

Mwaniki, D.L. and Guthua, S.W. (1992). Occupational exposure to glutaraldehyde in tropical climates. *Lancet* 340: 1476–1477.

Nagington, J., Sutehall, G.M., and Whipp, P. (1983). Tonometer disinfection and viruses. *Br. J. Ophthalmol.* 67: 674–676.

Najdovski, L., Dragas, A.Z., and Kotnik, V. (1991). The killing activity of microwaves on some non-sporogenic and sporogenic medically important bacterial strains. *J. Hosp. Infect.* 19: 239–247.

Narang, H.K. and Codd, A.A. (1983). Action of commonly used disinfectants against enteroviruses. *J. Hosp. Infect.* 4: 209–212.

National Research Council (1964). Postoperative wound infections – the influence of ultraviolet irradiation of the operating room and of various other factors. *Ann. Surg.* 160: 1–125.

National Toxicology Program External (n.d.). Current version of this document may differ from original.

Neely, A.N. (2000). A survey of gram-negative bacteria survival on hospital fabrics and plastics. *J. Burn Care Rehabil.* 21: 523–527.

Neely, A.N. and Maley, M.P. (1999). The 1999 Lindberg award. 3% hydrogen peroxide for the gram-positive disinfection of fabrics. *J. Burn Care Rehabil.* 20: 471–477.

Neely, A.N. and Maley, M.P. (2000). Survival of enterococci and staphylococci on hospital fabrics and plastic. *J. Clin. Microbiol.* 38: 724–726.

Nelson, D. (2000). Newer technologies for endoscope disinfection: electrolyzed acid water and disposable-component endoscope systems. *Gastrointest. Endosc. Clin. N. Am.* 10: 319–328.

Nelson, D.B. (2003). Infectious disease complications of GI endoscopy: Part II, exogenous infections. *Gastrointest. Endosc.* 57: 695–711.

Nelson, D.B. and Muscarella, L.F. (2006). Current issues in endoscope reprocessing and infection control during gastrointestinal endoscopy. *World J. Gastroenterol.* 12: 3953–3964.

Nelson, D.B., Jarvis, W.R., Rutala, W.A. et al. (2003). Multi-society guideline for reprocessing flexible gastrointestinal endoscopes. *Infect. Control Hosp. Epidemiol.* 24: 532–537.

Nethercott, J.R., Holness, D.L., and Page, E. (1988). Occupational contact dermatitis due to glutaraldehyde in health care workers. *Contact Derm.* 18: 193–196.

Newman, M.A. and Kachuba, J.B. (1992). Glutaraldehyde: a potential health risk to nurses. *Gastroenterol. Nurs.* 14: 296–300, discussion 300-1.

Newman, K.A., Tenney, J.H., Oken, H.A. et al. (1984). Persistent isolation of an unusual *Pseudomonas species* from a phenolic disinfectant system. *Infect. Control* 5: 219–222.

Ninemeier, J.D. (1998). *Central Service Technical Manual*. Chicago, IL: International Association of Healthcare Central Service Materiel Management.

NIOSH (1981). *Formaldehyde: Evidence of Carcinogenicity*. NIOSH Current Intelligence Bulletin 34. DHEW (NIOSH) Publication No. 81-111.

Nomura, K., Ogawa, M., Miyamoto, H. et al. (2004). Antibiotic susceptibility of glutaraldehyde-tolerant *Mycobacterium chelonae* from bronchoscope washing machines. *J. Hosp. Infect.* 32: 185–188.

Norback, D. (1988). Skin and respiratory symptoms from exposure to alkaline glutaraldehyde in medical services. *Scand. J. Work Environ. Health* 14: 366–371.

Noyce, J.O., Michels, H., and Keevil, C.W. (2006). Potential use of copper surfaces to reduce survival of epidemic methicillin-resistant *Staphylococcus aureus* in the healthcare environment. *J. Hosp. Infect.* 63: 289–297.

Nye, R.N. and Mallory, T.B. (1923). A note on the fallacy of using alcohol for the sterilization of surgical instruments. *Boston Med. Surg. J.* 189: 561–563.

Nystrom, B. (1981). Disinfection of surgical instruments. *J. Hosp. Infect.* 2: 363–368.

Nystrom, B. (1991). New technology for sterilization and disinfection. *Am. J. Med.* 91: 264S–266S.

O'Rourke, E., Runyan, D., O'Leary, J., and Stern, J. (2003). Contaminated iodophor in the operating room. *Am. J. Infect. Control* 31: 255–256.

Obee, P.C., Griffith, C.J., Cooper, R.A. et al. (2005). Real-time monitoring in managing the decontamination of flexible gastrointestinal endoscopes. *Am. J. Infect. Control* 33: 202–206.

Occupational Health and Safety Administration (2002). *OSHA Fact Sheet: Formaldehyde: Occupational Safety and Health Administration*. U.S. Department of Labor.

Occupational Health and Safety Administration (2002). *Chemical Sampling Information-Ethylene Chlorohydrin*. Occupational Safety and Health Administration.

Occupational Health and Safety Administration (n.d.). Hazard Communication Standard. 29 CFR 1910.1200, OSHA, Washington, DC.

Occupational Health and Safety Administration (n.d.). Ethylene Oxide Standard. Vol. 29 CFR 1910.1047, OSHA, Washington, DC.

Occupational Safety and Health Administration (1991). Occupational exposure to bloodborne pathogens; final rule. *Fed. Regist.* 56: 64003–64182.

Occupational Safety and Health Administration (1991). *OSHA Amends Formaldehyde Standard*, 1. Occupational Safety and Health News.

Occupational Safety and Health Administration (1992). *OSHA Instruction CPL 2-2.44C*. Washington, DC: Office of Health Compliance Assistance.

Occupational Safety and Health Administration (1993). Air contaminants final rule. *Fed. Regist.* 58: 35338–35351.

Occupational Safety and Health Administration (1997). *OSHA Memorandum from Stephen Mallinger*. Washington, DC: EPA-registered disinfectants for HIV/HBV.

Occupational Safety and Health Administration (2002). *Formaldehyde: OSHA Fact Sheet*. Occupational Safety and Health Administration.

Occupational Safety and Health Administration (2002). *Ethylene Oxide: OSHA Fact Sheet*. Occupational Safety and Health Administration.

Occupational Safety and Health Administration (n.d.). Formaldehyde Standard. Vol. 29 CFR 1910.1048, Washington, DC.

Kevin O'Donnell. *Cold Chain Chronicles: A Practitioners Outside-The-Box Perspectives on the Importance of Temperature-Sensitive Drug Stewardship*.

Odwin, C.S., Fleischer, A.C., Kepple, D.M., and Chiang, D.T. (1990). Probe covers and disinfectants for transvaginal transducers. *J. Diagn. Med. Sonogr.* 6: 130–135.

Ohara, T., Itoh, Y., and Itoh, K. (1998). Ultrasound instruments as possible vectors of staphylococcal infection. *J. Hosp. Infect.* 40: 73–77.

Oie, S. and Kamiya, A. (1996). Microbial contamination of antiseptics and disinfectants. *Am. J. Infect. Control* 24: 389–395.

Oie, S. and Kamiya, A. (2002). Assessment of and intervention for the misuse of aldehyde disinfectants in Japan. *Infect. Control Hosp. Epidemiol.* 23: 98–99.

Oie, S., Huang, Y., Kamiya, A. et al. (1996). Efficacy of disinfectants against biofilm cells of methicillin-resistant *Staphylococcus aureus*. *Microbios* 85: 223–230.

Okpara-Hofmann, J., Knoll, M., Durr, M. et al. (2005). Comparison of low-temperature hydrogen peroxide gas plasma sterilization for endoscopes using various Sterrad models. *J. Hosp. Infect.* 59: 280–285.

Omidbakhsh, N. and Sattar, S.A. (2006). Broad-spectrum microbicidal activity, toxicologic assessment, and materials compatibility of a new generation of accelerated hydrogen peroxide-based environmental surface disinfectant. *Am. J. Infect. Control* 34: 251–257.

Orsi, G.B., Filocamo, A., Di Stefano, L., and Tittobello, A. (1997). Italian national survey of digestive endoscopy disinfection procedures. *Endoscopy* 29: 732–738; quiz 739-40.

Overton, D., Burgess, J.O., Beck, B., and Matis, B. (1989). Glutaraldehyde test kits: evaluation for accuracy and range. *Gen. Dent.* 37: 126, 128.

Oxborrow, G.S. and Berube, R. (1991). Sterility testing-validation of sterilization processes, and sporicide testing. In: *Disinfection, Sterilization, and Preservation* (ed. S.S. Block), 1047–1057. Philadelphia, PA: Lea & Febiger.

Palenik, C.J. and Cumberlander, N.D. (1993). Effects of steam sterilization on the contents of sharps containers. *Am. J. Infect. Control* 21: 28–33.

Palmer, P.H. and Yeoman, D.M. (1972). A study to assess the value of disinfectants when washing ward floors. *Med. J. Aust.* 2: 1237–1239.

Pang, J., Perry, P., Ross, A., and Forbes, G.M. (2002). Bacteria-free rinse water for endoscope disinfection. *Gastrointest. Endosc.* 56: 402–406.

Parisi, A.N. and Young, W.E. (1991). Sterilization with ethylene oxide and other gases. In: *Disinfection, Sterilization, and Preservation* (ed. S.S. Block), 580–595. Philadelphia, PA: Lea & Febiger.

Parker, H.H. and Johnson, R.B. (1995). Effectiveness of ethylene oxide for sterilization of dental handpieces. *J. Dent.* 23: 113–115.

Parnes, C.A. (1997). Efficacy of sodium hypochlorite bleach and "alternative" products in preventing transfer of bacteria to and from inanimate surfaces. *Environ. Health* 59: 14–20.

Parrott, P.L., Terry, P.M., Whitworth, E.N. et al. (1982). *Pseudomonas aeruginosa* peritonitis associated with contaminated poloxamer-iodine solution. *Lancet* 2: 683–685.

Payan, C., Cottin, J., Lemarie, C., and Ramont, C. (2001). Inactivation of hepatitis B virus in plasma by hospital in-use chemical disinfectants assessed by a modified HepG2 cell culture. *J. Hosp. Infect.* 47: 282–287.

Payan, C., Pivert, A., Kampf, G. et al. (2004). Assessment of new chemical disinfectants for HBV virucidal activity in a cell culture model. *J. Hosp. Infect.* 56 (Suppl): S58–S63.

PDA Technical Series: Pharmaceutical Glass PDF Single user.

Penna, T.C., Ferraz, C.A., and Cassola, M.A. (1999). The presterilization microbial load on used medical devices and the effectiveness of hydrogen peroxide gas plasma against *Bacillus subtilis* spores. *Infect. Control Hosp. Epidemiol.* 20: 465–472.

Pentella, M.A., Fisher, T., Chandler, S. et al. (2000). Are disinfectants accurately prepared for use in hospital patient care areas? *Infect. Control Hosp. Epidemiol.* 21: 103.

Pepose, J.S., Linette, G., Lee, S.F., and MacRae, S. (1989). Disinfection of Goldmann tonometers against human immunodeficiency virus type 1. *Arch. Ophthalmol.* 107: 983–985.

Pepper, R.E. (1980). Comparison of the activities and stabilities of alkaline glutaraldehyde sterilizing solutions. *Infect. Control* 1: 90–92.

Perez, J., Springthorpe, S., and Sattar, S.A. (2005). Activity of selected oxidizing microbicides against spores of *Clostridium difficile*: relevance to environmental control. *Am. J. Infect. Control* 33: 320–325.

Perkins, J.J. (1969). *Principles and Methods of Sterilization in Health Sciences*. Charles C Thomas: Springfield, IL.

Perkins, R.E., Bodman, H.A., Kundsin, R.B., and Walter, C.W. (1981). Monitoring steam sterilization of surgical instruments: a dilemma. *Appl. Environ. Microbiol.* 42: 383–387.

Petersen, N.J., Carson, L.A., Doto, I.L. et al. (1982). Microbiologic evaluation of a new glutaraldehyde-based disinfectant for hemodialysis systems. *Trans. Am. Soc. Artif. Intern. Organs* 28: 287–290.

Petrocci, A.N. (1983). Surface active agents: quaternary ammonium compounds. In: *Disinfection, Sterilization, and Preservation* (ed. S.S. Block), 309–329. Philadelphia, PA: Lea & Febiger.

Pfeifer, M. (1998). Standardised test soil blood 1: composition, preparation, application. *Zentr. Steril.* 6: 381–385.

Pfeifer, M. (1998). Blood as a soil on surgical instruments: chemical profile, cleaning, detection. *Zentr. Steril.* 6: 304–310.

Phillips, J., Hulka, B., Hulka, J. et al. (1977). Laparoscopic procedures: The American Association of Gynecologic Laparoscopists' Membership Survey for 1975. *J. Reprod. Med.* 18: 227–232.

Platt, R., Lehr, J.L., Marino, S. et al. (1988). Safe and cost-effective cleaning of pressure-monitoring transducers. *Infect. Control Hosp. Epidemiol.* 9: 409–416.

Paul Pluta. *Cleaning and Cleaning Validation*, vols. 1 and 2.

Portner, D.M. and Hoffman, R.K. (1968). Sporicidal effect of peracetic acid vapor. *Appl. Microbiol.* 16: 1782–1785.

Power, E.G. and Russell, A.D. (1990). Sporicidal action of alkaline glutaraldehyde: factors influencing activity and a comparison with other aldehydes. *J. Appl. Bacteriol.* 69: 261–268.

Price, D. and Ahearn, D.G. (1988). Incidence and persistence of *Pseudomonas aeruginosa* in whirlpools. *J. Clin. Microbiol.* 26: 1650–1654.

Prince, D.L., Prince, R.N., and Prince, H.N. (1990). Inactivation of human immunodeficiency virus type 1 and herpes simplex virus type 2 by commercial hospital disinfectants. *Chem. Times Trends* 13: 13–16.

Prince, D.L., Prince, H.N., Thraenhart, O. et al. (1993). Methodological approaches to disinfection of human hepatitis B virus. *J. Clin. Microbiol.* 31: 3296–3304.

Prindle, R.F. (1983). Phenolic compounds. In: *Disinfection, Sterilization, and Preservation* (ed. S.S. Block), 197–224. Philadelphia, PA: Lea & Febiger.

Purohit, A., Kopferschmitt-Kubler, M.C., Moreau, C. et al. (2000). Quaternary ammonium compounds and occupational asthma. *Int. Arch. Occup. Environ. Health* 73: 423–427.

Pyle, B.H., Broadaway, S.C., and McFeters, G.A. (1992). Efficacy of copper and silver ions with iodine in the inactivation of *Pseudomonas cepacia*. *J. Appl. Bacteriol.* 72: 71–79.

Ransjo, U., Engstrom, L., Hakansson, P. et al. (2001). A test for cleaning and disinfection processes in a washer-disinfector. *APMIS* 109: 299–304.

Ray, A.J., Hoyen, C.K., Taub, T.F. et al. (2002). Nosocomial transmission of vancomycin-resistant enterococci from surfaces. *JAMA* 287: 1400–1401.

Reckitt & Colman (1997). *Sporox Sterilant and High-Level Disinfectant Technical Report*, 1–12. Montvale, NJ: Reckitt & Colman.

Reich, R.R., Fleming, W., and Burgess, D.J. (1996). Sterilization validation: it's not just for industry. *Infect. Control Steril. Technol.* 2.

Reichert, M. (1997). Preparation of supplies for terminal sterilization. In: *Sterilization Technology for the Health Care Facility* (eds. M. Reichert and J.H. Young), 36–50. Gaithersburg, MD: Aspen Publication.

Reichert, M. and Young, J.H. (1997). *Sterilization Technology for the Health Care Facility*, vol. 307. Gaithersburg, MD: Aspen Publication.

Reiss, I., Borkhardt, A., Fussle, R. et al. (2000). Disinfectant contaminated with *Klebsiella oxytoca* as a source of sepsis in babies. *Lancet* 356: 310.

Reisz, G.R. and Gammon, R.S. (1986). Toxic pneumonitis from mixing household cleaners. *Chest* 89: 49–52.

Rejchrt, S., Cermak, P., Pavlatova, L. et al. (2004). Bacteriologic testing of endoscopes after high-level disinfection. *Gastrointest. Endosc.* 60: 76–78.

Resnick, L., Veren, K., Salahuddin, S.Z. et al. (1986). Stability and inactivation of HTLV-III/LAV under clinical and laboratory environments. *JAMA* 255: 1887–1891.

Rey, J.F., Halfon, P., Feryn, J.M. et al. (1995). Risk of transmission of hepatitis C virus by digestive endoscopy. *Gastroenterol. Clin. Biol.* 19: 346–349.

Reynolds, C.D., Rhinehart, E., Dreyer, P., and Goldmann, D.A. (1992). Variability in reprocessing policies and procedures for flexible fiberoptic endoscopes in Massachusetts hospitals. *Am. J. Infect. Control* 20: 283–290.

Rheinbaben, F.V., Schunemann, S., Grob, T., and Wolff, M.H. (2000). Transmission of viruses via contact in a household setting: experiments using bacteriophage OX174 as a model virus. *J. Hosp. Infect.* 46: 61–66.

Rhodes, P., Zelner, L., and Laufman, H. (1982). A new disposable bowie-Dick-type test pack for prevacuum high-temperature sterilizers. *Med. Instrum.* 16: 117–120.

Rice, E.W., Clark, R.M., and Johnson, C.H. (1999). Chlorine inactivation of *Escherichia coli* O157:H7. *Emerg. Infect. Dis.* 5: 461–463.

Richburg, F.A., Reidy, J.J., Apple, D.J., and Olson, R.J. (1986). Sterile hypopyon secondary to ultrasonic cleaning solution. *J. Cataract Refract. Surg.* 12: 248–251.

Rideout, K., Teschke, K., Dimich-Ward, H., and Kennedy, S.M. (2005). Considering risks to healthcare workers from glutaraldehyde alternatives in high-level dinfection. *J. Hosp. Infect.* 59: 4–11.

Ries, M.D., Weaver, K., and Beals, N. (1996). Safety and efficacy of ethylene oxide sterilized polyethylene in total knee arthroplasty. *Clin. Orthop.*: 159–163.

Riley, R., Beanland, C., and Bos, H. (2002). Establishing the shelf life of flexible colonoscopes. *Gastroenterol. Nurs.* 25: 114–119.

Riney, S., Grimes, M., Khalife, K. et al. (1991). Diarrhea associated with disinfection of sigmoidoscopes. [abstract]. *Am. J. Infect. Control* 19: 109.

Roberts, C.G. (2001). Studies on the bioburden on medical devices and the importance of cleaning. In: *Disinfection, Sterilization and Antisepsis: Principles and Practices in Healthcare Facilities* (ed. W.A. Rutala), 63–69. Washington, DC: Association for Professional in Infection Control and Epidemiology.

Roberts, C. and Antonoplos, P. (1998). Inactivation of human immunodeficiency virus type 1, hepatitis A virus, respiratory syncytial virus, vaccinia virus, herpes simplex virus type 1, and poliovirus type 2 by hydrogen peroxide gas plasma sterilization. *Am. J. Infect. Control* 26: 94–101.

Roberts, C.G. and Chan-Myers, M. (1998). Mycobactericidal activity of dilute ortho-phthalaldehyde solutions. *Abstracts in Environmental and General Applied Microbiology, Q-265, ASM 98th General Meeting*, Atlanta, Georgia, USA, pp. 464–465.

Robison, R.A., Bodily, H.L., Robinson, D.F., and Christensen, R.P. (1988). A suspension method to determine reuse life of chemical disinfectants during clinical use. *Appl. Environ. Microbiol.* 54: 158–164.

Rodrigues, C., Mehta, A.C., Jha, U. et al. (2001). Nosocomial *Mycobacterium chelonae* infection in laparoscopic surgery. *Infect. Control Hosp. Epidemiol.* 22: 474–475.

Rohrer, M.D. and Bulard, R.A. (1985). Microwave sterilization. *J. Am. Dent. Assoc.* 110: 194–198.

Rohrer, M.D., Terry, M.A., Bulard, R.A. et al. (1986). Microwave sterilization of hydrophilic contact lenses. *Am. J. Ophthalmol.* 101: 49–57.

Roncoroni, A.J., Casewell, M.W., and Phillips, I. (1980). The disinfection of clinically contaminated Matburn suction pumps and baby incubators in an 'Aseptor' formalin cabinet. *J. Hosp. Infect.* 1: 251–259.

Rooks, V.J., Yancey, M.K., Elg, S.A., and Brueske, L. (1996). Comparison of probe sheaths for endovaginal sonography. *Obstet. Gynecol.* 87: 27–29.

Roosendaal, R., Kuipers, E.J., van den Brule, A.J. et al. (1993). Detection of *Helicobacter pylori* DNA by PCR in gastrointestinal equipment. *Lancet* 341: 900.

Rosaspina, S., Salvatorelli, G., Anzanel, D., and Bovolenta, R. (1994). Effect of microwave radiation on *Candida albicans*. *Microbios* 78: 55–59.

Rosaspina, S., Salvatorelli, G., and Anzanel, D. (1994). The bactericidal effect of microwaves on *Mycobacterium bovis* dried on scalpel blades. *J. Hosp. Infect.* 26: 45–50.

Rotter, M.L. (1997). Handwashing, hand disinfection, and skin disinfection. In: *Prevention and Control of Nosocomial Infections* (ed. R.P. Wenzel), 691–709. Baltimore, MD: Williams and Wilkins.

Rowland, A.S., Baird, D.D., Shore, D.L. et al. (1996). Ethylene oxide exposure may increase the risk of spontaneous abortion, preterm birth, and postterm birth. *Epidemiology* 7: 363–368.

Royce, A. and Bowler, C. (1961). Ethylene oxide sterilisation-some experiences and some practical limitations. *J. Pharm. Pharmacol.* 13: 87t–94t.

Rozen, P., Somjen, G.J., Baratz, M. et al. (1994). Endoscope-induced colitis: description, probable cause by glutaraldehyde, and prevention. *Gastrointest. Endosc.* 40: 547–553.

Rubbo, S.D., Gardner, J.F., and Webb, R.L. (1967). Biocidal activities of glutaraldehyde and related compounds. *J. Appl. Bacteriol.* 30: 78–87.

Ruden, H. and Daschner, F. (2002). Should we routinely disinfect floors? *J. Hosp. Infect.* 51: 309.

Rudolph, H. and Hilbert, M. (1997). Practical testing of the new plasma sterilizer "Sterrad 100S" in the Diakonkrankenhaus Rotenburg. *Zentr. Steril.* 5: 207–215.

Rusin, P., Orosz-Coughlin, P., and Gerba, C. (1998). Reduction of faecal coliform, coliform and heterotrophic plate count bacteria in the household kitchen and bathroom by disinfection with hypochlorite cleaners. *J. Appl. Microbiol.* 85: 819–828.

Rusin, P., Bright, K., and Gerba, C. (2003). Rapid reduction of *Legionella pneumophila* on stainless steel with zeolite coatings containing silver and zinc ions. *Lett. Appl. Microbiol.* 36: 69–72.

Russell, A.D. (1981). Neutralization procedures in the evaluation of bactericidal activity. In: *Disinfectants: Their Use and Evaluation of Effectiveness* (eds. C.H. Collins, M.C. Allwood, S.F. Bloomfield and A. Fox), 45–59. London: Academic Press.

Russell, A.D. (1990). Bacterial spores and chemical sporicidal agents. *Clin. Microbiol. Rev.* 3: 99–119.

Russell, A.D. (1994). Glutaraldehyde: current status and uses. *Infect. Control Hosp. Epidemiol.* 15: 724–733.

Russell, A.D. (1997). Plasmids and bacterial resistance to biocides. *J. Appl. Microbiol.* 83: 155–165.

Russell, A.D. (1998). Bacterial resistance to disinfectants: present knowledge and future problems. *J. Hosp. Infect.* 43: S57–S68.

Russell, A.D. (1999). Factors influencing the efficacy of antimicrobial agents. In: *Principles and Practice of Disinfection, Preservation and Sterilization* (eds. A.D. Russell, W.B. Hugo and G.A.J. Ayliffe), 95–123. Oxford: Blackwell Science.

Russell, A.D. (1999). Ultraviolet radiation. In: *Principles and Practices of Disinfection, Preservation and Sterilization* (eds. A.D. Russell, W.B. Hugo and G.A.J. Ayliffe), 688–702. Oxford: Blackwell Science.

Russell, A.D. (1999). Ionizing radiation. In: *Principles and Practices of Disinfection, Preservation and Sterilization* (eds. A.D. Russell, W.B. Hugo and G.A.J. Ayliffe), 675–687. Oxford: Blackwell Science.

Russell, A.D. (2001). Principles of antimicrobial activity and resistance. In: *Disinfection, Sterilization, and Preservation* (ed. S.S. Block), 31–55. Philadelphia, PA: Lippincott Williams & Wilkins.

Russell, A.D. (2004). Bacterial adaptation and resistance to antiseptics, disinfectants and preservatives is not a new phenomenon. *J. Hosp. Infect.* 57: 97–104.

Russell, A.D. (2004). Factors influencing the efficacy of germicides. In: *Disinfection, Sterilization and Antisepsis: Principles, Practices, Challenges, and New Research* (ed. W.A. Rutala), 162–170. Washington, DC: Association for Professionals in Infection Control and Epidemiology.

Russell, A.D. and Maillard, J.Y. (2000). Reaction and response-relationship between antibiotic resistance and resistance to antiseptics and disinfectants. *Am. J. Infect. Control* 28: 204–206.

Russell, A.D. and McDonnell, G. (2000). Concentration: a major factor in studying biocidal action. *J. Hosp. Infect.* 44: 1–3.

Russell, A.D. and Russell, N.J. (1995). Biocides: activity, action and resistance. In: *Fifty Years of Antimicrobials: Past Perspectives and Future Trends* (eds. P.A. Hunter, G.K. Darby and N.J. Russell), 327–365. England: Cambridge University Press.

Russell, A.D., Ahonkhai, I., and Rogers, D.T. (1979). Microbiological applications of the inactivation of antibiotics and other antimicrobial agents. *J. Appl. Bacteriol.* 46: 207–245.

Russell, A.D., Suller, M.T., and Maillard, J.Y. (1999). Do antiseptics and disinfectants select for antibiotic resistance? *J. Med. Microbiol.* 48: 613–615.

Russell, A.D., Hugo, W.B., and Ayliffe, G.A.J. (1999). *Principles and Practice of Disinfection, Preservation and Sterilization*. Oxford, England: Blackwell Scientific Publications.

Rutala, W.A. (1990). APIC guideline for selection and use of disinfectants. *Am. J. Infect. Control* 18: 99–117.

Rutala, W.A. (1991). Disinfection and flash sterilization in the operating room. *J. Ophthal. Nurs. Technol.* 10: 106–115.

Rutala, W.A. (1996). APIC guideline for selection and use of disinfectants. 1994, 1995, and 1996 APIC Guidelines Committee. Association for Professionals in Infection Control and Epidemiology, Inc. *Am. J. Infect. Control* 24: 313–342.

Rutala, W.A. (1996). Disinfection and sterilization of patient-care items. *Infect. Control Hosp. Epidemiol.* 17: 377–384.

Rutala, W.A. (1997). Disinfection, sterilization and waste disposal. In: *Prevention and Control of Nosocomial Infections* (ed. R.P. Wenzel), 539–593. Baltimore, MD: Williams and Wilkins.

Rutala, W.A. (1999). Selection and use of disinfectants in healthcare. In: *Infection Control and Hospital Epidemiology* (ed. C.G. Mayhall), 1161–1187. Philadelphia, PA: Lippincott Williams & Wilkins.

Rutala, W.A. and Cole, E.C. (1984). Antiseptics and disinfectants–safe and effective? *Infect. Control* 5: 215–218.

Rutala, W.A. and Cole, E.C. (1987). Ineffectiveness of hospital disinfectants against bacteria: a collaborative study. *Infect. Control* 8: 501–506.

Rutala, W.A. and Hamory, B.H. (1989). Expanding role of hospital epidemiology: employee health–chemical exposure in the health care setting. *Infect. Control Hosp. Epidemiol.* 10: 261–266.

Rutala, W.A. and Shafer, K.M. (1996). General information on cleaning, disinfection, and sterilization. In: *APIC Infection Control and Applied Epidemiology: Principles and Practice* (ed. J.A. Pfeiffer), 15.1–15.17. St. Louis, MO: Mosby.

Rutala, W.A. and Weber, D.J. (1995). FDA labeling requirements for disinfection of endoscopes: a counterpoint. *Infect. Control Hosp. Epidemiol.* 16: 231–235.

Rutala, W.A. and Weber, D.J. (1996). Low-temperature sterilization technology: do we need to redefine sterilization? *Infect. Control Hosp. Epidemiol.* 17: 89–91.

Rutala, W.A. and Weber, D.J. (1997). Water as a reservoir of nosocomial pathogens. *Infect. Control Hosp. Epidemiol.* 18: 609–616.

Rutala, W.A. and Weber, D.J. (1997). Uses of inorganic hypochlorite (bleach) in health-care facilities. *Clin. Microbiol. Rev.* 10: 597–610.

Rutala, W.A. and Weber, D.J. (1998). Principles of disinfecting patient-care items. In: *Disinfection, Sterilization, and Antisepsis in Healthcare* (ed. W.A. Rutala), 133–149. Champlain, NY: Polyscience Publications.

Rutala, W.A. and Weber, D.J. (1998). Clinical effectiveness of low-temperature sterilization technologies. *Infect. Control Hosp. Epidemiol.* 19: 798–804.

Rutala, W.A. and Weber, D.J. (1999). Disinfection of endoscopes: review of new chemical sterilants used for high-level disinfection. *Infect. Control Hosp. Epidemiol.* 20: 69–76.

Rutala, W.A. and Weber, D.J. (1999). Importance of lumen flow in liquid chemical sterilization. *Am. J. Infect. Control* 20: 458–459.

Rutala, W.A. and Weber, D.J. (1999). Infection control: the role of disinfection and sterilization. *J. Hosp. Infect.* 43: S43–S55.

Rutala, W.A. and Weber, D.J. (2000). Choosing a sterilization wrap. *Infect. Control Today* 4: 64,70.

Rutala, W.A. and Weber, D.J. (2001). Surface disinfection: should we do it? *J. Hosp. Infect.* 48 (Suppl. A): S64–S68.

Rutala, W.A. and Weber, D.J. (2001). New disinfection and sterilization methods. *Emerg. Inf. Dis.* 7: 348–353.

Rutala, W.A. and Weber, D.J. (2001). A review of the use of gowns and drapes (single use and reusable) in healthcare. *Infect. Control Hosp. Epidemiol.* 22: 248–257.

Rutala, W.A. and Weber, D.J. (2002). Should we routinely disinfect floors? Reply to Professor F. Daschner. *J. Hosp. Infect.* 51: 309–311.

Rutala, W.A. and Weber, D.J. (2004). Reprocessing endoscopes: United States perspective. *J. Hosp. Infect.* 56: S27–S39.

Rutala, W.A. and Weber, D.J. (2004). The benefits of surface disinfection. *Am. J. Infect. Control* 32: 226–231.

Rutala, W.A., Stiegel, M.M., and Sarubbi, F.A. Jr. (1982). Decontamination of laboratory microbiological waste by steam sterilization. *Appl. Environ. Microbiol.* 43: 1311–1316.

Rutala, W.A., Clontz, E.P., Weber, D.J., and Hoffmann, K.K. (1991). Disinfection practices for endoscopes and other semicritical items. *Infect. Control Hosp. Epidemiol.* 12: 282–288.

Rutala, D.R., Rutala, W.A., Weber, D.J., and Thomann, C.A. (1991). Infection risks associated with spirometry. *Infect. Control Hosp. Epidemiol.* 12: 89–92.

Rutala, W.A., Cole, E.C., Wannamaker, N.S., and Weber, D.J. (1991). Inactivation of *Mycobacterium tuberculosis* and *Mycobacterium bovis* by 14 hospital disinfectants. *Am. J. Med.* 91: 267S–271S.

Rutala, W.A., Gergen, M.F., and Weber, D.J. (1993). Sporicidal activity of chemical sterilants used in hospitals. *Infect. Control Hosp. Epidemiol.* 14: 713–718.

Rutala, W.A., Gergen, M.F., and Weber, D.J. (1993). Inactivation of *Clostridium difficile* spores by disinfectants. *Infect. Control Hosp. Epidemiol.* 14: 36–39.

Rutala, W.A., Gergen, M.F., and Weber, D.J. (1993). Evaluation of a rapid readout biological indicator for flash sterilization with three biological indicators and three chemical indicators. *Infect. Control Hosp. Epidemiol.* 14: 390–394.

Rutala, W.A., Jones, S.M., and Weber, D.J. (1996). Comparison of a rapid readout biological indicator for steam sterilization with four conventional biological indicators and five chemical indicators. *Infect. Control Hosp. Epidemiol.* 17: 423–428.

Rutala, W.A., Stiegel, M.M., Sarubbi, F.A., and Weber, D.J. (1997). Susceptibility of antibiotic-susceptible and antibiotic-resistant hospital bacteria to disinfectants. *Infect. Control Hosp. Epidemiol.* 18: 417–421.

Rutala, W.A., Cole, E.C., Thomann, C.A., and Weber, D.J. (1998). Stability and bactericidal activity of chlorine solutions. *Infect. Control Hosp. Epidemiol.* 19: 323–327.

Rutala, W.A., Gergen, M.F., and Weber, D.J. (1998). Comparative evaluation of the sporicidal activity of new low-temperature sterilization technologies: ethylene oxide, 2 plasma sterilization systems, and liquid peracetic acid. *Am. J. Infect. Control* 26: 393–398.

Rutala, W.A., Gergen, M.F., Jones, J.F., and Weber, D.J. (1998). Levels of microbial contamination on surgical instruments. *Am. J. Infect. Control* 26: 143–145.

Rutala, W.A., Weber, D.J., and Chappell, K.J. (1999). Patient injury from flash-sterilized instruments. *Infect. Control Hosp. Epidemiol.* 20: 458.

Rutala, W.A., Gergen, M.F., and Weber, D.J. (1999). Sporicidal activity of a new low-temperature sterilization technology: the Sterrad 50 sterilizer. *Infect. Control Hosp. Epidemiol.* 20: 514–516.

Rutala, W.A., Barbee, S.L., Aguiar, N.C. et al. (2000). Antimicrobial activity of home disinfectants and natural products against potential human pathogens. *Infect. Control Hosp. Epidemiol.* 21: 33–38.

Rutala, W.A., Weber, D.J., and Gergen, M.F. (2000). Studies on the disinfection of VRE-contaminated surfaces. *Infect. Control Hosp. Epidemiol.* 21: 548.

Rutala, W.A., Weber, D.J., Gergen, M.F., and Gratta, A.R. (2000). Efficacy of a washer-pasteurizer for disinfection of respiratory-care equipment. *Infect. Control Hosp. Epidemiol.* 21: 333–336.

Rutala, W.A., White, M.S., Gergen, M.F., and Weber, D.J. (2006). Bacterial contamination of keyboards: efficacy and functional impact of disinfectants. *Infect. Control Hosp. Epidemiol.* 27: 372–377.

Rutala, W.A., Peacock, J.E., Gergen, M.F. et al. (2006). Efficacy of hospital germicides against adenovirus 8, a common cause of epidemic keratoconjunctivitis in health care facilities. *Antimicrob. Agents Chemother.* 50: 1419–1424.

Ryan, C.K. and Potter, G.D. (1995). Disinfectant colitis. Rinse as well as you wash. *J. Clin. Gastroenterol.* 21: 6–9.

Sagripanti, J.L. and Bonifacino, A. (1996). Comparative sporicidal effect of liquid chemical germicides on three medical devices contaminated with spores of *Bacillus subtilis*. *Am. J. Infect. Control* 24: 364–371.

Sagripanti, J.L. and Bonifacino, A. (1996). Comparative sporicidal effects of liquid chemical agents. *Appl. Environ. Microbiol.* 62: 545–551.

Sagripanti, J.L. and Bonifacino, A. (1997). Effects of salt and serum on the sporicidal activity of liquid disinfectants. *J. AOAC Int.* 80: 1198–1207.

Sagripanti, J.L., Eklund, C.A., Trost, P.A. et al. (1997). Comparative sensitivity of 13 species of pathogenic bacteria to seven chemical germicides. *Am. J. Infect. Control* 25: 335–339.

Sakagami, Y. and Kajimura, K. (2002). Bactericidal activities of disinfectants against vancomycin-resistant enterococci. *J. Hosp. Infect.* 50: 140–144.

Saknimit, M., Inatsuki, I., Sugiyama, Y., and Yagami, K. (1988). Virucidal efficacy of physico-chemical treatments against coronaviruses and parvoviruses of laboratory animals. *Jikken Dobutsu.* 37: 341–345.

Salinas, E., Sasich, L., Hall, D.H. et al. (1981). Acute ethylene oxide intoxication. *Drug Intell. Clin. Pharm.* 15: 384–386.

Sampson, M.N. and Muir, A.V. (2002). Not all super-oxidized waters are the same. *J. Hosp. Infect.* 52: 227–228.

Sanborn, M.R., Wan, S.K., and Bulard, R. (1982). Microwave sterilization of plastic tissue culture vessels for reuse. *Appl. Environ. Microbiol.* 44: 960–964.

Sanders, F.T. (2001). Environmental protection agency's role in the regulation of antimicrobial pesticides in the United States. In: *Disinfection, Sterilization and Antisepsis: Principles and Practices in Healthcare Facilities* (ed. W.A. Rutala), 28–40. Washington, DC: Association for Professional in Infection Control and Epidemiology.

Sanders, F.T. and Morrow, M.S. (2004). The EPA's role in the regulation of antimicrobial pesticides in the United States. In: *Disinfection, Sterilization and Antisepsis: Principles, Practices, Challenges, and New Research* (ed. W.A. Rutala), 29–41. Washington, DC: Association for Professionals in Infection Control and Epidemiology.

Tim Sandle. *Microbiological Culture Media: A Complete Guide for Pharmaceutical and Healthcare Manufacturers.*

Tim Sandle. *Recalls of Pharmaceutical Products: Eliminating Contamination and Adulteration Causes.*

Sarin, P.S., Scheer, D.I., and Kross, R.D. (1985). Inactivation of human T-cell lymphotropic retrovirus (HTLV-III) by LD. *N. Engl. J. Med.* 313: 1416.

Sarin, P.S., Scheer, D.I., and Kross, R.D. (1990). Inactivation of human T-cell lymphotropic retrovirus. *Environ. Microbiol* 56: 1423–1428.

Sartor, C., Charrel, R.N., de Lamballerie, X. et al. (1999). Evaluation of a disinfection procedure for hysteroscopes contaminated by hepatitis C virus. *Infect. Control Hosp. Epidemiol.* 20: 434–436.

Sartor, C., Jacomo, V., Duvivier, C. et al. (2000). Nosocomial *Serratia marcescens* infections associated with extrinsic contamination of a liquid nonmedicated soap. *Infect. Control Hosp. Epidemiol.* 21: 196–199.

Sattar, S.A. (1997). Effect of liquid chemical germicides on mycobacteria including multi-drug resistant isolates of *Mycobacteria tuberculosis. Abstracts of the 37th Interscience Conference on Antimicrobial Agents of Chemotherapy* (September 28-October 1 1997); Toronto, Ontario, Canada; E166, 1997.

Sattar, S.A. (1998). Microbicidal testing of germicides: an update. In: *Disinfection, Sterilization, and Antisepsis in Healthcare* (ed. W.A. Rutala), 227–240. Champlain, NY: Polyscience Publications.

Sattar, S.A. (2004). Microbicides and the environmental control of nosocomial viral infections. *J. Hosp. Infect.* 56 (Suppl): S64–S69.

Sattar, S.A. and Springthorpe, V.S. (1991). Survival and disinfectant inactivation of the human immunodeficiency virus: a critical review. *Rev. Infect. Dis.* 13: 430–447.

Sattar, S.A. and Springthorpe, V.S. (2001). New methods for efficacy testing of disinfectants and antiseptics. In: *Disinfection, Sterilization and Antisepsis: Principles and Practices in Healthcare Facilities* (ed. W.A. Rutala), 174–186. Washington, DC: Association for Professional in Infection Control and Epidemiology.

Sattar, S.A. and Springthorpe, V.S. (2004). Recent developments in methods for testing the germicidal activity of disinfectants and antiseptics. In: *Disinfection, Sterilization and Antisepsis: Principles, Practices, Challenges, and New Research* (ed. W.A. Rutala), 180–188. Washington, DC: Association for Professionals in Infection Control and Epidemiology.

Sattar, S.A., Raphael, R.A., Lochnan, H., and Springthorpe, V.S. (1983). Rotavirus inactivation by chemical disinfectants and antiseptics used in hospitals. *Can. J. Microbiol.* 29: 1464–1469.

Sattar, S.A., Lloyd-Evans, N., Springthorpe, V.S., and Nair, R.C. (1986). Institutional outbreaks of rotavirus diarrhoea: potential role of fomites and environmental surfaces as vehicles for virus transmission. *J. Hyg. (Lond).* 96: 277–289.

Sattar, S.A., Springthorpe, V.S., Karim, Y., and Loro, P. (1989). Chemical disinfection of non-porous inanimate surfaces experimentally contaminated with four human pathogenic viruses. *Epidemiol. Infect.* 102: 493–505.

Sattar, S.A., Jacobsen, H., Springthorpe, V.S. et al. (1993). Chemical disinfection to interrupt transfer of rhinovirus type 14 from environmental surfaces to hands. *Appl. Environ. Microbiol.* 59: 1579–1585.

Sattar, S.A., Jacobsen, H., Rahman, H. et al. (1994). Interruption of rotavirus spread through chemical disinfection. *Infect. Control Hosp. Epidemiol.* 15: 751–756.

Sattar, S.A., Springthorpe, V.S., Conway, B., and Xu, Y. (1994). Inactivation of the human immunodeficiency virus: an update. *Rev. Med. Microbiol.* 5: 139–150.

Sattar, S.A., Taylor, Y.E., Paquette, M., and Rubino, J. (1996). In-hospital evaluation of 7.5% hydrogen peroxide as a disinfectant for flexible endoscopes. *Can. J. Infect. Control* 11: 51–54.

Sattar, S.A., Springthorpe, V.S., and Rochon, M. (1998). A product based on accelerated and stabilized hydrogen peroxide: evidence for broad-spectrum germicidal activity. *Can. J. Infect. Control*: 123–130.

Sattar, S.A., Adegbunrin, O., and Ramirez, J. (2002). Combined application of simulated reuse and quantitative carrier test to assess high-level disinfection: experiments with an accelerated hydrogen peroxide-based formulation. *Am. J. Infect. Control* 30: 449–457.

Sattar, S.A., Springthorpe, V.S., and Adegbunrin, O. (2004). Is *Bacillus subtilis* (ATCC 19659) a suitable surrogate for evaluating microbicides against *Bacillus anthracis*.

Association for Official Analytical Chemists International Annual Meeting, St. Louis, Missouri, USA.
Saurina, G., Landman, D., and Quale, J.M. (1997). Activity of disinfectants against vancomycin-resistant *Enterococcus faecium*. *Infect. Control Hosp. Epidemiol.* 18: 345–347.
Schaeffer, A.J., Jones, J.M., and Amundsen, S.K. (1980). Bacterial effect of hydrogen peroxide on urinary tract pathogens. *Appl. Environ. Microbiol.* 40: 337–340.
Schembre, D.B. (2000). Infectious complications associated with gastrointestinal endoscopy. *Gastrointest. Endosc. Clin. N. Am.* 10: 215–232.
Schmitt, S. (ed.). *Good Distribution Practice: A Handbook for Healthcare Manufacturers and Suppliers*, vol. 1.
Schmitt, S. (ed.). *Good Distribution Practice: A Handbook for Healthcare Manufacturers and Suppliers*, vol. 2.
Schneider, P.M. (1994). Low-temperature sterilization alternatives in the 1990s. *Tappi J.* 77: 115–119.
Schneider, P.M. (1998). Emerging low temperature sterilization technologies (non-FDA approved). In: *Disinfection, Sterilization, and Antisepsis in Healthcare* (ed. W.A. Rutala), 79–92. Champlain, NY: Polyscience Publications.
Schneider, P.M. (2004). New technologies for disinfection and sterilization. In: *Disinfection, Sterilization and Antisepsis: Principles, Practices, Challenges, and New Research* (ed. W.A. Rutala), 127–139. Washington, DC: Association for Professionals in Infection Control and Epidemiology.
Schneider, P.M., Reich, R.R., Kirckof, S.S., and Foltz, W.G. (2004). Perfomance of various steam sterilization indicators under optimum and sub-optimum exposure conditions. In: *Disinfection, Sterilization and Antisepsis: Principles, Practices, Challenges, and New Research* (ed. W.A. Rutala), 200–223. Washington DC: Association for Professionals in Infection Control and Epidemiology.
Schnuch, A., Uter, W., Geier, J. et al. (1998). Contact allergies in healthcare workers. Results from the IVDK. *Acta Derm. Venereol.* 78: 358–363.
Schultz, J.K. (1990). Decontamination alternative. *Infect. Control Hosp. Epidemiol.* 11: 8–9.
Schultz, J.K. (1997). Decontamination: recommended practices. In: *Sterilization Technology for the Health Care Facility* (eds. M. Reichert and J.H. Young), 10–20. Gaithersburg, MD: Aspen Publication.
Sciortino, C.V., Xia, E.L., and Mozee, A. (2004). Assessment of a novel approach to evaluate the outcome of endoscope reprocessing. *Infect. Control Hosp. Epidemiol.* 25: 284–290.
Scott, E. and Bloomfield, S.F. (1990). The survival and transfer of microbial contamination via cloths, hand and utensils. *J. Appl. Bacteriol.* 68: 271–278.

Scott, E. and Bloomfield, S.F. (1990). Investigations of the effectiveness of detergent washing, drying and chemical disinfection on contamination of cleaning cloths. *J. Appl. Bacteriol.* 68: 279–283.

Scott, E.M. and Gorman, S.P. (1991). Glutaraldehyde. In: *Disinfection, Sterilization, and Preservation* (ed. S.S. Block), 596–616. Philadelphia, PA: Lea & Febiger.

Scott, E.M. and Gorman, S.P. (2001). Glutaraldehyde. In: *Disinfection, Sterilization, and Preservation* (ed. S.S. Block), 361–381. Philadelphia, PA: Lippincott Williams & Wilkins.

Scott, G.H. and Williams, J.C. (1990). Susceptibility of *Coxiella burnetii* to chemical disinfectants. *Ann. N.Y. Acad. Sci.* 590: 291–296.

Scott, E., Bloomfield, S.F., and Barlow, C.G. (1982). An investigation of microbial contamination in the home. *J. Hyg. (Lond).* 89: 279–293.

Seballos, R.J., Walsh, A.L., and Mehta, A.C. (1995). Clinical evaluation of a liquid chemical sterilization system for flexible bronchoscopes. *J. Bronch.* 2: 192–199.

Sehulster, L.M. and Anderson, R.L. (1998). Susceptibility of glycopeptide-intermediate resistant *Staphylococcus aureus* (GISA) to surface disinfectants, hand washing chemicals, and a skin antiseptic. *Abstract Y-3. 98th General Meeting of American Society for Microbiology*, May, 1998, p. 547.

Sehulster, L., Chinn, R.Y.W., and Healthcare Infection Control Practices Advisory Committee (2003). Guidelines for environmental infection control in health-care facilities. *MMWR* 52: 1–44.

Selkon, J.B., Babb, J.R., and Morris, R. (1999). Evaluation of the antimicrobial activity of a new super-oxidized water, Sterilox®, for the disinfection of endoscopes. *J. Hosp. Infect.* 41: 59–70.

Sensakovic, J.W. and Smith, L.G. (1982). Nosocomial ultraviolet keratoconjunctivitis. *Infect. Control* 3: 475–476.

Shaham, J., Levi, Z., Gurvich, R. et al. (2000). Hematological changes in hospital workers due to chronic exposure to low levels of ethylene oxide. *J. Occup. Environ. Med.* 42: 843–850.

Shapshak, P., McCoy, C.B., Rivers, J.E. et al. (1993). Inactivation of human immunodeficiency virus-1 at short time intervals using undiluted bleach. *J. Acquir. Immune Defic. Syndr.* 6: 218–219.

Shapshak, P., McCoy, C.B., Shah, S.M. et al. (1994). Preliminary laboratory studies of inactivation of HIV-1 in needles and syringes containing infected blood using undiluted household bleach. *J. Acquir. Immune Defic. Syndr.* 7: 754–759.

Shechmeister, I.L. (1991). Sterilization by ultraviolet irradiation. In: *Disinfection, Sterilization, and Preservation* (ed. S.S. Block), 553–565. Philadelphia, PA: Lea & Febiger.

Shere, L. (1948). Some comparisons of the disinfecting properties of hypochlorites and quaternary ammonium compounds. *Milk Plant Mon.*: 66–69.

Shetty, N., Srinivasan, S., Holton, J., and Ridgway, G.L. (1999). Evaluation of microbicidal activity of a new disinfectant: Sterilox® 2500 against *Clostridium difficile* spores, *Helicobacter pylori*, vancomycin resistant *Enterococcus species, Candida albicans* and several *Mycobacterium species. J. Hosp. Infect.* 41: 101–105.

Shickman, M.D., Guze, L.B., and Pearce, M.L. (1959). Bacteremia following cardiac catheterization. *N. Engl. J. Med.* 260: 1164–1166.

Shimada, T., Terano, A., Ota, S. et al. (1996). Risk of iatrogenic transmission of *Helicobacter pylori* by gastroscopes. *Lancet* 347: 1342–1343.

Silberman, H.D. (2001). Non-inflatable sterile sheath for introduction of the flexible nasopharyngolaryngoscope. *Ann Otol, Rhinol, Laryngol* 110: 385–387.

Silvany, R.E., Dougherty, J.M., McCulley, J.P. et al. (1990). The effect of currently available contact lens disinfection systems on *Acanthamoeba castellanii* and *Acanthamoeba polyphaga. Ophthalmology* 97: 286–290.

Silverman, J., Vazquez, J.A., Sobel, J.D., and Zervos, M.J. (1999). Comparative in vitro activity of antiseptics and disinfectants versus clinical isolates of *Candida species. Infect. Control Hosp. Epidemiol.* 20: 676–684.

Silverstone, S.E. and Hill, D.E. (1999). Evaluation of sterilization of dental handpieces by heating in synthetic compressor lubricant. *Gen. Dent.* 47: 158–160.

Simmons, B.P. (1983). CDC guidelines for the prevention and control of nosocomial infections. Guideline for hospital environmental control. *Am. J. Infect. Control* 11: 97–120.

Simons, C., Walsh, S.E., Maillard, J.Y., and Russell, A.D. (2000). A note: ortho-phthalaldehyde: proposed mechanism of action of a new antimicrobial agent. *Lett. Appl. Microbiol.* 31: 299–302.

Singh, S. and Schaaf, N.G. (1989). Dynamic sterilization of titanium implants with ultraviolet light. *Int. J. Oral Maxillofac. Implants* 4: 139–146.

Singh, J., Bhatia, R., Gandhi, J.C. et al. (1998). Outbreak of viral hepatitis B in a rural community in India linked to inadequately sterilized needles and syringes. *Bull. World Health Organ.* 76: 93–98.

Singh, D., Kaur, H., Gardner, W.G., and Treen, L.B. (2002). Bacterial contamination of hospital pagers. *Infect. Control Hosp. Epidemiol.* 23: 274–276.

Sisco, V., Winters, L.L., Zange, L.L., and Brennan, P.C. (1988). Efficacy of various methods of sterilization of acupuncture needles. *J. Manip. Physiol. Therap.* 11: 94–97.

Sizun, J., Yu, M.W., and Talbot, P.J. (2000). Survival of human coronaviruses 229E and OC43 in suspension and after drying onsurfaces: a possible source ofhospital-acquired infections. *J. Hosp. Infect.* 46: 55–60.

Skoutelis, A.T., Westenfelder, G.O., Beckerdite, M., and Phair, J.P. (1994). Hospital carpeting and epidemiology of *Clostridium difficile. Am. J. Infect. Control* 22: 212–217.

Smeets, E., Koonman, J., van der Sande, F. et al. (2003). Prevention of biofilm formation in dialysis water treatment systems. *Kidney Int.* 63: 1574–1576.

Smith, C.R. (1947). Alcohol as a disinfectant against the tubercle bacillus. *Public Health Rep.* 62: 1285–1295.

Smith, G.E. (1986). Glass bead sterilization of orthodontic bands. *Am. J. Orthod. Dentofacial Orthop.* 90: 243–249.

Smith, C.R., Nishihara, H., Golden, F. et al. (1950). The bactericidal effect of surface-active agents on tubercle bacilli. *Public Health Rep.* 48: 1588–1600.

Smith, T.L., Iwen, P.C., Olson, S.B., and Rupp, M.E. (1998). Environmental contamination with vancomycin-resistant enterococci in an outpatient setting. *Infect. Control Hosp. Epidemiol.* 19: 515–518.

Smith, A., Dickson, M., Aitken, J., and Bagg, J. (2002). Contaminated dental instruments. *J. Hosp. Infect.* 51: 233–235.

Snydman, D.R. (1976). Hepatitis B infection from medical personnel. *JAMA* 236: 1009.

Society of Gastroenterology Nurses and Associates (2000). Standards for infection control and reprocessing of flexible gastrointestinal endoscopes. *Gastroenterol. Nurs.* 23: 172–179.

Society of Gastroenterology Nurses and Associates (2000). Guideline for the use of high-level disinfectants and sterilants for reprocessing of flexible gastrointestinal endoscopes. *Gastroenterol. Nurs.* 23: 180–187.

Society of Gastroenterology Nurses and Associates (2006). Standards of infection control in reprocessing of flexible gastrointestinal endoscopes. *Gastroenterol. Nurs.* 29: 142–148.

Society of Gastroenterology Nurses and Associates (n.d.). Guideline for the use of high-level disnfectants and sterilants in reprocessing of flexible gastrointestinal endoscopes.

Sokol, W.N. (2004). Nine episodes of anaphylaxis following cytoscopy caused by Cidex OPA (ortho-phthalaldehyde) high-level disinfectant in 4 patients after cystoscopy. *J. Allergy Clin. Immunol.* 114: 392–397.

Sommermeyer, L. and Frobisher, M. (1973). Laboratory studies on disinfection of rectal thermometers. *Nurs. Res.* 2: 85–89.

Sorin, M., Segal-Maurer, S., Urban, C. et al. (2001). Nosocomial transmission of imipenem-resistant *Pseudomonas aeruginosa* following bronchoscopy associated with improper connection to the STERIS System 1 Processor. *Infect. Control Hosp. Epidemiol.* 22: 409–413.

Soukiasian, S.H., Asdourian, G.K., Weiss, J.S., and Kachadoorian, H.A. (1988). A complication from alcohol-swabbed tonometer tips. *Am. J. Ophthalmol.* 105: 424–425.

Spach, D.H., Silverstein, F.E., and Stamm, W.E. (1993). Transmission of infection by gastrointestinal endoscopy and bronchoscopy. *Ann. Intern. Med.* 118: 117–128.

Spaulding, E.H. (1964). Alcohol as a surgical disinfectant. *AORN J.* 2: 67–71.

Spaulding, E.H. (1968). Chemical disinfection of medical and surgical materials. In: *Disinfection, Sterilization, and Preservation* (eds. C. Lawrence and S.S. Block), 517–531. Philadelphia, PA: Lea & Febiger.

Spire, B., Barre-Sinoussi, F., Montagnier, L., and Chermann, J.C. (1984). Inactivation of lymphadenopathy associated virus by chemical disinfectants. *Lancet* 2: 899–901.

Springthorpe, V.S., Grenier, J.L., Lloyd-Evans, N., and Sattar, S.A. (1986). Chemical disinfection of human rotaviruses: efficacy of commercially-available products in suspension tests. *J. Hyg. (Lond).* 97: 139–161.

Srinivasan, A., Wolfenden, L.L., Song, X. et al. (2003). An outbreak of *Pseudomonas aeruginosa* infections associated with flexible bronchoscopes. *N. Engl. J. Med.* 348: 221–227.

Srinivasan, A., Bova, G., Ross, T. et al. (2003). A 17-month evaluation of a chlorine dioxide water treatment system to control *Legionella species* in a hospital water supply. *Infect. Control Hosp. Epidemiol.* 24: 575–579.

Stanley, P. (1998). Destruction of a glutaraldehyde-resistant mycobacterium by a per-oxygen disinfectant. (Abstract). *Am. J. Infect. Control* 26: 185.

Steve, L., Goodhart, P., and Alexander, J. (1979). Hydrotherapy burn treatment: use of chloramine-T against resistant microorganisms. *Arch. Phys. Med. Rehabilit.* 60: 301–303.

Stonehill, A.A., Krop, S., and Borick, P.M. (1963). Buffered glutaraldehyde – a new chemical sterilizing solution. *Am. J. Hosp. Pharm.* 20: 458–465.

Storment, J.M., Monga, M., and Blanco, J.D. (1997). Ineffectiveness of latex condoms in preventing contamination of the transvaginal ultrasound transducer head. *South. Med. J.* 90: 206–208.

Stout, J.E. and Yu, V.L. (2003). Experience of the first 16 hospitals using copper-silver ionization for *Legionella* control: implications for the evaluation of other disinfection modalities. *Infect Control Hosp Epidemiol* 24: 563–568.

Stout, J.E., Lin, Y.S., Goetz, A.M., and Muder, R.R. (1998). Controlling *Legionella* in hospital water systems: experience with the superheat-and-flush method and copper-silver ionization. *Infect. Control Hosp. Epidemiol.* 19: 911–914.

Streckenbach, S.C. and Alston, T.A. (2003). Perioral stains after ortho-phthalaldehyde disinfectioon of echo probes. *Anesthesiol* 99: 1032.

Strzelecki, L.R. and Nelson, J.H. (1989). Evaluation of closed container flash sterilization system. *Orthoped. Nurs.* 8: 21–24.

Sykes, G. (1939). The influence of germicides on the dehydrogenases of Bact. coli. Part I. The succinic acid dehydrogenase of *Bact. coli. J. Hyg. (Camb).* 39: 463–469.

Sykes, G. (1965). *Disinfection and Sterilization*. London: E & FN Spon Ltd.

Takashina, M. (2001). Application of a bioluminescent method for checking cleaning results. *Zentr. Steril.* 9: 248–258.

Talbot, G.H., Skros, M., and Provencher, M. (1985). 70% alcohol disinfection of transducer heads: experimental trials. *Infect. Control* 6: 237–239.

Tan, J.A. and Schnagl, R.D. (1981). Inactivation of a rotavirus by disinfectants. *Med. J. Aust.* 1: 19–23.

Tanaka, H., Hirakata, Y., Kaku, M. et al. (1996). Antimicrobial activity of superoxidized water. *J. Hosp. Infect.* 34: 43–49.

Tanaka, N., Fujisawa, T., Daimon, T. et al. (2000). The use of electrolyzed solutions for the cleaning and disinfecting of dialyzers. *Artif. Organs* 24: 921–928.

Taormina, P.J. and Beuchat, L.R. (1999). Behavior of enterohemorrhagic *Escherichia coli* O157:H7 on alfalfa sprouts during the sprouting process as influenced by treatments with various chemicals. *J. Food Prot.* 62: 850–856.

Taormina, P.J. and Beuchat, L.R. (1999). Comparison of chemical treatments to eliminate enterohemorrhagic *Escherichia coli* O157:H7 on alfalfa seeds. *J. Food Prot.* 62: 318–324.

Taurasi, R. (2004). Comfortable PPE? Maximum tray weight? *Healthc. Purch. News* 48.

Taylor, D.M. (1999). Inactivation of prions by physical and chemical means. *J. Hosp. Infect.* 43 (Suppl): S69–S76.

Taylor, E.W., Mehtar, S., Cowan, R.E., and Feneley, R.C. (1994). Endoscopy: disinfectants and health. Report of a meeting held at the Royal College of Surgeons of England, February 1993. *J. Hosp. Infect.* 28: 5–14.

Tennent, J.M., Lyon, B.R., Gillespie, M.T. et al. (1985). Cloning and expression of *Staphylococcus aureus* plasmid-mediated quaternary ammonium resistance in *Escherichia coli*. *Antimicrob. Agents Chemother.* 27: 79–83.

Tennent, J.M., Lyon, B.R., Midgley, M. et al. (1989). Physical and biochemical characterization of the qacA gene encoding antiseptic and disinfectant resistance in *Staphylococcus aureus*. *J. Gen. Microbiol.* 135: 1–10.

Terleckyj, B. and Axler, D.A. (1987). Quantitative neutralization assay of fungicidal activity of disinfectants. *Antimicrob. Agents Chemother.* 31: 794–798.

Thompson, R.L., Haley, C.E., Searcy, M.A. et al. (1984). Catheter-associated bacteriuria. Failure to reduce attack rates using periodic instillations of a disinfectant into urinary drainage systems. *JAMA* 251: 747–751.

Tiller, J.C., Liao, C.J., Lewis, K., and Klibanov, A.M. (2001). Designing surfaces that kill bacteria on contact. *Proc. Natl. Acad. Sci. U.S.A.* 98: 5981–5985.

Tilley, F.W. and Schaffer, J.M. (1926). Relation between the chemical constitution and germicidal activity of the monohydric alcohols and phenols. *J. Bacteriol.* 12: 303–309.

Timm, D. and Gonzales, D. (1997). Effect of sterilization on microstructure and function of microsurgical scissors. *Surg. Serv. Manage.* 3: 47–49.

Tokars, J.I., Miller, E.R., Alter, M.J., and Arduino, M.J. (1998). National surveillance of dialysis associated diseases in the United States, 1995. *ASAIO J.* 44: 98–107.

Tokars, J.I., Miller, E.R., Alter, M.J., and Arduino, M.J. (2000). National surveillance of dialysis-associated diseases in the United States, 1997. *Semin. Dial.* 13: 75–85.

Townsend, T.R., Wee, S.B., and Koblin, B. (1982). An efficacy evaluation of a synergized glutaraldehyde-phenate solution in disinfecting respiratory therapy equipment contaminated during patient use. *Infect. Control* 3: 240–244.

Townsend, D.E., Ashdown, N., Greed, L.C., and Grubb, W.B. (1984). Transposition of gentamicin resistance to staphylococcal plasmids encoding resistance to cationic agents. *J. Antimicrob. Chemother.* 14: 115–124.

Tsuji, S., Kawano, S., Oshita, M. et al. (1999). Endoscope disinfection using acidic electrolytic water. *Endoscopy* 31: 528–535.

Tucker, R.C., Lestini, B.J., and Marchant, R.E. (1996). Surface analysis of clinically used expanded PTFE endoscopic tubing treated by the STERIS PROCESS. *ASAIO J.* 42: 306–313.

Tulis, J.J. (1972). Formaldehyde as a gas. In: *Industrial Sterilization* (eds. G.B. Phillips and W.S. Miller), 209–238. Durham: Duke University Press.

Tunuguntla, A. and Sullivan, M.J. (2004). Monitoring quality of flexible endoscopic disinfection by microbiologic surveillance cultures. *Tenn. Med.*: 453–456.

Turner, F.J. (1983). Hydrogen peroxide and other oxidant disinfectants. In: *Disinfection, Sterilization, and Preservation* (ed. S.S. Block), 240–250. Philadelphia, PA: Lea & Febiger.

Turner, A.G., Higgins, M.M., and Craddock, J.G. (1974). Disinfection of immersion tanks (Hubbard) in a hospital burn unit. *Arch. Environ. Health* 28: 101–104.

Turner, N.A., Russell, A.D., Furr, J.R., and Lloyd, D. (1999). *Acanthamoeba* spp., antimicrobial agents and contact lenses. *Sci. Prog.* 82: 1–8.

Tyler, R. and Ayliffe, G.A. (1987). A surface test for virucidal activity of disinfectants: preliminary study with herpes virus. *J. Hosp. Infect.* 9: 22–29.

Tyler, R., Ayliffe, G.A., and Bradley, C. (1990). Virucidal activity of disinfectants: studies with the poliovirus. *J. Hosp. Infect.* 15: 339–345.

Ulatowski, T.A. (1998). Current activities concerning the premarket evaluation of infection control devices at the Food and Drug Administration. In: *Disinfection, Sterilization, and Antisepsis in Healthcare* (ed. W.A. Rutala), 1–7. Champlain, NY: Polyscience Publications.

Ulatowski, T.A. (2004). FDA: Reuse of single-use devices. In: *Disinfection, Sterilization and Antisepsis: Principles, Practices, Challenges, and New Research* (ed. W.A. Rutala), 15–23. Washington, DC: Association for Professionals in Infection Control and Epidemiology.

United States General Accounting Office (1990). *Disinfectants: EPA Lacks Assurance They Work*. United States General Accounting Office.

Urata, M., Isomot, H., Murase, K. et al. (2003). Comparison of the microbicidal activities of superoxidized and ozonated water in the disinfection of endoscopes. *J. Intern. Med. Res.* 31: 299–306.

Urayama, S., Kozarek, R.A., Sumida, S. et al. (1996). Mycobacteria and glutaraldehyde: is high-level disinfection of endoscopes possible? *Gastrointest. Endosc.* 43: 451–456.

Uttley, A.H. and Simpson, R.A. (1994). Audit of bronchoscope disinfection: a survey of procedures in England and Wales and incidents of mycobacterial contamination. *J. Hosp. Infect.* 26: 301–308.

Van Bueren, J., Simpson, R.A., Salman, H. et al. (1995). Inactivation of HIV-1 by chemical disinfectants: sodium hypochlorite. *Epidemiol. Infect.* 115: 567–579.

Vassal, S., Favennec, L., Ballet, J.J., and Brasseur, P. (1998). Hydrogen peroxide gas plasma sterilization is effective against *Cryptosporidium parvum* oocysts. *Am. J. Infect. Control* 26: 136–138.

Velandia, M., Fridkin, S.K., Cardenas, V. et al. (1995). Transmission of HIV in dialysis centre. *Lancet* 345: 1417–1422.

Venkitanarayanan, K.S., Ezeike, G.O., Hung, Y.C., and Doyle, M.P. (1999). Inactivation of *Escherichia coli* O157:H7 and *Listeria monocytogenes* on plastic kitchen cutting boards by electrolyzed oxidizing water. *J. Food Prot.* 62: 857–860.

Ventura, L.M. and Dix, R.D. (1987). Viability of herpes simplex virus type 1 on the applanation tonometer. *Am. J. Ophthalmol.* 103: 48–52.

Vesley, D., Norlien, K.G., Nelson, B. et al. (1992). Significant factors in the disinfection and sterilization of flexible endoscopes. *Am. J. Infect. Control* 20: 291–300.

Vesley, D., Langholz, A.C., Rohlfing, S.R., and Foltz, W.E. (1992). Fluorimetric detection of a *Bacillus stearothermophilus* spore-bound enzyme, α-D-glucosidase, for rapid identification of flash sterilization failure. *Appl. Environ. Microbiol.* 58: 717–719.

Vesley, D., Nellis, M.A., and Allwood, P.B. (1995). Evaluation of a rapid readout biological indicator for 121°C gravity and 132°C vacuum-assisted steam sterilization cycles. *Infect. Control Hosp. Epidemiol.* 16: 281–286.

Vesley, D., Melson, J., and Stanley, P. (1999). Microbial bioburden in endoscope reprocessing and an in-use evaluation of the high-level disinfection capabilities of Cidex* PA. *Gastroenterol. Nurs.* 22: 63–68.

James Vesper. *Risk Assessment and Risk Management in the Pharmaceutical Industry: Clear and Simple.*

Vickery, K., Deva, A.K., Zou, J. et al. (1999). Inactivation of duck hepatitis B virus by a hydrogen peroxide gas plasma sterilization system: laboratory and 'in use' testing. *J. Hosp. Infect.* 41: 317–322.

Vickery, K., Pajkos, A., and Cossart, Y. (2004). Removal of biofilm from endoscopes: evaluation of detergent efficiency. *Am. J. Infect. Control* 32: 170–176.

Vizcaino-Alcaide, M.J., Herruzo-Cabrera, R., and Fernandez-Acenero, M.J. (2003). Comparison of the disinfectant efficacy of Persafe and 2% glutaraldehyde in in vitro tests. *J. Hosp. Infect.* 53: 124–128.

Wallace, R.J. Jr., Brown, B.A., and Driffith, D.E. (1998). Nosocomial outbreaks/pseudo-outbreaks caused by nontuberculous mycobacteria. *Annu. Rev. Microbiol.* 52: 453–490.

Wallace, C.G., Agee, P.M., and Demicco, D.D. (1995). Liquid chemical sterilization using peracetic acid. An alternative approach to endoscope processing. *ASAIO J.* 41: 151–154.

Wallbank, A.M., Drulak, M., Poffenroth, L. et al. (1978). Wescodyne: lack of activity against poliovirus in the presence of organic matter. *Health Lab. Sci.* 15: 133–137.

Wallhausser, K.H. (1979). Is the removal of microorganisms by filtration really a sterilization method? *J. Parenter. Drug Assoc.* 33: 156–170.

Walsh, S.E., Maillard, J.Y., Simons, C., and Russell, A.D. (1999). Studies on the mechanisms of the antibacterial action of ortho-phthalaldehyde. *J. Appl. Microbiol.* 87: 702–710.

Walsh, S.E., Maillard, J.Y., and Russell, A.D. (1999). Ortho-phthalaldehyde: a possible alternative to glutaraldehyde for high level disinfection. *J. Appl. Microbiol.* 86: 1039–1046.

Walsh, S.E., Maillard, J.-Y., Russell, A.D., and Hann, A.C. (2001). Possible mechanisms for the relative efficiacies of ortho-phthalaldehyde and glutaraldehyde aganst glutaradehyde-resistant *Mycobacterium chelonae*. *J. Appl. Microbiol.* 91: 80–92.

Wang, C.-Y., Wu, H.-D., Lee, L.-N. et al. (2006). Pasteurization is effective against multidrug-resistant bacteria. *Am. J. Infect. Control* 34: 320–322.

Ward, M.J. and Routledge, P.A. (1988). Hypernatraemia and hyperchloraemic acidosis after bleach ingestion. *Hum. Toxicol.* 7: 37–38.

Ward, R.L., Bernstein, D.I., Knowlton, D.R. et al. (1991). Prevention of surface-to-human transmission of rotaviruses by treatment with disinfectant spray. *J. Clin. Microbiol.* 29: 1991–1996.

Wardle, E. and Jones, D. (2003). Determination of rinsing volumes following manual endoscope disinfection with ortho-phthalaldehyde (OPA). *J. Gastroenterol. Nurses Coll. Aust.*: 7–9.

Wardle, M.D. and Renninger, G.M. (1975). Bactericidal effect of hydrogen peroxide on spacecraft isolates. *Appl. Microbiol.* 30: 710–711.

Webb, B.C., Thomas, C.J., Harty, D.W., and Willcox, M.D. (1998). Effectiveness of two methods of denture sterilization. *J. Oral Rehabil.* 25: 416–423.

Weber, D.J. and Rutala, W.A. (1991). Environmental issues and nosocomial infections. In: *Prevention and Control of Nosocomial Infections* (ed. R.P. Wenzel), 1042–1064. Baltimore, MD: Williams and Wilkins.

Weber, D.J. and Rutala, W.A. (1995). Use of metals and microbicides in the prevention of nosocomial infections. In: *Disinfection, Sterilization, and Antisepsis in Healthcare* (ed. W. Rutala), 271–285. Champlain, NY: Polyscience Publications.

Weber, D.J. and Rutala, W.A. (1997). Environmental issues and nosocomial infections. In: *Prevention and Control of Nosocomial Infections* (ed. R.P. Wenzel), 491–514. Baltimore, MD: Williams and Wilkins.

Weber, D.J. and Rutala, W.A. (1997). Role of environmental contamination in the transmission of vancomycin-resistant enterococci. *Infect. Control Hosp. Epidemiol.* 18: 306–309.

Weber, D.J. and Rutala, W.A. (1998). Occupational risks associated with the use of selected disinfectants and sterilants. In: *Disinfection, Sterilization, and Antisepsis in Healthcare* (ed. W.A. Rutala), 211–226. Champlain, NY: Polyscience Publications.

Weber, D.J. and Rutala, W.A. (1999). Nosocomial ocular infections. In: *Infection Control and Hospital Epidemiology* (ed. C.G. Mayhall), 287–299. Philadelphia, PA: Lippincott Williams & Wilkins.

Weber, D.J. and Rutala, W.A. (2001). Lessons from outbreaks associated with bronchoscopy. *Infect. Control Hosp. Epidemiol.* 22: 403–408.

Weber, D.J. and Rutala, W.A. (2001). Risks and prevention of nosocomial transmission of rare zoonotic diseases. *Clin. Infect. Dis.* 32: 446–456.

Weber, D.J. and Rutala, W.A. (2001). Use of metals as microbicides in preventing infections in healthcare. In: *Disinfection, Sterilization, and Preservation* (ed. S.S. Block), 415–430. Philadelphia, PA: Lippincott Williams & Wilkins.

Weber, D.J. and Rutala, W.A. (2004). Disinfection and sterilization of potential bioterrorism agents. In: *Disinfection, Sterilization and Antisepsis: Principles, Practices and New Research* (ed. W.A. Rutala), 86–103. Washington, DC: Association for Professionals in Infection Control and Epidemiology.

Weber, D.J., Wilson, M.B., Rutala, W.A., and Thomann, C.A. (1990). Manual ventilation bags as a source for bacterial colonization of intubated patients. *Am. Rev. Respir. Dis.* 142: 892–894.

Weber, D.J., Barbee, S.L., Sobsey, M.D., and Rutala, W.A. (1999). The effect of blood on the antiviral activity of sodium hypochlorite, a phenolic, and a quaternary ammonium compound. *Infect. Control Hosp. Epidemiol.* 20: 821–827.

Weber, D.J., Rutala, W.A., and DiMarino, A.J. Jr. (2002). The prevention of infection following gastrointestinal endoscopy: the importance of prophylaxis and reprocessing. In: *Gastrointestinal Diseases: An Endoscopic Approach* (eds. A.J. DiMarino Jr. and S.B. Benjamin), 87–106. Thorofare, NJ: Slack Inc.

Webster, E., Ribner, B., Streed, L.L., and Hutton, N. (1996). Microbial contamination of activated 2% glutaraldehyde used in high-level disinfection of endoscopes (abstract). *Am. J. Infect. Control* 24: 153.

Webster, J., Lloyd, W., Ho, P. et al. (2003). Rethinking sterilization practices: evidence for event-related outdating. *Infect Control Hosp Epidemiol* 24: 622–624.

Wellons, S.L., Trawick, E.G., Stowers, M.F. et al. (1998). Laboratory and hospital evaluation of four personal monitoring methods for glutaraldehyde in ambient air. *Am. Ind. Hyg. Assoc. J.* 59: 96–103.

Welt, B.A., Tong, C.H., Rossen, J.L., and Lund, D.B. (1994). Effect of microwave radiation on inactivation of *Clostridium sporogenes* (PA 3679) spores. *Appl. Environ. Microbiol.* 60: 482–488.

Wendt, C. and Herwaldt, L.A. (1997). Epidemics: identification and Management. In: *Prevention and Control of Nosocomial Infections* (ed. R.P. Wenzel), 175–214. Baltimore, MD: Williams & Wilkins.

Wendt, C., Wiensenthal, B., Dietz, E., and Ruden, H. (1998). Survival of enterococci on dry surfaces. *J. Clin. Microbiol.* 36: 3734–3736.

West, A.B., Kuan, S.F., Bennick, M., and Lagarde, S. (1995). Glutaraldehyde colitis following endoscopy: clinical and pathological features and investigation of an outbreak. *Gastroenterology* 108: 1250–1255.

Westwood, J.C., Mitchell, M.A., and Legace, S. (1971). Hospital sanitation: the massive bacterial contamination of the wet mop. *Appl. Microbiol.* 21: 693–697.

Wheeler, P.W., Lancaster, D., and Kaiser, A.B. (1989). Bronchopulmonary cross-colonization and infection related to mycobacterial contamination of suction valves of bronchoscopes. *J. Infect. Dis.* 159: 954–958.

Whitby, J.L. and Rampling, A. (1972). Pseudomonas aeruginosa contamination in domestic and hospital environments. *Lancet* 1: 15–17.

Whitney, E.A.S., Beatty, M.E., Taylor, T.H. Jr. et al. (2003). Inactivation of *Bacillus anthracis* spores. *Emerg. Infect. Dis.* 9: 623–627.

Widmer, A.F. and Frei, R. (2003). Antimicrobial activity of glucoprotamin: a clinical study of a new disinfectant for instruments. *Infect. Control Hosp. Epidemiol.* 24: 762–764.

Widmer, A.F., Houston, A., Bollinger, E., and Wenzel, R.P. (1992). A new standard for sterility testing for autoclaved surgical trays. *J. Hosp. Infect.* 21: 253–260.

Wiggins, P., McCurdy, S.A., and Zeidenberg, W. (1989). Epistaxis due to glutaraldehyde exposure. *J. Occup. Med.* 31: 854–856.

Wilcox, M.H. and Fawley, W.N. (2000). Hospital disinfectants and spore formation by *Clostridium difficile*. *Lancet* 356: 1324.

Wilcox, M.H., Fawley, W.N., Wigglesworth, N. et al. (2003). Comparison of the effect of detergent versus hypochlorite cleaning on environmental contamination and incidence of *Clostridium difficile* infection. *J. Hosp. Infect.* 54: 109–114.

Williams, N.D. and Russell, A.D. (1991). The effects of some halogen-containing compounds on *Bacillus subtilis* endospores. *J. Appl. Bacteriol.* 70: 427–436.

Williams, I.R., Mayor, M.B., and Collier, J.P. (1998). The impact of sterilization method on wear in knee arthroplasty. *Clin. Orthop.*: 170–180.

Willis, C. (2005). Bacteria-free endoscopy rinse water – a realistic aim? *Epidemiol. Infect.* 134: 279–284.

Wilson, M. (2003). Light-activated antimicrobial coating for the continuous disinfection of surfaces. *Infect. Control Hosp. Epidemiol.* 24: 782–784.

Wilson, J.A. and Margolin, A.B. (1999). The efficacy of three common hospital liquid germicides to inactivate *Cryptosporidium parvum* oocysts. *J. Hosp. Infect.* 42: 231–237.

Windebank, A.J. and Blexrud, M.D. (1989). Residual ethylene oxide in hollow fiber hemodialysis units is neurotoxic in vitro. *Ann. Neurol.* 26: 63–68.

Wright, E.P., Collins, C.H., and Yates, M.D. (1985). *Mycobacterium xenopi* and *Mycobacterium kansasii* in a hospital water supply. *J. Hosp. Infect.* 6: 175–178.

Wu, M.S., Wang, J.T., Yang, J.C. et al. (1996). Effective reduction of *Helicobacter pylori* infection after upper gastrointestinal endoscopy by mechanical washing of the endoscope. *Hepatogastroenterology* 43: 1660–1664.

Wullt, M., Odenholt, I., and Walder, M. (2003). Activity of three disinfectants and acidified nitrite against *Clostridium difficile* spores. *Infect. Control Hosp. Epidemiol.* 24: 765–768.

Wysowski, D.K., Flynt, J.W. Jr., Goldfield, M. et al. (1978). Epidemic neonatal hyperbilirubinemia and use of a phenolic disinfectant detergent. *Pediatrics* 61: 165–170.

Yahya, M.T., Landeen, L.K., Messina, M.C. et al. (1990). Disinfection of bacteria in water systems by using electrolytically generated copper:silver and reduced levels of free chlorine. *Can. J. Microbiol.* 36: 109–116.

Young, J.H. (1993). Sterilization with steam under pressure. In: *Sterilization Technology: A Practical Guide for Manufacturers and Users of Health Care Product* (eds. R.F. Morrissey and G.B. Phillips), 81–119. New York: Van Nostrand Reinhold.

Young, J.H. (1997). Steam sterilization: scientific principles. In: *Sterilization Technology for the Health Care Facility* (eds. M. Reichert and J.H. Young), 123–144. Gaithersburg, MD: Aspen Publication.

Young, M. (2004). Sterilization process monitoring. *Manag. Infect. Control*: 70–76.

Zachary, K.C., Bayne, P.S., Morrison, V.J. et al. (2001). Contamination of gowns, gloves, and stethoscopes with vancomycin-resistant enterococci. *Infect. Control Hosp. Epidemiol.* 22: 560–564.

Zaidi, M., Angulo, M., and Sifuentes-Osornio, J. (1995). Disinfection and sterilization practices in Mexico. *J. Hosp. Infect.* 31: 25–32.

Zuhlsdorf, B.E.M., Floss, H., and Martiny, H. (2002). Cleaning efficacy of nine different cleaners in a washer-disinfector designed for flexible endoscopes. *J. Hosp. Infect.* 52: 206–211.

Zuhlsdorf, B., Floss, H., and Martiny, H. (2004). Efficacy of 10 different cleaning processes in a washer-disinfector for flexible endoscopes. *J. Hosp. Infect.* 56: 305–311.

Index

Sterile Processing of Pharmaceutical Products: Engineering Practice, Validation, and Compliance in Regulated Environments, First Edition. Sam A. Hout.

Back Cover

Dr. Sam A. Hout, PhD, MBA is an expert in pharmaceutical, medical, and health-care systems compliance. His experience spans over three decades in design, construction, and validation of cGMP manufacturing of drugs, medical devices, and delivery systems in clinical and commercial applications. As a senior consultant affiliated with Lachman, he provided cGMP inspection services both in the US domestic and international markets, including PAI on new drug introductions. Previously, he was a Senior Director of Engineering, Project Management, and Process Technology Transfer at Siegfried, Irvine, CA. He held engineering and operations management positions at Teva Pharmaceuticals, Phenomenex, Johnson & Johnson, and US government process chemistry research. In addition, his practice covered microbiology and chemistry laboratory operations, including high-potency drugs for oncology treatments with emphasis on data integrity. For the last 20 years, his focus has been to design, validate, and operate sterile manufacturing facilities with automated aseptic and lyophilization capabilities.